FORD MUSTANG
OWNERS WORKSHOP MANUAL
STRAIGHT SIX
200 CID & 250 CID
1965-1972

ISBN: 9781783181100

Models covered:
Ford Mustang Straight Six 200 CID 1965-1970
Ford Mustang Straight Six 250 CID 1968-1972

BROOKLANDS BOOKS LTD.
P.O. BOX 904, AMERSHAM,
BUCKS. HP6 9JA. UK

MU6WH

www.brooklandsbooks.com

12W5/2553

INTRODUCTION

This do-it-yourself Workshop Manual has been specially written for the owner who wishes to maintain his car in first class condition and to carry out his own servicing and repairs. Considerable savings on garage charges can be made, and one can drive in safety and confidence knowing the work has been done properly.

Comprehensive step-by-step instructions and illustrations are given on all dismantling, overhauling and assembling operations. Certain assemblies require the use of expensive special tools, the purchase of which would be unjustified. In these cases information is included but the reader is recommended to hand the unit to the agent for attention.

Throughout the Manual hints and tips are included which will be found to be invaluable, and there is an easy to follow fault diagnosis at the end of each chapter.

Whilst every care has been taken to ensure correctness of information it is obviously not possible to guarantee complete freedom from errors or to accept liability arising from such errors or omissions.

Instructions may refer to the right-hand or left-hand sides of the vehicle or the components. These are the same as the right-hand or left-hand of an observer standing behind the car and looking forward.

By the Autobooks Team of Technical Writers

Brooklands Books Ltd., PO Box 904, Amersham, Bucks, England. HP6 9JA
brooklandsbooks.com

CONTENTS

HINTS ON MAINTENANCE AND OVERHAUL

There are few things more rewarding than the restoration of a vehicle's original peak of efficiency and smooth performance.

The following notes are intended to help the owner to reach that state of perfection. Providing that he possesses the basic manual skills he should have no difficulty in performing most of the operations detailed in this manual. It must be stressed, however, that where recommended in the manual, highly-skilled operations ought to be entrusted to experts, who have the necessary equipment, to carry out the work satisfactorily.

Quality of workmanship:
The hazardous driving conditions on the roads today demand that vehicles should be as nearly perfect, mechanically, as possible. It is therefore most important that amateur work be carried out with care, bearing in mind the often inadequate working conditions, and also the inferior tools which may have to be used. It is easy to counsel perfection in all things, and we recognize that it may be setting an impossibly high standard. We do, however, suggest that every care should be taken to ensure that a vehicle is as safe to take on the road as it is humanly possible to make it.

Safe working conditions:
Even though a vehicle may be stationary, it is still potentially dangerous if certain sensible precautions are not taken when working on it while it is supported on jacks or blocks. It is indeed preferable not to use jacks alone, but to supplement them with carefully placed blocks, so that there will be plenty of support if the car rolls off the jacks during a strenuous manoeuvres. Axle stands are an excellent way of providing a rigid base which is not readily disturbed. Piles of bricks are a dangerous substitute. Be careful not to get under heavy loads on lifting tackle, the load could fall. It is preferable not to work alone when lifting an engine, or when working underneath a vehicle which is supported well off the ground. To be trapped, particularly under the vehicle, may have unpleasant results if help is not quickly forthcoming. Make some provision, however humble, to deal with fires. Always disconnect a battery if there is a likelihood of electrical shorts. These may start a fire if there is leaking fuel about. This applies particularly to leads which can carry a heavy current, like those in the starter circuit. While on the subject of electricity, we must also stress the danger of using equipment which is run off the mains and which has no earth or has faulty wiring or connections. So many workshops have damp floors, and electrical shocks are of such a nature that it is sometimes impossible to let go of a live lead or piece of equipment due to the muscular spasms which take place.

Work demanding special care:
This involves the servicing of braking, steering and suspension systems. On the road, failure of the braking system may be disastrous. Make quite sure that there can be no possibility of failure through the bursting of rusty brake pipes or rotten hoses, nor to a sudden loss of pressure due to defective seals or valves.

Problems:
The chief problems which may face an operator are:

1 External dirt.
2 Difficulty in undoing tight fixings
3 Dismantling unfamiliar mechanisms.
4 Deciding in what respect parts are defective.
5 Confusion about the correct order for reassembly.
6 Adjusting running clearances.
7 Road testing.
8 Final tuning.

Practical suggestions to solve the problems:

1. Preliminary cleaning of large parts - engines, transmissions, steering, suspensions, etc., should be carried out before removal from the car. Where road dirt and mud alone are present, wash clean with a high pressure water jet, brushing to remove stubborn adhesions, and allow to drain and dry. Where oil or grease is also present, wash down with a proprietary compound (Gunk, Teepol etc.,) applying with a stiff brush - an old paint brush is suitable, into all crevices. Cover the distributor and ignition coils with a polythene bag and then apply a strong water jet to clear the loosened deposits. Allow to drain and dry. The assemblies will then be sufficiently clean to remove and transfer to the bench for the next stage.

   On the bench, further cleaning can be carried out, first wiping the parts as free as possible from grease with old newspaper. Avoid using rag or cotton waste which can leave clogging fibres behind. Any remaining grease can be removed with a brush dipped in paraffin. Avoid using paraffin or petrol in large quantities for cleaning in enclosed areas, such as garages, on account of the high fire risk.

When all exteriors have been cleaned, and not before, dismantling can be commenced. This ensures that dirt will not enter into interiors and orifices revealed by dismantling. In the next phases, where components have to be cleaned and keep the containers covered except when in use. After the components have been cleaned, plug small holes with tapered hard wood plugs cut to size and blank off larger orifices with greaseproof paper and masking tape. Do not use soft wood plugs or matchsticks as they may break.

2. It is not advisable to hammer on the end of a screw thread, but if it must be done, first screw on a nut to protect the thread, and use a lead hammer. This applies particularly to the removal of tapered cotters. Nuts and bolts seem to 'grow' together, especially in exhaust systems. If penetrating oil does not work, try the judicious application of heat, but be careful of starting a fire.

   Tight bushes or pieces of tail-pipe rusted into a silencer can be removed by splitting them with an open-ended hacksaw. Tight screws can sometimes be started by a tap from a hammer on the end of a suitable screwdriver. Many tight fittings will yield to the judicious use of a hammer, but it must be a soft-faced hammer if damage is to be avoided, use a heavy block on the opposite side to absorb shock. Any parts of the steering system which have been damaged should be renewed, as attempts to repair them may lead to cracking and subsequent failure, and steering ball joints should be disconnected using a recommended tool to prevent damage.

3. It often happens that an owner is baffled when trying to dismantle an unfamiliar piece of equipment. So many modern devices are pressed together or assembled by spinning-over flanges, that they must be sawn apart. The intention is that the whole assembly must be renewed. However, parts which appear to be in one piece to the naked eye, may reveal close-fitting joint lines when inspected with a magnifying glass, and, this may provide the necessary clue to dismantling. Left-handed screw threads are used where rotational forces would tend to unscrew a right handed screw thread.

   Be very careful when dismantling mechanisms which may come apart suddenly. Work in an enclosed space where the parts will be contained, and drape a piece of cloth over the device if springs are likely to fly in all directions. Mark everything which might be reassembled in the wrong position, scratched symbols may be used on unstressed parts, or a sequence of tiny dots from a centre punch can be useful. Stressed parts should never be scratched or centre-popped as this may lead to cracking under working conditions. Store parts which look alike in the correct order for reassembly. Never rely upon memory to assist in the assembly of complicated mechanisms, especially when they will be dismantled for a long time, but make notes, and drawings to supplement the diagrams in the manual, and put labels on detached wires. Rust stains may indicate unlubricated wear. This can sometimes be seen round the outside edge of a bearing cup in a universal joint. Look for bright rubbing marks on parts which normally should not make heavy contact. These might prove that something is bent or running out of truth. For example, there might be bright marks on one side of a piston, at the top near the ring grooves, and others at the bottom of the skirt on the other side. This could well be the clue to a bent connecting rod. Suspected cracks can be proved by heating the component in a light oil to approximately 100°C, removing, drying off, and dusting with French chalk, if a crack is present the oil retained in the crack will stain the French chalk.

4. In determining wear, and the degree, against the permissible limits set in the manual, accurate measurement can only be achieved by the use of a micrometer. In many cases the wear is given to the fourth place of decimals; that is in ten-thousandths of an inch. This can be read by the vernier scale on the barrel of a good micrometer. Bore diameters are more difficult to determine. If, however, the matching shaft is accurately measured, the degree of play in the bore can be felt as a guide to its suitability. In other cases, the shank of a twist drill of known diameter is a handy check.

   Many methods have been devised for determining the clearance between bearing surfaces. Today the best and simplest is by the use of Plastigage, obtainable from most garages. A thin plastic thread is laid between the two surfaces and the bearing is tightened, flattening the thread. On removal, the width of the thread is compared with a scale supplied with the thread and the clearance is read off directly. Sometimes joint faces leak persistently, even after gasket renewal. The fault will then be traceable to distortion, dirt or burrs. Studs which are screwed into soft metal frequently raise burrs at the point of entry. A quick cure for this is to chamfer the edge of the hole in the part which fits over the stud.

5. Always check a replacement part with the original one before it is fitted.

If parts are not marked, and the order for reassembly is not known, a little detective work will help. Look for marks which are due to wear to see if they can be mated. Joint faces may not be identical due to manufacturing errors, and parts which overlap may be stained, giving a clue to the correct position. Most fixings leave identifying marks especially if they were painted over on assembly. It is then easier to decide whether a nut, for instance, has a plain, a spring, or a shake-proof washer under it. All running surfaces become 'bedded' together after long spells of work and tiny imperfections on one part will be found to have left corresponding marks on the other. This is particularly true of shafts and bearings and even a score on a cylinder wall will show on the piston.

6. Checking end float or rocker clearances by feeler gauge may not always give accurate results because of wear. For instance, the rocker tip which bears on a valve stem may be deeply pitted, in which case the feeler will simply be bridging a depression. Thrust washers may also wear depressions in opposing faces to make accurate measurement difficult. End float is then easier to check by using a dial gauge. It is common practice to adjust end play in bearing assemblies, like front hubs with taper rollers, by doing up the axle nut until the hub becomes stiff to turn and then backing it off a little. Do not use this method with ball-bearing hubs as the assembly is often pre-loaded by tightening the axle nut to its fullest extent. If the split-pin hole will not line up, file the base of the nut a little.

Steering assemblies often wear in the straight-ahead position. If any part is adjusted, make sure that it remains free when moved from lock to lock. Do not be surprised if an assembly like a steering gearbox, which is known to be carefully adjusted outside the car, becomes stiff when it is bolted in place. This will be due to distortion of the case by the pull of the mounting bolts, particularly if the mounting points are not all touching together. This problem may be met in other equipment and is cured by careful attention to the alignment of mounting points.

When a spanner is stamped with a size and A/F it means that the dimension is the width between the jaws and has no connection with ANF, which is the designation for the American National Fine thread. Coarse threads like Whitworth are rarely used on cars to-day except for studs which screw into soft aluminium or cast iron. For this reason it might be found that the top end of a cylinder head stud has a fine thread and the lower end a coarse thread to screw into the cylinder block. If the car has mainly UNF threads then it is likely that any coarse threads will be UNC, which are not the same as Whitworth. Small sizes have the same number of threads in Whitworth and UNC, but in the 1/4 inch size for example, there are twelve threads to the inch in the former and thirteen in the latter.

7. After a major overhaul, particularly if a great deal of work has been done on the braking, steering and suspension systems, it is advisable to approach the problem of testing with care. If the braking system has been overhauled, apply heavy pressure to the brake pedal and get a second operator to check every possible source of leakage. The brakes may work extremely well, but a leak could cause complete failure after a few miles.

Do not fit the hub caps until every wheel nut has been checked for tightness, and make sure the tyre pressures are correct. Check the levels of coolant, lubricants and hydraulic fluids. Being satisfied that all is well, take the car on the road and test the brakes at once. Check the steering and the action of the handbrake. Do all this at moderate speeds on quiet roads, and make sure there is no other vehicle behind you when you try a rapid stop.

Finally, remember that many parts settle down after a time, so check for tightness of all fixings after the car has been on the road for a hundred miles or so.

8. It is useless to tune an engine which has not reached its normal running temperature. In the same way, the tune of an engine which is stiff after a re-bore will be different when the engine is again running free. Remember too, that rocker clearances on push-rod operated valve gear will change when the cylinder head nuts are tightened after an initial period of running with a new head gasket. Trouble may not always be due to what seems the obvious cause. Ignition, carburation and mechanical condition are interdependent and spitting back through the carburetter, which might be attributed to a weak mixture, can be caused by a sticking inlet valve. For one final hint on tuning, never adjust more than one thing at a time or it will be impossible to tell which adjustment produced the desired result.

GLOSSARY OF TERMS

Allen key Cranked wrench of hexagonal section for use with socket head screws.

Alternator Electrical generator producing alternating current. Rectified to direct current for battery charging.

Ambient temperature Surrounding atmospheric temperature.

Annulus Used in engineering to indicate the outer ring gear of an epicyclic gear train.

Armature The shaft carrying the windings, which rotates in the magnetic field of a generator or starter motor. That part of a solenoid or relay which is activated by the magnetic field.

Axial In line with, or pertaining to, an axis.

Backlash Play in meshing gears.

Balance lever A bar where force applied at the centre is equally divided between connections at the ends.

Banjo axle Axle casing with large diameter housing for the crownwheel and differential.

Bar Standard unit of pressure equal to 14.5lb/sq in.

Bendix pinion A self-engaging and self-disengaging drive on a starter motor shaft.

Bevel pinion A conical shaped gearwheel, designed to mesh with a similar gear with an axis usually at 90° to its own.

bhp Brake horse power. Now superseded by kW.

bmep Brake mean effective pressure. Average pressure on a piston during the working stroke.

Brake cylinder Cylinder with hydraulically operated piston(s) acting on brake shoes or pads.

Brake regulator Control valve fitted in hydraulic braking system which limits brake pressure to rear brakes during heavy braking to prevent rear wheel locking.

Camber Angle at which a wheel is tilted from the vertical.

Capacitor Modern term for an electrical condenser. Part of distributor assembly, connected across contact breaker points, acts as an interference suppressor.

Castellated Top face of a nut, slotted across the flats to take a locking splitpin.

Castor Angle at which the kingpin or swivel pin is tilted when viewed from the side.

cc or cm^3 Cubic centimetres. Engine capacity is arrived at by multiplying the area of the bore in sq cm by the stroke in cm by the number of cylinders.

Clevis U-shaped forked connector used with a clevis pin, usually at handbrake connections.

Collet A type of collar, usually split and located in a groove in a shaft, and held in place by a retainer. The arrangement used to retain the spring(s) on a valve stem in most cases.

Commutator Rotating segmented current distributor between armature windings and brushes.

Compression ratio The ratio, or quantitative relation, of the total volume (piston at bottom of stroke) to the unswept volume (piston at top of stroke) in an engine cylinder.

Condenser See 'Capacitor'.

Core plug Plug for blanking off a manufacturing hole.

Crownwheel Large bevel gear in rear axle, driven by a bevel pinion attached to the propeller shaft.

'C'-spanner Like a 'C' with a handle. For use on screwed collars without flats, but with slots or holes.

Damper Modern term for shock absorber, used in vehicle suspension systems to damp out spring oscillations.

Depression The lowering of atmospheric pressure as in the inlet manifold and carburetter.

Dowel Close tolerance pin, peg, tube or bolt, which accurately locates mating parts.

Drag link Rod connecting steering box drop arm (pitman arm) to nearest front wheel steering arm in certain types of steering systems.

Dry liner Thinwall tube pressed into cylinder bore.

Dry sump Lubrication system where all oil is scavenged from the sump, and returned to a separate tank.

Dynamo Electrical generator producing direct current.

Electrode Terminal part of an electrical component, such as the points or 'Electrodes' of a sparking plug.

Electrolyte In lead-acid car batteries a solution of sulphuric acid and distilled water.

End float The axial movement between associated parts, end play.

EP Extreme pressure. In lubricants, special grades for heavily loaded bearing surfaces, such as gear teeth in a gearbox, or crownwheel and pinion in a rear axle.

Fade Of brakes. Reduced efficiency due to overheating.

Field coils Windings on the polepieces of motors and generators.

Fillets Narrow finishing strips usually applied to interior bodywork.

First motion shaft Input shaft from clutch to gearbox.

Fullflow filter Filters in which all the oil is pumped to the engine. If the element becomes clogged, a bypass valve operates to pass unfiltered oil to the engine.

FWD Front wheel drive.

Gear pump Two meshing gears in a close fitting casing. Oil is carried from the inlet round the outside of both gears in the spaces between the gear teeth and casing to the outlet, the meshing gear teeth prevent oil passing back to the inlet, and the oil is forced through the outlet port.

Generator An alternator or a dynamo.

Grommet A ring of protective or sealing material. Can be used to protect pipes or leads passing through bulkheads.

Grubscrew Fully threaded headless screw with screwdriver slot. Used for locking or alignment purposes.

Gudgeon pin Shaft which connects a piston to its connecting rod. Sometimes called 'wrist pin' or 'piston pin'.

Halfshaft One of a pair transmitting drive from the differential.

Helical In spiral form. The teeth of helical gears are cut at a spiral angle to the side faces of the gearwheel.

Hot spot Hot area that assists vapourisation of fuel on its way to cylinders. Often provided by close contact between inlet and exhaust manifolds.

HT High Tension. Applied to electrical current produced by the ignition coil for the sparking plugs.

Hydrometer A device for checking specific gravity of liquids. Used to check specific gravity of electrolyte.

Hypoid bevel gears A form of bevel gear used in the rear axle drive gears. The bevel pinion meshes below the centre line of the crownwheel, giving a lower propeller shaft line.

Idler A device for passing on movement. A free running gear between driving and driven gears. A lever transmitting track rod movement to a side rod in steering gear.

Impeller A centrifugal pumping element. Used in water pumps to stimulate flow.

Journals Those parts of a shaft that are in contact with the bearings.

kW Standard unit of power, equal to 1.34102hp

Kingpin The main vertical pin which carries the front wheel spindle, and permits steering movement. May be called 'steering pin' or 'swivel pin'.

Layshaft The shaft which carries the laygear in the gearbox. The laygear is driven by the first motion shaft and drives the third motion shaft according to the gear selected. Sometimes called the 'countershaft' or 'second motion shaft'.

lbf ft A measure of twist or torque. A pull of 10lb at a radius of 1ft is a torque of 10lbf ft.

lb/sq in Pounds per square inch.

Little-end The small, or piston end of a connecting rod. Sometimes called the 'small-end'.

LT Low Tension. The current output from the battery.

Mandrel Accurately manufactured bar or rod used for test or centring purposes.

Manifold A pipe, duct or chamber, with several branches.

Nm Standard unit of torque equal to 0.738lbf ft.

Needle rollers Bearing rollers with a length many times their diameter.

Oil bath Reservoir which lubricates parts by immersion. In air filters, a separate oil supply for wetting a wire mesh element to hold the dust.

Oil wetted In air filters, a wire mesh element lightly oiled to trap and hold airborne dust.

Overlap Period during which inlet and exhaust valves are open together.

Panhard rod Bar connected between fixed point on chassis and another on axle to control sideways movement.

Pawl Pivoted catch which engages in the teeth of a ratchet to permit movement in one direction only.

Peg spanner Tool with pegs, or pins, to engage in holes or slots in the part to be turned.

Pendant pedals Pedals with levers pivoted at the top.

Phillips screwdriver A cross-point screwdriver for use with the cross-slotted heads of Phillips screws.

Pinion A small gear, usually in relation to another gear.

Piston-type damper Shock absorber in which damping is controlled by a piston working in a closed oil-filled cylinder.

Preloading Preset static pressure on ball or roller bearings not due to working loads.

Radial Radiating from a centre, like the spokes of a wheel.

Radius rod Pivoted arm confining movement of a part to an arc of fixed radius.

Ratchet Toothed wheel or rack which can move in one direction only, movement in the other being prevented by a pawl.

Ring gear A gear tooth ring attached to outer periphery of flywheel. Starter pinion engages with it during starting.

Runout Amount by which rotating part is out of true.

Semi-floating axle Outer end of rear axle halfshaft is carried on bearing inside axle casing. Wheel hub is secured to end of shaft.

Servo A hydraulic or pneumatic system for assisting, or, augmenting a physical effort. See 'Vacuum Servo'.

Setscrew Fastener threaded the full length of the shank.

Shackle A coupling link, used in the form of two parallel pins connected by side plates to secure the end of the master suspension spring and absorb the effects of deflection.

Shell bearing Thinwalled steel shell lined with antifriction metal. Usually semi-circular and used in pairs for main and big-end bearings.

Shock absorber See 'Damper'.

Silentbloc Rubber bush bonded to inner and outer metal sleeves.

Socket-head screw Screw with hexagonal socket for an Allen key.

Solenoid A coil of wire creating a magnetic field when electric current passes through it. Used with a soft iron core to operate contacts or a mechanical device.

Spur gear A gear with teeth cut axially.

Stub axle Short axle fixed at one end only.

Tachometer An instrument for accurate measurement of rotating speed. Usually indicates in revolutions per minute.

TDC Top Dead Centre. The highest point reached by a piston in a cylinder, with the crank and connecting rod in line.

Thermostat Automatic device for regulating temperature. Used in vehicle coolant systems to open a valve which restricts circulation at low temperature.

Third motion shaft Output shaft of gearbox.

Threequarter floating axle Outer end of rear axle halfshaft flanged and bolted to wheel hub, which runs on bearing mounted on outside of axle casing. Vehicle weight is not carried by the axle shaft.

Thrust bearing or washer Used to reduce friction in rotating parts subject to axial loads.

Torque Turning or twisting effort. See 'lbf ft'.

Track rod The bar(s) across the vehicle which connect the steering arms and maintain the front wheel alignment.

UJ Universal joint. A coupling between shafts which permits angular movement.

UNF Unified National Fine screw thread.

Vacuum servo Device used in brake system, using difference between atmospheric pressure and inlet manifold depression to operate a piston which acts to augment brake pressure as required. See 'Servo'.

Venturi A restriction or 'choke' in a tube, as in a carburetter, used to increase velocity to obtain a reduction in pressure.

Vernier A sliding scale for obtaining fractional readings of the graduations of an adjacent scale.

Welch plug A domed thin metal disc which is partially flattened to lock in a recess. Used to plug core holes in castings.

Wet liner Removable cylinder barrel, sealed against coolant leakage, where the coolant is in direct contact with the outer surface.

Wet sump A reservoir attached to the crankcase to hold the lubricating oil.

CHAPTER 1

THE ENGINE

1 :1 Description

The Mustang models covered by this manual all have straight-six engines. Two versions can be fitted, very similar in design, with capacities of either 200 or 250 cubic inches. Details of bore, stroke and other dimensions are given in **Technical Data** in the **Appendix** to this manual.

A sectional view of a 200 engine is shown in **FIG 1 :1**. Both the crankshaft and camshaft rotate in renewable bearing inserts in the crankcase. Split inserts are used on the main bearings and they are held in place by bearing caps attaching directly to the crankcase. The camshaft is driven, at half engine speed, by two sprockets and a timing chain from the front of the crankshaft. On the very front of the crankshaft is installed a damper (to damp out torsional vibrations in the crankshaft) and the pulleys for driving the engine accessories. Hydraulic valve lifters operate in bores in the crankcase and they are raised by the cams on the camshaft. The valve lifters automatically take up the lash in the valve train and the motion is transmitted to the rocker arms by pushrods. The rocker arms pivot around a rocker arm shaft, running the length of the engine, and their opposite ends act directly onto the tops of the valve stems to open the valves. The valves are closed by the combined pressures of the gases in the cylinder and the valve return springs. Gears on the camshaft also drive the distributor and oil pump. The connecting rods are an interference fit with the piston pins at the top end but the lower end is split, using a cap, and renewable bearing inserts are used for rotation about the crankshaft.

Lubrication:

A typical lubrication system is shown in **FIG 1 :2**. The oil runs down inside the engine to the sump of the oil pan, where it is cooled by the airflow over the engine. From the sump the oil is drawn into the oil pump through a gauze strainer. The oil pump is fitted with a relief valve which ensures that the maximum oil pressure is limited and the surplus oil is returned directly to the inlet of the pump. From the pump the oil flows through a fullflow oil filter. The filter contains a bypass valve which opens under the differential pressure if the filter element is clogged. This is a safety precaution to ensure that the engine will not be starved of oil but if the valve does operate, the oil to the engine will be unfiltered. The filter also contains an anti-drain back diaphragm which prevents oil from draining down out of the system when

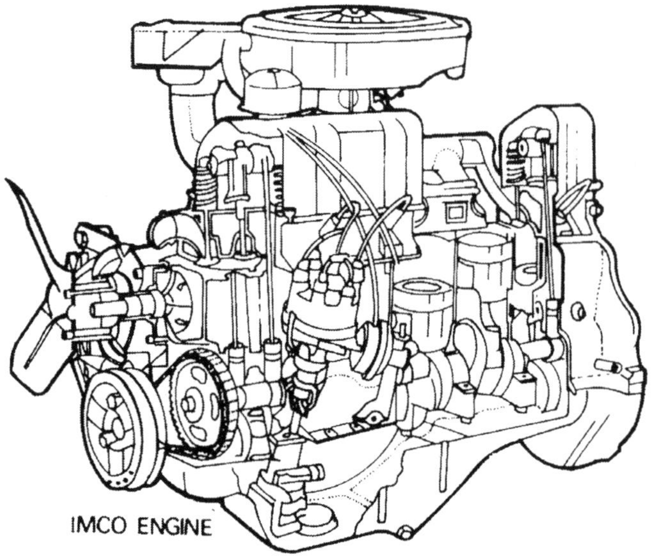

FIG 1:1 Three-quarter sectional view of engine

the engine is stationary. From the filter the oil passes to the engine main oil gallery. Internal passages feed oil directly from the main gallery to the crankshaft main bearings and hydraulic valve lifters. Other internal passages then lead the oil from the main bearings to the camshaft bearings. The rear camshaft journal has a groove machined in it and this groove allows a metered quantity of oil to pass from the rear camshaft bearing up through passages to the rocker shaft running along the cylinder head. From the rocker shaft the oil flows out through drilled holes to lubricate the rocker arms.

The crankshaft has internal oilways drilled in it and these lead oil from the main bearings to the connecting rod bearings for their lubrication. The connecting rods each have a squirt hole machined in them and when this hole indexes with the oil hole in the crankshaft a jet of oil lubricates and cools the cylinder walls. Oil flingers and oil seals are fitted at either end of the crankshaft to prevent the oil from leaking out.

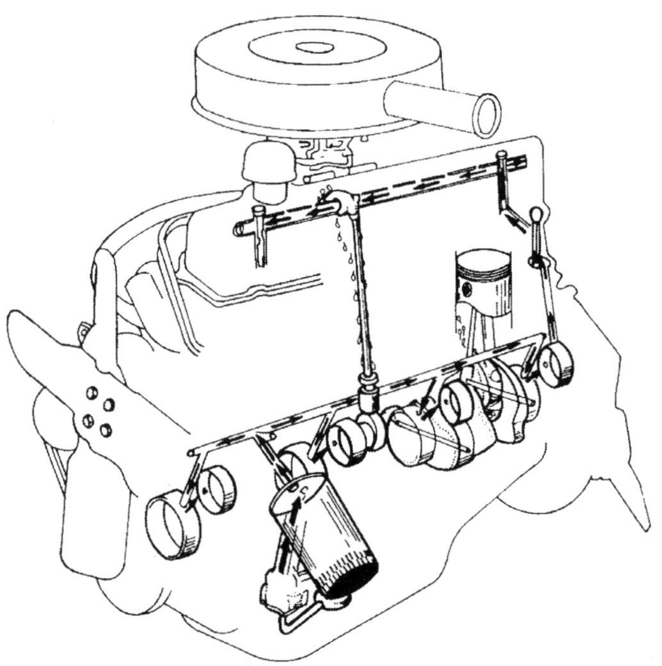

FIG 1:2 Typical lubrication system

The oil to the valve lifters lubricates their bores and also supplies the oil for their operation. A reservoir is fitted beside each valve lifter to ensure that each lifter will have an adequate supply of oil immediately on start-up.

The remainder of the parts are lubricated by the oil mist in the engine and splash thrown from the bearings.

Crankcase ventilation:

A typical ventilation system is shown in **FIG 1:3**. Air is drawn in from under the hood through the inlet duct to the carburetor air cleaner and it then passes through a hose connected to the oil filler cap on the rocker cover. The air for crankcase ventilation does not pass through the carburetor air cleaner element but a filter is fitted into the filler cap to ensure that the air entering the engine crankcase is clean. The rotation of the crankshaft sweeps the air around the engine from the front to rear and it then passes up through the pushrod tubes at the rear and passes out of the rocker cover through a regulator valve. The regulator valve is shown in **FIG 1:4**. During normal running, the return spring holds the plunger in the open position providing an unrestricted flow for the air. When the throttle is closed the vacuum in the manifold increases and if full flow were allowed to take place the extra air would be sufficient to alter the slow-running. At high vacuum the plunger in the valve is pulled across against the pressure of the spring making the flow through the valve restricted. The valve can neither be dismantled or cleaned and when it is time-expired or defective a new valve must be fitted in its place. After having passed through the regulator valve the air and fumes are drawn into the intake manifold to mix with the ingoing fuel/air and be burnt in the combustion chambers.

Emission control:

Usually the Imco system is used on models fitted with automatic transmission and this system relies on specially made and adjusted carburetors to give efficient combustion and reduce emissions. On models fitted with a manual-shift transmission the Thermactor system is usually used. This system uses an air pump, driven by a belt from the engine, to inject excess cold air into the exhaust manifold where it mixes with the very hot exhaust gases to burn all surplus hydrocarbons. The Thermactor system is fully described in **Section 1:12**.

1:2 Engine removal and installation

The only operations for which it is essential to remove the engine from the vehicle are removing the crankshaft or renewing the camshaft bearings. All other operations, including crankshaft main bearing renewal, can be carried out with the engine installed in the vehicle. However, if major repair work is to be carried out it will often be found more convenient to have the engine out and on the bench.

If the owner is not a skilled automobile mechanic it is suggested that he reads the 'Hints on Maintenance and Overhaul' at the end of this manual as much useful information is contained there.

This section only refers to ancilliaries and components connected to the engine without giving full details and for complete information details the relevant chapter should be consulted. The cooling system and drive belts are fully dealt with in **Chapter 4**, the drive shaft in **Chapter 8** while the manual-shift transmission is in **Chapter 6** and the automatic transmission in **Chapter 7**. Ignition and distributor is dealt with in **Chapter 3**.

Removal:

1 Disconnect the battery. Drain the cooling system and crankcase. Remove the hood. Remove the radiator, cooling fan and hub after disconnecting the water hoses. On a vehicle fitted with automatic transmission it will be necessary to disconnect the oil cooler lines before removing the radiator. The oil cooler and lines should be blanked off to prevent the entry of dirt. Slacken all the drive belts.

2 Disconnect the automatic choke air tubes and the crankcase ventilation system hoses from the air cleaner and remove the air cleaner assembly. Disconnect all vacuum hoses from the intake manifold and carburetor.

3 Models fitted with power steering; free the pump from the engine and hang it safely out of the way so that fluid is not spilled from the reservoir. Models fitted with Thermactor; remove the air cleaner if it is not mounted on the pump or engine. Models fitted with air-conditioning; remove the compressor from the engine but leave the lines connected (see **Chapter 13**).

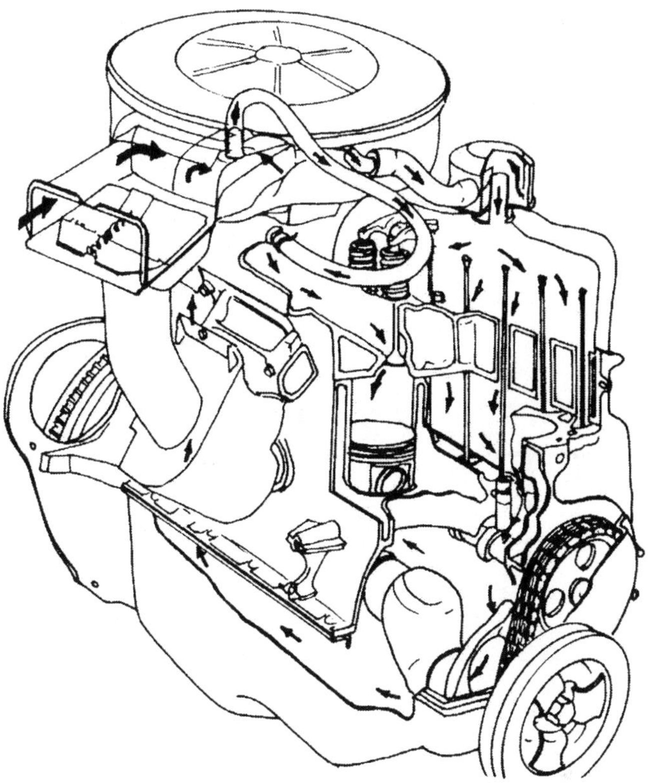

FIG 1:3 Positive closed-circuit crankcase ventilation system

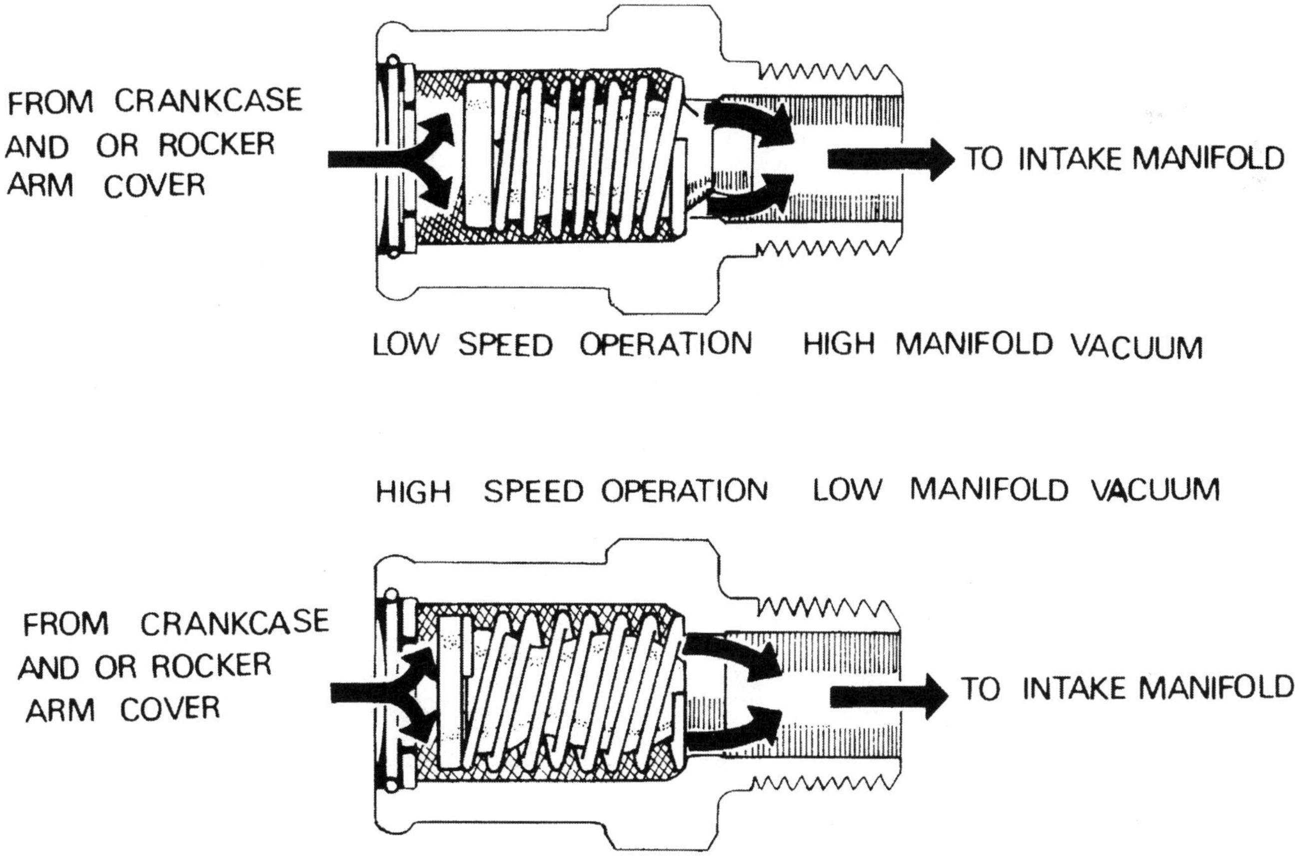

FIG 1:4 The crankcase ventilation regulator valve

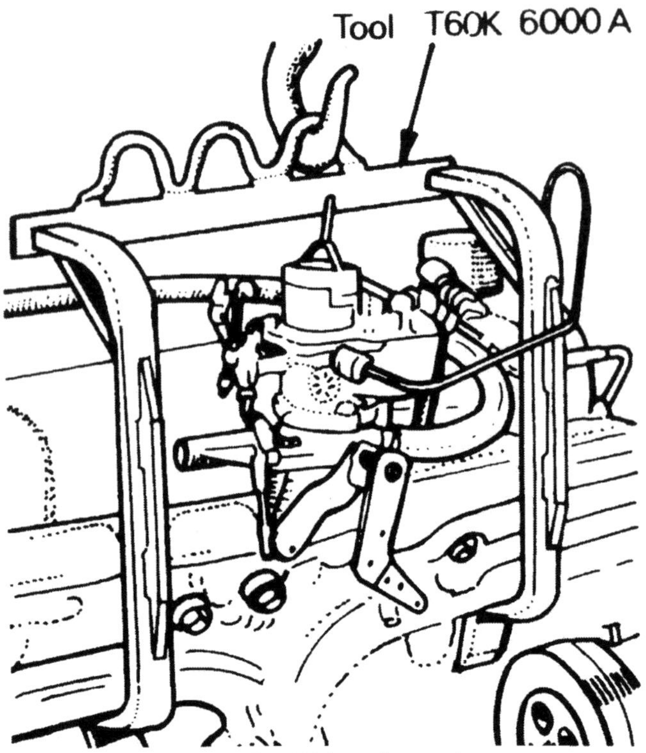

FIG 1 : 5 Lifting the engine

4 Disconnect the heater hoses. Disconnect all electrical wires and cables connecting the frame to the engine. These include the alternator wires, ignition coil, temperature and oil pressure sender units and the starter motor. If the colors have faded on the wires it is advisable to label them to ensure correct installation.

5 Disconnect the main fuel feed to the fuel pump and blank it off to prevent fuel from siphoning through. Disconnect the throttle rod and choke operating cable from the carburetor, as well as the automatic transmission rod if fitted.

6 Raise the vehicle and place it securely on stands. Remove the starter motor and its dust shield. On models with manual-shift transmission, disconnect the clutch operating linkage parts from the release lever and housing and then remove the bolts that secure the flywheel housing to the engine. On models fitted with automatic transmission, remove the inspection cover so that the converter can be disconnected from the engine drive plate and then remove the bolts that secure the converter housing to the engine. On all models disconnect the muffler inlet pipe from the exhaust manifold and free the clip that secures the pipe to the engine.

7 Lower the vehicle back to the ground and support the transmission with a jack and pad of wood underneath it. Disconnect the front engine mountings at the brackets on the underbody. **Check that there are no connections left holding the engine to the frame or body**. Attach an engine lifting hook as shown in **FIG 1 : 5**. Use the hook in conjunction with a suitable hoist and raise the engine slightly so that it clears the front mountings. Pull the engine forward and then lift it out of the vehicle. **On models fitted with manual-shift transmission, make sure that the transmission input shaft is well clear of the clutch before lifting out the engine. On models with automatic transmission, make sure that the converter does not fall out of the transmission when the engine is free.**

Installation:

The engine is installed in the reverse order of removal. Before installing the engine, check that the mounting insulators are in good condition and renew them if they are defective, distorted or perished.

The main point to watch when installing the engine is the mating to the transmission. **Do not attempt to force the engine back in place.** Make sure that the engine and transmission are accurately aligned. Rotate the crankshaft, using the damper and pulley until the splines on the input shaft slide through the clutch and into the spigot bearing in the crankshaft. **Do not allow the weight of the engine to hang on the input shaft of the transmission otherwise the clutch may be severely damaged.**

Make sure that all connections are tight enough to prevent leaks or faults. Fill the crankcase to the correct level with fresh oil and check the transmission fluid level. Fill and bleed the cooling system and adjust all drive belt tensions correctly. Check the clutch adjustment.

Check for leaks both before and after starting the engine. Allow the engine to run at a fast idle until it has reached its normal operating temperature and make final adjustments to carburetor and ignition.

1 :3 Cylinder head removal

Normally the cylinder head should only need removal for decarbonizing or attention to the pistons and cylinder bores. It should be noted that it is possible to renew the valve springs and valve stem without removing the cylinder head from the engine. Air under pressure is injected through the spark plug hole, using a suitable adaptor, and this keeps the valves closed while the rocker assembly and valve spring parts are removed. The engine may turn until the piston is at BDC under the action of the air pressure. **If the air supply is cut off it is essential that the valves are held up using elastic bands or adhesive tape to prevent them from falling through into the engine.**

To prevent distortion the cylinder head should not be removed while the engine is hot.

Removal:

1 Disconnect the battery and drain the cooling system. Remove the air cleaner after disconnecting all hoses to it. Remove the crankcase ventilation system and, on models fitted with Thermactor, disconnect the air hose at the check valve and remove the intermediate exhaust air supply tube assembly.

2 Disconnect the vacuum lines between carburetor and distributor at the carburetor and identify them as necessary. Disconnect the choke and throttle controls from the carburetor and on models fitted with automatic transmission, also disconnect the kickdown rod. Disconnect the upper radiator hose at the engine. Disconnect the fuel line at the line filter. It is advisable to disconnect and remove the distributor vacuum lines and the carburetor fuel line as an assembly so that they will not become damaged or distorted.

3 Label the spark plug leads and disconnect them from the spark plugs, **by pulling and twisting on the moulded caps only.** Disconnect the wire from the temperature sender unit.

4 Remove the valve rocker cover. In order, slacken each rocker shaft holding pedestal bolt by two turns until all the bolts are slack and can be removed. Lift off the rocker parts as an assembly and withdraw the pushrods. The pushrods must be stored in the correct order as they must be installed back into their original positions.

5 Evenly slacken and remove the remaining cylinder head holding bolts, preferably in the sequence shown in **FIG 1 :6**. Remove the cylinder head. **If the head sticks, do not pry between the faces as this may damage them**. Try tapping the head up using a wooden drift and if this fails, crank the engine over on the starter motor with the plugs in place to break the seal.

Cleanliness:

Service the cylinder head as described in the next section. When all the parts have been checked and serviced, make sure that they are scrupulously clean before assembling them. Before installing the head, make sure that all the gasket surfaces are clean and have not picked up specks of dirt.

Distortion:

If the cylinder head gasket has failed, the cylinder head should be checked for distortion, as shown in **FIG 1 :7**. Distortion must not exceed .003 inch in any 6 inches or .007 inch overall. If distortion is excessive the head may be machined down provided that not more than .010 inch of stock is removed.

Gaskets:

All gaskets and seals should be discarded and new ones fitted on installation. The rocker cover gasket should be cemented into the cover. Lay the cover on a flat surface with weights on it until the cement has dried.

No gasket is required on the exhaust manifold but the face and threads should be smeared with graphite grease. **Do not use any other form of grease as it will bake hard with the heat and lock the nuts in place.**

On the 200 engine a composition gasket is generally used and this must not have any sealer applied to it. On the 250 engine which is fitted with an ordinary gasket, a light smear of sealer should be applied to both sides of the gasket. On all engines, guide studs should be screwed into the block and the gasket and head laid into place over these studs.

Installation:

Make up two guide studs, as shown in **FIG 1 :8**, and fit them to the block.

1 Lay the gasket back into place, making sure that any markings are facing in the correct direction. **Do not use sealer on composition gaskets**. Lower the cylinder head into place down the studs and secure it with two end bolts fingertight. Remove the guide studs and install the remainder of the cylinder head bolts fingertight.

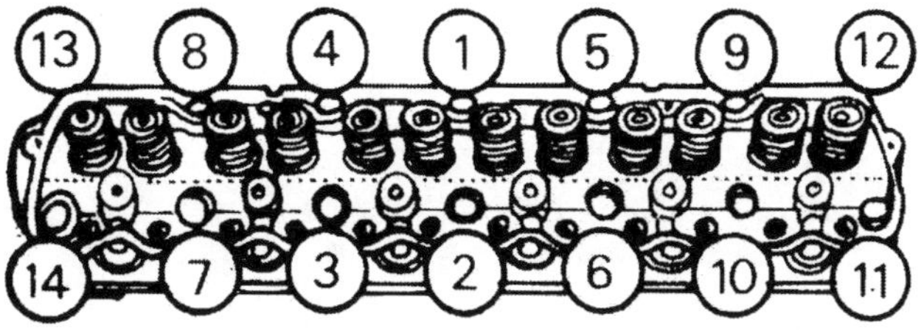

FIG 1 :6 Cylinder head bolt torque sequence

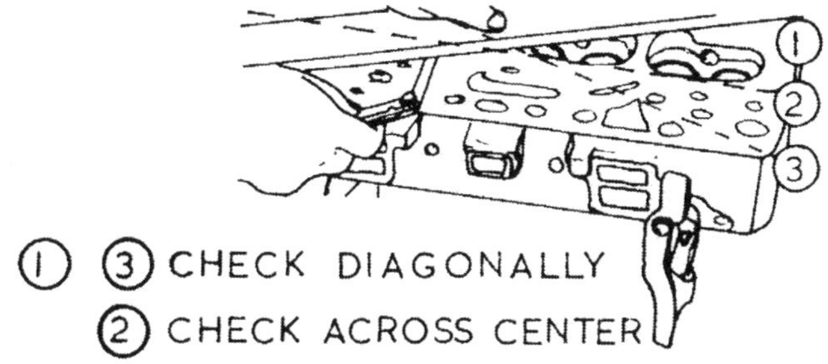

FIG 1 :7 Checking the cylinder head for distortion

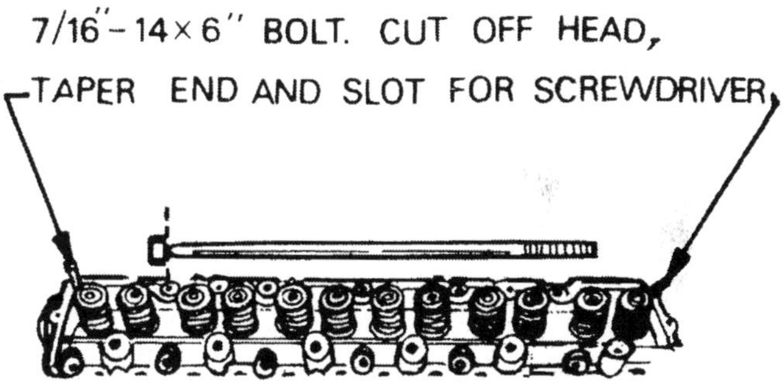

FIG 1 :8 Guide studs for installing the cylinder head

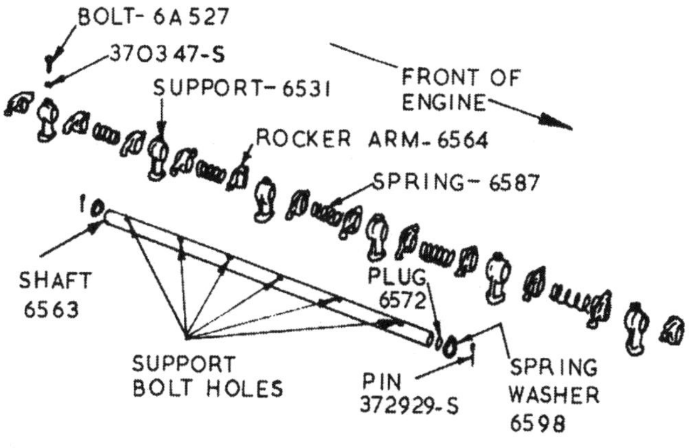

FIG 1 :9 Valve rocker arm shaft assembly

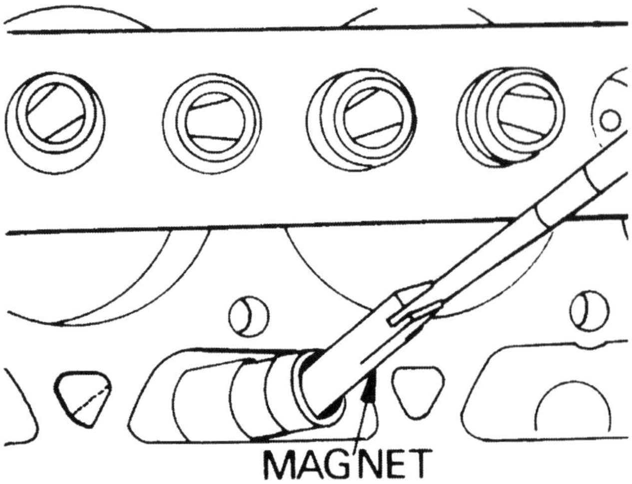

FIG 1:10 Removing the valve lifters

2 Tighten the cylinder head bolts progressively in the sequence shown in **FIG 1:6** to a torque load of 55 lb ft and then tighten them again in sequence to a torque load of 65 lb ft. Finally tighten the bolts, again in sequence to a torque load of 70 to 75 lb ft. The bolts will not require torque loading again but it may be carried out on the composition gaskets if required.
3 Insert all the pushrods into their original positions after lubricating their ends with Lubriplate. Install the rocker assembly and make sure that all the pushrods are correctly located on the rocker arms and in the valve lifters. Tighten the rocker pedestal bolts in order and two turns at a time until they are all at 30 to 35 lb ft. The valve clearances are checked as described in **Section 1:5**.
4 The remainder of the parts are then installed in the reverse order of disassembly. Fill and bleed the cooling system. Check for leaks both before and after starting the engine. Set all adjustments after the engine has reached its normal operating temperature.

1:4 Servicing the cylinder head

Remove the cylinder head as described in the previous section.

Decarbonizing:

Normally decarbonizing will only be required after the vehicle has been used for at least 30,000 miles. When the cylinder head has been removed, a check should be made on the wear in the bores.

Scrape away the carbon from the combustion chambers using a sharpened piece of perspex or hardwood and then polish the chambers using old emerycloth dipped in kerosene. On cast iron components, a rotary wire brush will make the removal of the carbon much easier but such treatment must not be given to light alloy parts. Remove the deposits from the combustion chambers before removing the valves, as the valves will protect the seats in the cylinder head.

Remove the valves as described later and take off the manifolds. Clean out the ports in the cylinder head and manifolds, taking great care not to score or damage the valve seats in the cylinder head. Blow away all loose dust and dirt with compressed air or a tire pump. Clean the valve guides in the cylinder head. Special wire rotary brushes are made for this purpose but if these are not available, pull strips of rag soaked in gasoline through them. Wash all the parts in a suitable solvent to remove all dirt, grease and oil. On models fitted with Thermactor, check the air nozzles and clean them with a suitable brush. Renew any nozzles that are eroded or damaged.

Plug all water- and oil-ways in the top of the block with suitable pieces of rag. **The pieces must be large enough not to be pushed through into the block or forgotten on reassembly.** Smear a little grease around the top of each bore. Crank the engine until a pair of pistons is at TDC. Remove the carbon from the piston crowns with a sharpened piece of hardwood or perspex. Do not use any form of abrasive on the piston crowns as particles cannot be effectively removed. Blow away loose dust and dirt. Crank the engine until the next pair of pistons is at TDC and repeat the procedure. When all the piston crowns have been cleaned, crank the engine for a few rotations and wipe away the grease and dirt that the pistons have swept to the upper portion of the bores. Remove all the pieces of rag from the water- and oil-ways.

The valve rocker assembly:

The components are shown in **FIG 1:9**. The assembly is removed from the cylinder head after evenly and progressively slackening the support bolts (two turns at a time).

To disassemble the parts, remove the cotterpins from either end and slide the parts off. The rear support is locked to the shaft by a bolt. The parts must be stored in the correct order as they are slid off so that they will be refitted into their original positions. Wash the parts in suitable solvent and make sure that all the oil holes are clear. If it is necessary to remove the plugs from the ends of the shaft, drill a hole in one and use a suitable hook to draw it out and drive out the other plug with a suitable piece of rod. New plugs must be driven back into place if the old ones have been removed.

Check the shaft for wear or fret marks and check the fit of the rocker arms on the shaft. Renew parts as required if they are defective or worn. Check the pads on the ends of the rocker arms for wear. Light scoring or wear can be smoothed down with an oilstone but if the wear is deep, new rocker arms must be fitted. Lubricate the parts liberally with fresh oil as they are reassembled and store them so that the assembly does not pick up dust or dirt.

Pushrods:

Remove the pushrods after the rocker assembly has been taken off. Store the pushrods in the correct order for installation.

Wash the pushrods with a suitable solvent and dry them with compressed air. Check the ends for wear, damage or scoring.

Check the pushrods for bends. The best method is to lay each rod onto two V-blocks, one at either end, and rotate the pushrod while measuring the runout with a dial gauge at the center. Renew any pushrods which have damaged ends or where the runout exceeds .025 inch. The pushrods can also be checked by rolling them along a flat surface.

Valve lifters:

These will be more fully dealt with in the next section. The method of removing the valve lifters, after the cylinder head has been taken off, is shown in **FIG 1:10**. Note that the lifters must slide freely in their bores and that they must be stored in the correct order.

Valve removal:

The components of the valve and its spring are shown in **FIG 1:11**. A spring compressor is required to remove or install the valves. Remove the cylinder head and use the tool to compress the valve spring, by squeezing down the retainer and holding the head of the valve up in the combustion chamber. It may be necessary to drive down the sleeve with a suitable piece of tube and hammer. Remove the locks and release the spring compressor slowly. The sleeve, retainer and spring can then be removed. Take off the valve stem oil seal and discard it. Slide the valve out through the combustion chamber. The valves must be stored in the correct order for installation, so use a suitable rack or even a cardboard box with twelve holes punched in it for the valve stems.

Valve springs:

Special rigs are made for testing the valve springs. The data is given in Technical Data and the complete set should be renewed if any springs fall below tolerance.

The springs can also be roughly tested, provided that one new spring is available. Mount the old spring end on to the new one and compress the pair between the jaws of a vice. The spring is defective if it is 10 per cent shorter than the new one, as both have the same pressure applied by the vice.

Before installing a spring, stand it with the closed coils downwards on a flat surface and check spring with a square. Rotate the spring and if the gap, shown in **FIG 1:12**, exceeds $\frac{5}{64}$ inch then the spring should be rejected.

Valves:

The critical dimensions to be checked on each valve are shown in **FIG 1:13**. Use a steel straightedge to check the stem for bends. Renew worn or burnt valves and any that have scored stems.

Remove carbon from the valve using a suitable scraper and worn emerycloth. **Take great care not to damage the seat face or the ground portion of the valve stem.** If the tip of the valve stem is damaged or bruised it may be ground down flat, provided that the minimum dimension is not reached.

Valve guides:

These are machined directly into the cylinder head. Wear can be checked using a special dial gauge but an estimate can also be made by partially inserting a new valve and noting the amount of rock on the head.

Valves with oversized diameter stems are available and these are used when the guides are worn. Before fitting new valves with oversized stems, ream out the guide to the correct diameter using the special piloted reamers made for the purpose. If the wear is such that the next oversize stem cannot be used, the guides must still be reamed out using the reamers in sequence until the correct

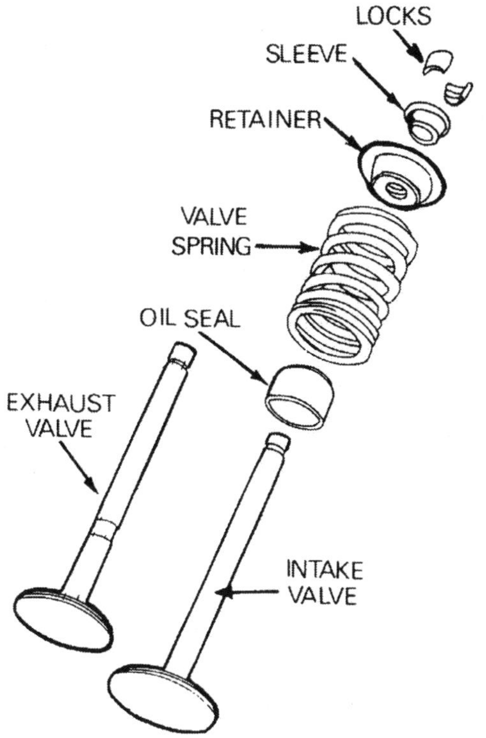

FIG 1:11 A typical valve assembly

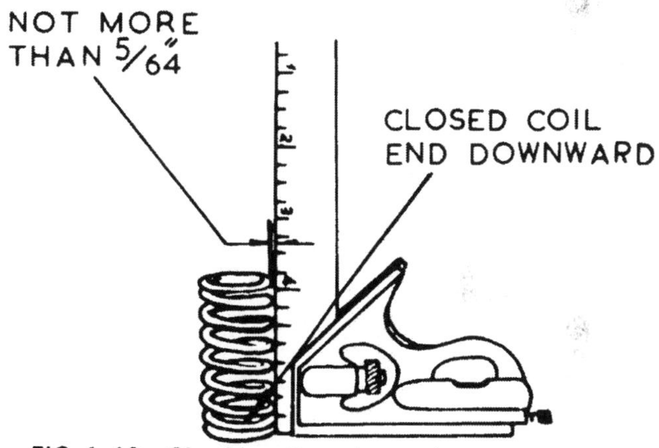

FIG 1:12 Checking the valve spring squareness

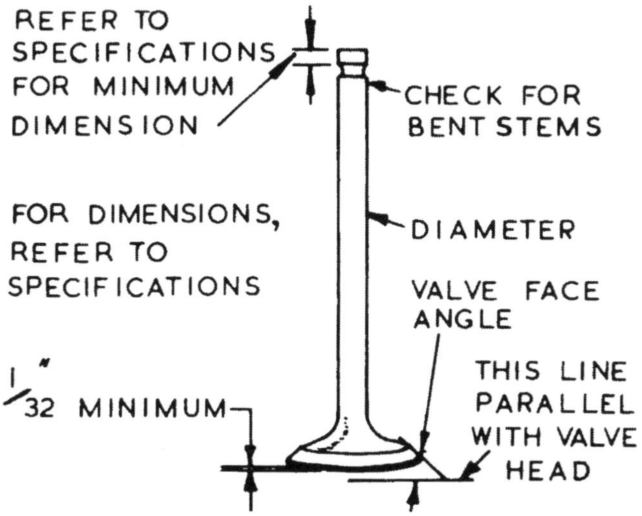

FIG 1:13 Critical valve dimensions

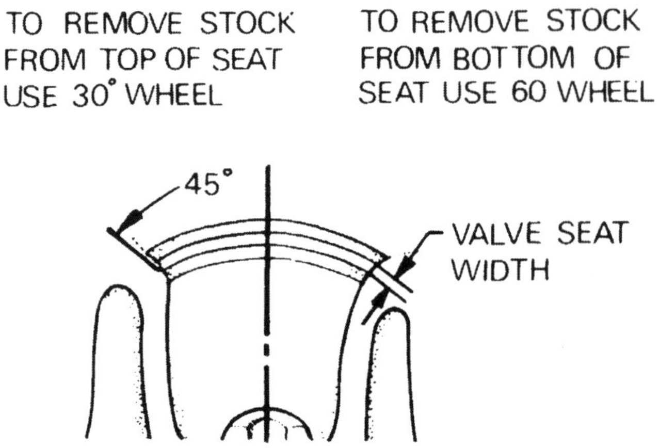

FIG 1:14 **The valve seat width in the cylinder head**

FIG 1:15 **The valve spring assembled height**

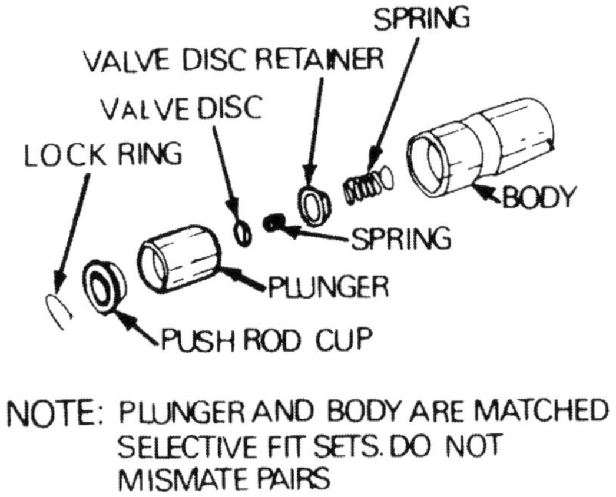

FIG 1:16 **Valve lifter components**

diameter is reached. Do not ream out more than one size at a time. **When the guides have been reamed, the valve seats must be recut to ensure that they are concentric with the guides.**

Valve seats:

Check the efficiency of the seats, with the valve installed and held closed by its spring, by making a dam of plasticene around the seat area. Invert the head and fill the dam with gasoline and leaks will be immediately apparent.

Leaking or pitted seats should be recut using garage equipment. The seats on the valves are cut to a 44 deg. angle while those in the cylinder head are cut to 45 deg.

When the seats in the cylinder head have been recut, they should be checked for width as shown in **FIG 1:14**. On inlet valves the seat width should be .040 to .060 inch and on exhaust valves it should be .070 to .090 inch. Reduce the width with suitable cutters if it is too wide. Use Prussian blue to ensure that the head face meets the valve face centrally, adjusting the position by the use of the cutters.

The use of grinding paste for lapping-in the valves is not recommended.

Assembly and installation:

Clean all the parts in suitable solvent and blow away dust, dirt and chippings with compressed air. Take great care not to damage the gasket surface of the cylinder head.

Install the valves, into their correct positions, in the reverse order of disassembly and using new stem seals. Lubricate the stems liberally with engine oil before inserting them into the valve guides.

When the valves are in place, tap the ends of the stems lightly to ensure that the locks are fully seated. Use dividers to measure the installed height of the valve spring, as shown in **FIG 1:15,** and transfer the measurement to a steel ruler for checking the actual dimension. If the height is not within limits, install .030 inch thick spacers between the cylinder head and spring to bring the height within limits. **Spacers must not be installed unnecessarily** otherwise the valve spring will be overstressed, leading to failure or excessive wear in the valve train.

Lubricate the valve lifters with clean engine oil and install them back into their correct positions. Install the cylinder head followed by the pushrods. Note that both ends of the pushrods should be smeared with Lubriplate and they must seat fully into the valve lifters. Smear the tips of the valve stems with Lubriplate and install the rocker gear assembly. Check the valve lash as described in the next section. When installing the rocker cover using a new gasket, tighten the attachment bolts to the correct torque and then wait two minutes before tightening them again to the correct torque.

1:5 The hydraulic valve lifters

These can be removed from the engine after removing the cylinder head as shown in **FIG 1:10**. Store the lifters in a rack in the correct order for installation.

The components of a valve lifter are shown in **FIG 1:16** and the operation in **FIG 1:17**. When the valve is closed the lifter is on the base circle of the cam and the plunger

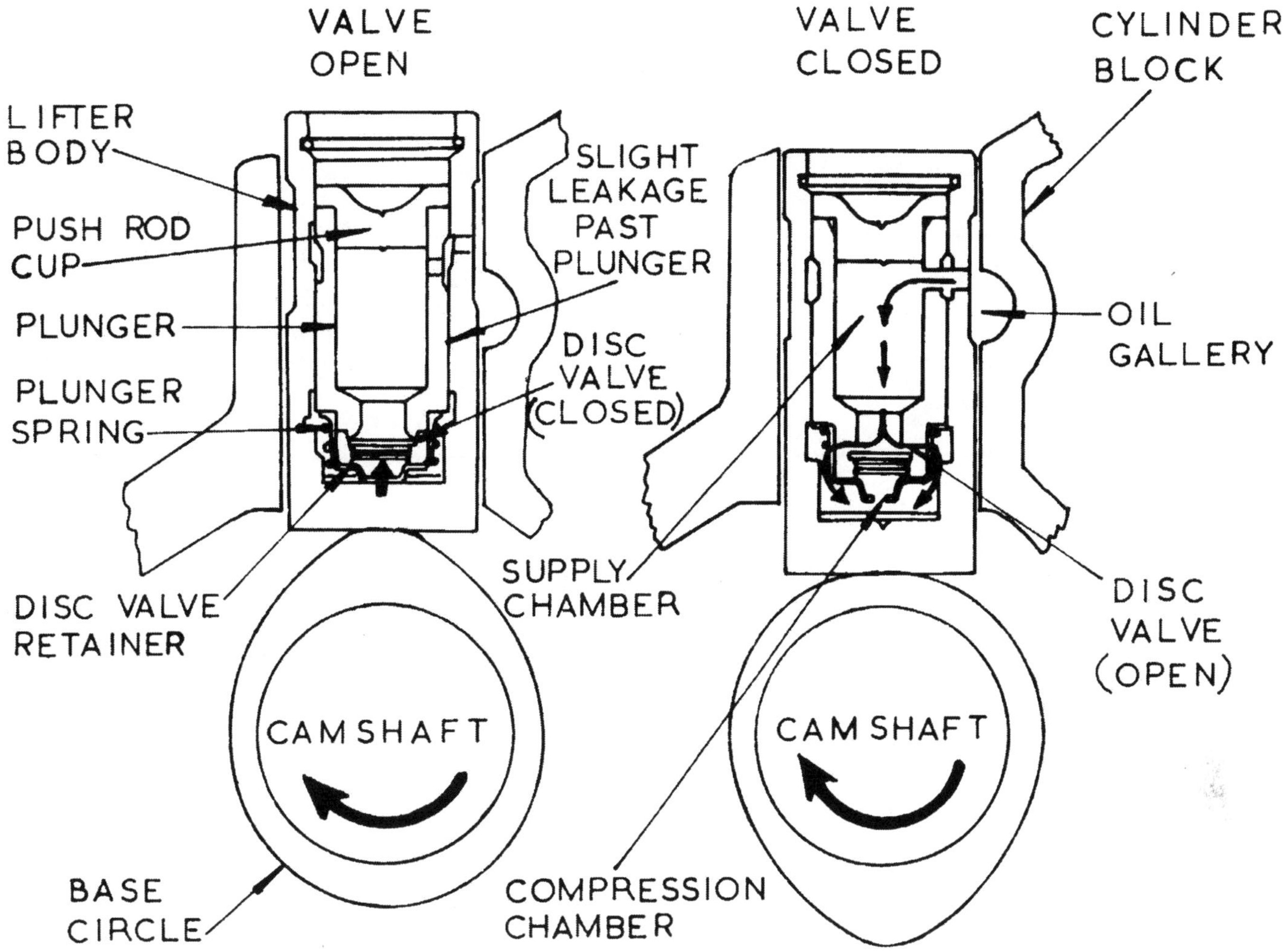

FIG 1:17 Valve lifter operation

spring moves the plunger upwards to take up all lash in the train. When the plunger moves upwards there is a reduced pressure in the compression chamber and the check valve opens allowing oil from the gallery to fill the compression chamber. As the cam rotates and raises the lifter, the pressure in the compression chamber rises and the check valve closes. The oil being non-compressible makes the lifter act as a hydraulic ram transferring the lift directly to the pushrod without any lash. A slight amount of leakage is allowed as this ensures that the lifter will not overcompensate and cause the valve to be held open. The leakage is replenished on the next stroke.

Noisy operation:

When the engine has stood stationary for some time, such as overnight, the oil will drain down from the lifters—particularly those which are holding valves open. On start-up there will be excessive lash and the lifters will sound noisy. As the compression chambers fill, the noise will die away. This noisy operation on start-up is perfectly normal and no action need be taken, unless the noise does not die away after a few minutes of running.

Intermittent clickings are mainly caused by incorrect initial adjustment or a piece of dirt momentarily holding the check valve open.

General noise through the valve train is caused by incorrect initial settings or a lack of oil to the valve lifters. If the crankcase is overfilled, the connecting rods thrashing in the oil will make foam, and this foam instead of oil may be drawn into the lifters. As air is compressible, all the lash will not be taken up and the valve train will sound noisy.

Moderate rapping is caused by a worn valve lifter or incorrect initial adjustment. Heavy rapping noises are caused by the plunger sticking in the body of the lifter.

If the operation is noisy and continues to be noisy, first check the oil level in the crankcase and top up (or drain) until it is at the correct level. If the level was correct on filling but it is found to rise with use, and the coolant level drops, there is an internal leak allowing coolant into the crankcase. If the oil level is correct, check the initial adjustment of the valve train. If the adjustments are satisfactory it will be necessary to remove the cylinder head and withdraw the lifters for cleaning and inspection. Use a magnet to extract them from the cylinder block. If they are stuck it may be necessary to use a claw type tool to rotate the lifter back and forth until any gummy deposits are removed.

Disassembly:

Refer to **FIG 1:16**. If any part of a valve lifter is defective the complete assembly must be renewed as the parts are matched on manufacture. To ensure that parts are not accidentally interchanged, work on one lifter at a time

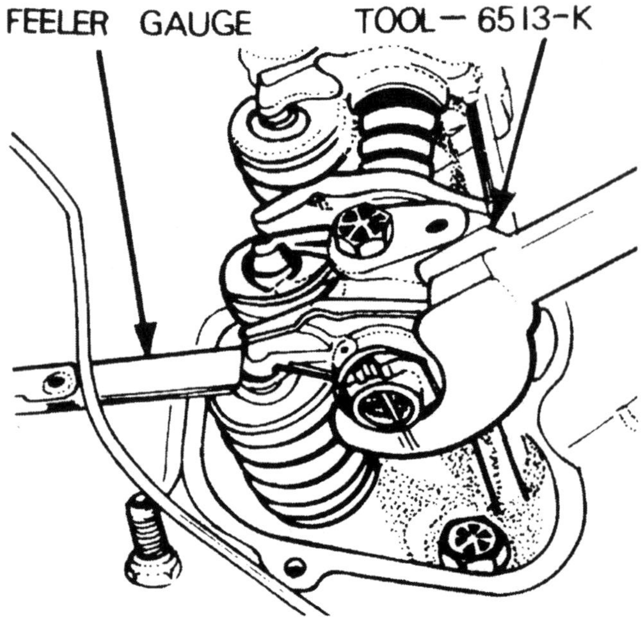

FIG 1:18 Checking the valve clearance

and complete it before starting the next. Discard any lifters that have scored or chipped bodies.

Remove the lock ring using a pair of needle-nosed pliers. It may be necessary to partially depress the plunger when removing the lock ring. Remove the pushrod cup, plunger and spring. Invert the plunger and carefully remove the check valve retainer by prying it up with a screwdriver. Shake out the check valve disc and its spring.

Wash all the parts in a suitable harsh solvent, such as trichlorethylene, to remove oil and gum. Check the parts for wear, scoring or distortion and make sure that the plunger slides freely in the body. **Renew the lifter complete if defects are found.**

Brush down the bench and lay clean paper on it to prevent the parts from picking up dirt. Invert the plunger on the bench. Place the valve disc in position over the oil hole in the plunger and position the spring over it. Insert the check valve retainer into the plunger and press it down into position. Place the plunger spring followed by the plunger assembly into the body and install the pushrod cup. Secure the parts in place using the lock ring.

If the lifter is to be installed directly into the engine, immerse it in a bath of engine oil and pump the plunger up and down to fill the unit with oil. With the compression chamber is full, extremely heavy pressure will be required to move the plunger.

If the lifter is suspect and requires testing, it should be taken to a service garage without filling with oil. The leak down rate is checked on a special test machine after the unit has been filled with special test fluid. **Any engine oil in the lifter would negate the test.** On test, the machine applies a constant load to the plunger and the time is noted for the plunger to move through its full range of movement.

Adjustment:

Correctly install the valve lifters, cylinder head, push-rods and valve rocker gear. Crank the engine until it is at TDC after the compression stroke on No. 1 (front cylinder). The timing marks on the cylinder front cover and crankshaft damper will give the TDC position for Nos. 1 and 6 cylinders. The compression stroke can be found by removing the spark plug and blocking the hole with a thumb as the engine is cranked. A rise in pressure in the cylinder will indicate that the piston is rising on the compression stroke. The correct position can also be found by removing the distributor cap and cranking the engine until the timing marks are aligned and the brass end of the rotor arm is pointing to the position normally occupied by the electrode in the cap connected to the correct cylinder sparking plug. With the engine set at TDC after the compression stroke on No. 1 (front cylinder) check the following valves:

No. 1 Intake
No. 1 Exhaust
No. 2 Intake
No. 3 Exhaust
No. 4 Intake
No. 5 Exhaust

When these valves have been checked, crank the engine through one complete turn until it is at TDC after the compression stroke on No. 6 (rear) cylinder, and check the following valves:

No. 2 Exhaust
No. 3 Intake
No. 4 Exhaust
No. 5 Intake
No. 6 Intake
No. 6 Exhaust

The method of checking the lash is shown in **FIG 1:18**. Use the tool shown to fully collapse the valve lifter and measure the gap between the rocker arm and tip of the valve stem with feeler gauges. On the 200 engine the limits are .079 to .209 inch and the desirable setting is .144 inch. On the 250 engines the limits should be between .095 and .195, with a desirable setting of .145.

If the clearance is incorrect the pushrod will have to be changed. Pushrods .060 inch longer than standard are available to reduce excessive clearance and pushrods .060 inch shorter than standard are available to increase the clearance if it is too small. The rocker assembly does not need to be removed in order to change the pushrod. **Turn the engine past TDC so that there is no danger of the valve hitting the piston crown** and use the special tool to compress the valve spring so that the old pushrod can be removed and a new one of the correct length can be fitted. Make sure that both ends of the pushrod are correctly seated and lubricated with Lubriplate.

1:6 The oil filter

The engine oil and filter element should be renewed at intervals not exceeding 6000 miles. If the vehicle is used in dusty or adverse conditions, used for short-mileage journeys only or kept at curb idle for long periods the interval between oil changes should be reduced.

The oil does not lose its lubrication properties but it does collect sludge from the engine and combustion products that pass by the piston rings. The filter element will remove the larger particles from the oil and therefore help to keep it clean. If the engine never reaches its full operating temperature, the combustion products will rapidly build up in the oil and cause wear and errosion of the metal parts. When the oil is at its full normal operating temperature for long periods, many of the combustion products are boiled out of the oil and taken away by the air ventilating the crankcase. It is for this reason that cars used under adverse conditions must have their oil changed at shorter intervals.

Before draining the oil from the crankcase take the vehicle for a short run to warm the oil and stir dirt up into suspension. **If the oil is drained after a long run, take great care as the oil will be hot enough to cause scalding**. Stop the engine and slide a container under the oil pan. Remove the drain plug from under the oil pan and allow the old oil to drain out into the container. Leave the oil to drain while changing the oil filter, but do not forget to fit the drain plug before filling the engine with fresh oil.

Slide the drip tray or container under the oil filter, on the lefthand side of the crankcase, so that any spillage will be caught. Unscrew the complete cartridge from the engine and discard it. Clean the area where the cartridge fits. Check that the gasket on the new cartridge is in good condition and smear it lightly with a little engine oil. Screw the new cartridge back into place until its gasket just touches the crankcase. Hand tighten the cartridge by a further half turn.

Install the drain plug in the oil pan. Fill the engine with fresh oil through the filler cap on the rocker cover. Approximately 4½ quarts of oil will be required to fill the crankcase and oil filter, allowing for the quart held by the oil filter. The level may be allowed to go over the top mark on the dipstick in this case. Stand the car on level ground and start the engine. Check for oil leaks and then switch off the engine. Leave the vehicle standing for at least five minutes, to allow the oil to run down inside the engine, and again check the level on the dipstick. Top up as required but **do not overfill the engine.**

The correct grades of oil to be used in various ambient temperatures are given in the following table:

Multi-viscosity oils:

When outside temperature is consistently	*Use SAE Viscosity Number*
Below +32°F	5W-30
—10°F to +90°F	10W-30, 10W-40
Above +32°F	20W-40

Single viscosity oils:

When outside temperature is consistently	*Use SAE Viscosity Number*
—10°F to +10°F	10W
+10°F to +32°F	20-20W
+32°F to +90°F	30
Above 90°F	40

It should be noted that an oil strainer is fitted to the inlet to the oil pump. If regular oil changes are carried out the strainer should not block. Whenever the oil pan is removed, for whatever reason, the oil pump strainer should be removed, cleaned and inspected before installing it using a new gasket.

1:7 The oil pan and oil pump

Both the oil pan and oil pump can be removed with the engine installed in the vehicle. Before removing the oil pan drain the oil from the crankcase by removing the drain plug, using a $\frac{7}{8}$ inch socket wrench.

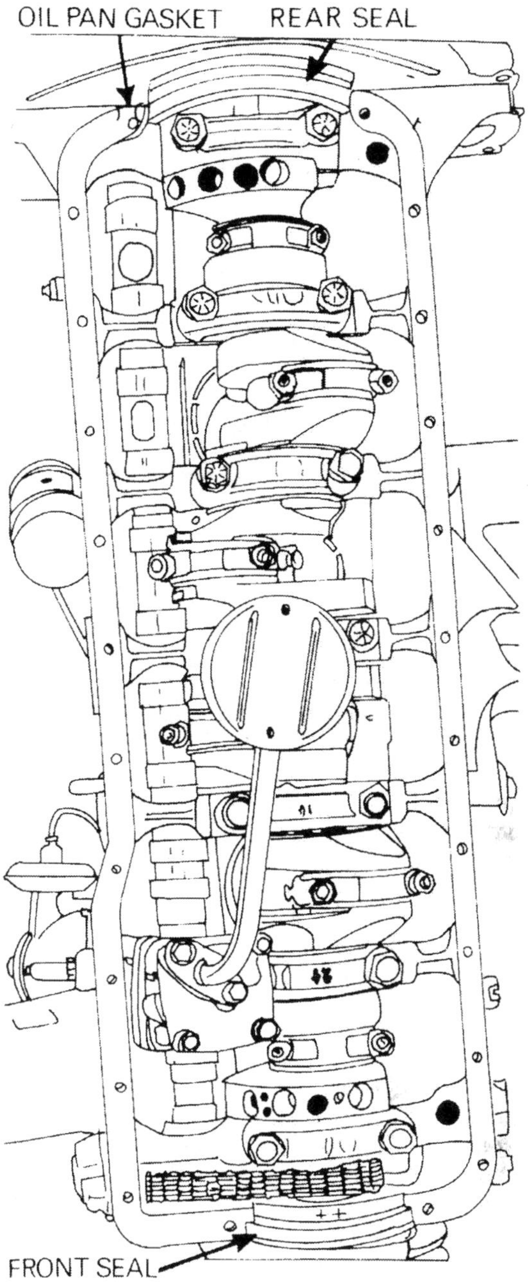

FIG 1:19 Oil pan seals and gaskets

Oil pan removal:

With the oil drained from the crankcase, remove the flywheel housing inspection cover. Remove the dipstick. Disconnect the stabilizer bar and swing it down to clear the oil pan. Remove one bolt and loosen the other on the No. 2 crossmember and lower it out of the way. Take out the screws securing the oil pan and lower it from the engine. Remove the old gaskets and seals. Remove the oil pump inlet strainer.

Oil pan installation:

Clean the oil pump inlet strainer and install it using a new gasket. The oil pan seals and gaskets are shown in **FIG 1:19**. Remove all traces of old seal, gasket, or sealing compound from the mounting faces, making sure that the grooves for the oil seals are also cleaned.

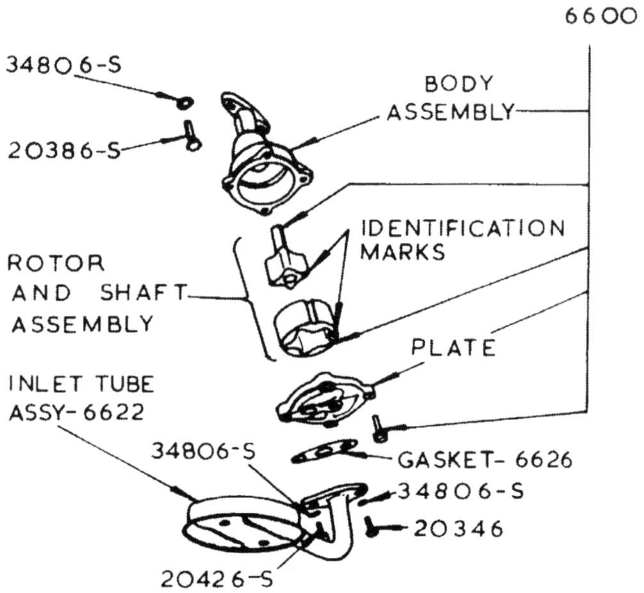

FIG 1:20 Oil pump components

Smear the gasket surfaces of the block and pan with oil-resistant sealer. Install the two-piece gasket set into place on the cylinder block. Install the front oil seal to the cylinder front cover and the rear seal to the rear main bearing cap, making sure that the tabs on the seals are over the oil pan gaskets.

Raise the oil pan into position and secure it with one screw on either side. Fit the remainder of the screws without disturbing the positions of the gaskets and then tighten all the screws to a torque of 7 to 9 lb ft, working from the center outwards.

Reposition the No. 2 crossmember and tighten the bolts to a torque of 45 to 60 lb ft. Refit the stabilizer bar and tighten the nuts and bolt holding the brackets to the appropriate torque: 17 to 25 lb ft for models up to 1970 inclusive, or 6 to 12 lb ft for later models.

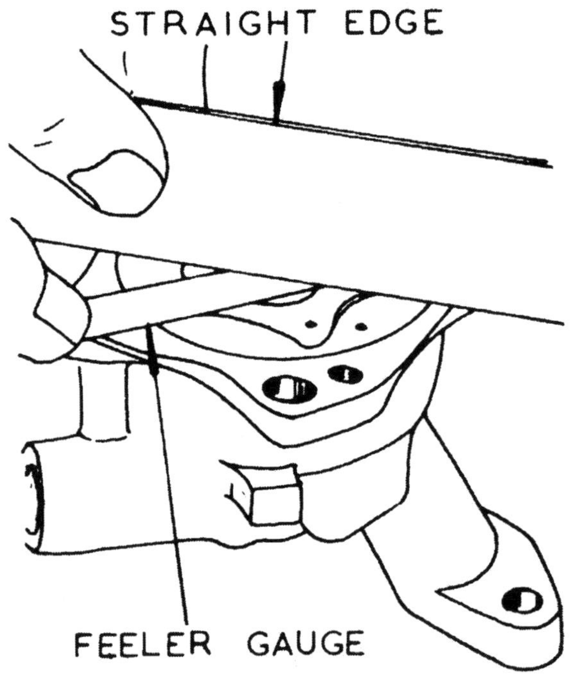

FIG 1:21 Checking the end play of the oil pump

Install the drain plug and tighten it to a torque of 15 to 20 lb ft. Fill the engine with fresh oil through the filler cap. **Do not overfill the engine** as the oil filter will be full and will not take any surplus oil. Start the engine and check for leaks. Top up to the correct level after the engine has stood stationary for at least five minutes.

Oil pump removal:

Remove the oil pan as described earlier. The components of a typical oil pump are shown in **FIG 1:20**. Take out the pump attachment bolts and withdraw the pump complete with the intermediate drive shaft.

Oil pump servicing:

Remove the inlet tube assembly and discard the gasket. Take the coverplate securing screws and remove the plate. Slide out the inner rotor and shaft assembly followed by the outer rotor.

Insert a self-threading screw into the cap for the oil relief valve and use a pair of grips on the screw to draw out the cap. The spring and plunger can then be removed from the body of the pump.

Wash all the parts in suitable solvent and examine them for obvious wear or scoring. If the rotors are worn or scored they must be renewed as a set as they are mated on manufacture.

Install the rotors back into the body assembly, making sure that the identification marks are outwards. Insert feeler gauges between the outer rotor side and the body assembly to measure the clearance. If the clearance is outside the limits of .006 to .013 inch, new rotors must be fitted and if the clearance is still not within limits the complete pump must be renewed.

Lay a straightedge across the pump, as shown in **FIG 1:21**, and use feeler gauges to measure the end play of the rotors. If the end play is outside the limits of .0011 to .0041 inch, try fitting new rotors and if this does not cure the end play then a new pump must be fitted.

Check the end plate for scores and install a new plate if the old one is worn or scored.

Examine the relief valve plunger and bore for fretting or wear. Check the spring and renew it if it is weak. If the spring is satisfactory it will take a load between 9.0 to 10.1 lbs to compress it to a length of 1.078 inches.

Assemble the pump in the reverse order of disassembly. Inject engine oil through the inlet or outlet port and rotate the pump to spread the oil evenly. Check that the pump rotates freely.

Oil pump installation:

Position the intermediate drive shaft in the distributor socket and place a new gasket onto the oil pump body. Fit the pump back into place, without trying to force it. If the pump sticks, turn the intermediate shaft to a new position to allow the hex to align. Secure the pump in place with the attachment bolts and fit the inlet tube assembly to the pump, using a new gasket. Install the oil pan and fill the crankcase to the correct level with fresh oil. If possible, crank the engine over on the starter motor without allowing it to start for a short while to make sure that the oil pump and oil system are fully primed. The engine can be prevented from starting by disconnecting the ignition low-tension circuit or by pumping the accelerator pedal 8 to 10 times to flood the engine.

1 :8 The clutch and flywheel

These are only accessible after the transmission has been removed from the vehicle (see **Chapter 6**) or the engine has been taken out.

This section deals only with the removal and installation of the clutch. All details of clutch maintenance or servicing are dealt with in **Chapter 5**.

Clutch removal:

Remove the transmission and flywheel housing as described in **Chapter 6**. The attachments of the clutch are shown in **FIG 1 :22**. The clutch arbor shown in the figure is not an integral part of the clutch and is used for aligning the clutch disc when installing the clutch.

If the clutch is to be used again, make aligning marks on the cover and flywheel to ensure that the cover assembly will be installed back into its original position, and the balance not lost. **When handling the clutch components, particularly the disc, avoid using dirty or greasy hands and prevent oil or grease from coming into contact with the disc friction linings.**

Evenly and in a diagonal sequence slacken the six bolts that secure the cover and when they are all free remove them, while holding the cover in place. Remove the cover and collect the clutch disc without allowing it to drop.

Install the clutch in the reverse order of removal. Fit the arbor through the disc and into the pilot service bearing in the crankshaft so that the disc is held centrally in place. If the disc is not centralized it will be impossible to fit the transmission input shaft back through the clutch. Install the cover so that the previously made marks are again in alignment. Secure the cover by progressively tightening the attachment bolts to a torque of 12 to 20 lb ft. The bolts must be progressively tightened in a diagonal sequence to prevent the pressure of the clutch springs from distorting the cover.

Clutch pilot bushing:

Whenever the clutch is removed, the bushing should be examined and measured for wear or fretting. If the bushing is worn, extract it using the tools shown in **FIG 1 :23**. If the special tools are not available, the bushing can still be removed provided that a snug fitting piece of rod can be found or turned. Pack the recess in the crankshaft with grease so that there are no air pockets. Insert the rod into the bushing and drive it in with a hammer. The hydraulic action of the grease will force the old bushing out. Drive the new bushing back into place, preferably with the special drift No. T65L-7600-A.

The flywheel:

This can be removed, after the clutch has been taken off, by undoing the bolts that secure it to the crankshaft and then drawing it back off the crankshaft, taking care not to drop it as it comes free.

Before installing the flywheel, clean the mating surfaces of the flywheel and crankshaft and make sure that there are no specks of dirt which could cause excessive run-out. Install the flywheel and progressively tighten the attachment bolts to a torque of 75 to 85 lb ft.

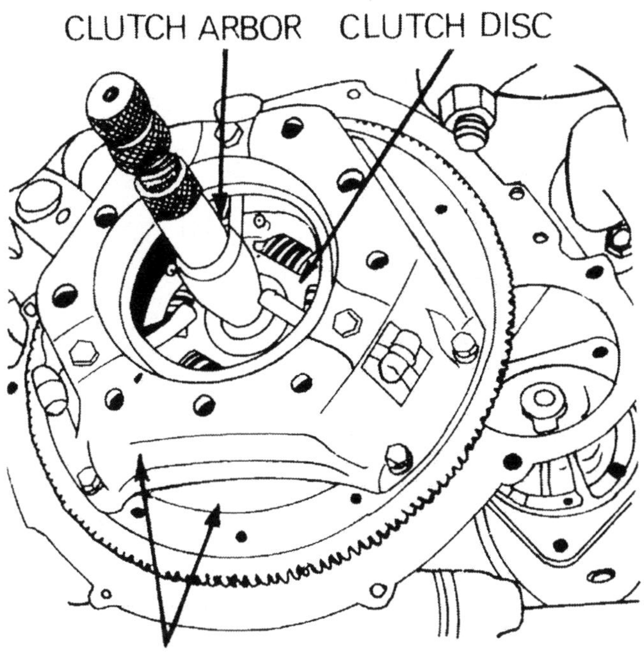

FIG 1 :22 Clutch attachments, showing the arbor in place

Use a dial gauge mounted on the crankcase to check the run-out of the flywheel friction surface and the periphery of the flywheel.

If the flywheel friction face is scored or worn, up to .045 inch of stock may be removed from the friction surface in order to remove the wear or damage, or to cure excessive run-out. If the surface cannot be restored by machining to the maximum depth a new flywheel must be fitted.

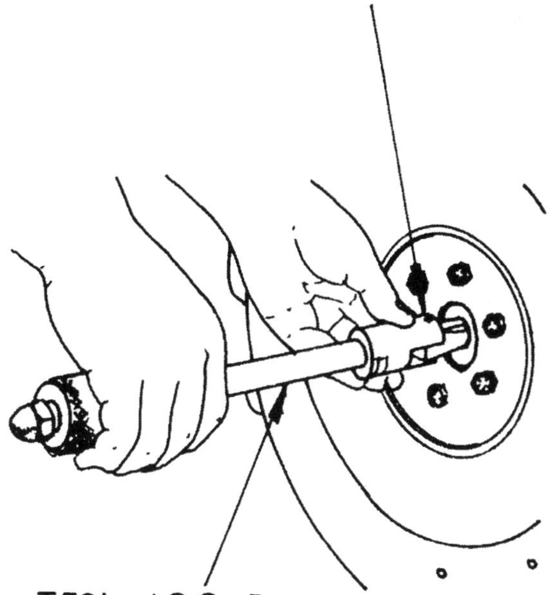

FIG 1 :23 Removing the clutch pilot bushing

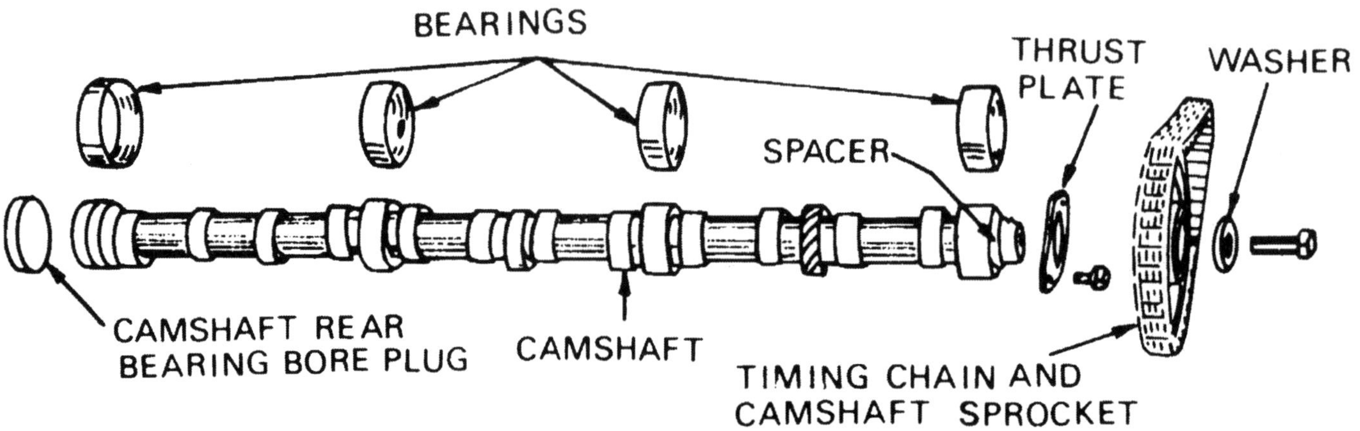

FIG 1:24 Camshaft and related parts

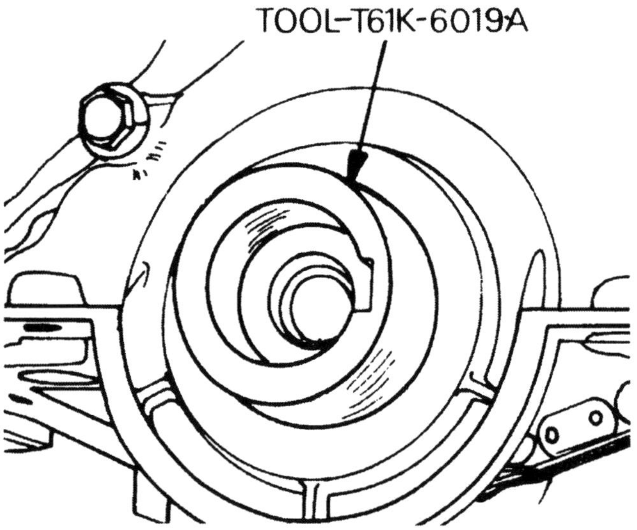

FIG 1:25 Aligning the cylinder front cover

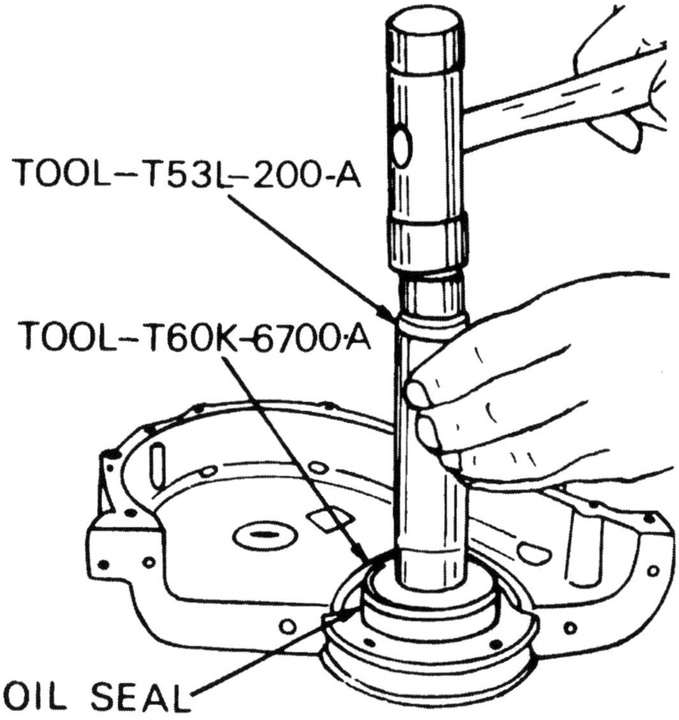

FIG 1:26 Installing a crankshaft front oil seal

Flywheel ring gear:

If the teeth on the ring are worn or damaged a new ring should be fitted. To remove the old ring, partially drill through at the root of a tooth and then split the ring at the weak spot using a cold chisel. **Take great care not to damage or mark the flywheel itself.**

Thoroughly clean the periphery of the flywheel, using a wire brush to remove dirt and scale and using emery-cloth to remove burrs and high-spots.

Heat the new ring evenly to a maximum temperature of 500°F. **If this temperature is exceeded the temper of the metal will be destroyed**. When the ring is hot enough and has expanded, use tongs to lay it into place on the flywheel. Tap the ring fully home using drifts made of soft-metal. Leave the flywheel and ring to cool naturally to room temperature before moving the flywheel.

1:9 The valve timing gear and camshaft

All the parts, except the camshaft bearings, can be removed and installed with the engine still fitted to the vehicle.

A typical camshaft and its associated parts are shown in **FIG 1:24**.

Cylinder front cover:

1 Drain the crankcase and cooling system. Disconnect the battery. Remove the radiator (see **Chapter 4**) noting that on models fitted with automatic transmission it will be necessary to disconnect the oil cooler lines. Remove all drive belts, the cooling fan and water pump pulley.

2 On models fitted with air-conditioning, free the compressor from the engine but do not disconnect the refrigerant lines. On models fitted with power-steering, free the pump from the engine and hang it safely out of the way so that fluid does not spill from the reservoir.

3 Remove the bolts that secure the front cover, not forgetting the bolts that secure the oil pan to the front cover. On the 250 engines also remove the oil pan. Remove the crankshaft damper and pulley, preferably using the special tool T58P-6316-B.

4 Pry the front cover slightly away from the top of the block and use a thin sharp knife to cut the oil pan gasket flush with the front face of the cylinder block. Remove the front cover.

After carrying out servicing as required, the front cover is installed in the reverse order of removal. It is most advisable to renew the oil seal in the cover. Clean all the parts and remove old gasket or sealer. Coat the surface on the block with sealer. Cut a new gasket to the required shape and attach it to the block using the sealer. Install pan seal as required. On a 200 engine smear the gasket surfaces of the cover with sealer and install it back into position, taking care not to disturb the gasket. On the 250 engine, if the timing chain rubber has been removed, refit it. Install a pilot tool on the crankshaft, as shown in **FIG 1:25**, to centralize the cover about the crankshaft. Downward pressure may have to be applied to the cover in order to align it correctly. No adjustment is necessary to the chain snubber on the 250 engine as this is automatically taken up during the front cover alignment. Secure the cover in place by tightening the attachment bolts to a torque of 7 to 9 lb ft. Remove the pilot and install the damper, preferably using the tool T52L-6306-AEE, after lubricating the hub with Lupriplate to prevent scuffing.

The remainder of the parts are then installed in the reverse order of disassembly. Set all drive belt tensions correctly. Fill the cooling system. **If coolant has run down into the engine when removing the front cover, it will be necessary to drain the crankcase and refill it with fresh oil** ($3\frac{1}{2}$ quarts only as the oil filter need not be changed).

Front oil seal:

When the front cylinder cover has been removed, drive out the old oil seal with a suitable pin punch, taking care not to damage the cover itself. Coat the new seal with grease and drive it in, as shown in **FIG 1:26**, until it is fully in its recess. Check that the spring is correctly located and the lips face into the engine.

Timing chain and sprockets:

Remove the cylinder front cover as described earlier. Crank the engine until the timing marks, shown in **FIG 1:27**, are in alignment. Turn the engine until the lefthand side (righthand side when viewed from in front of the car) of the chain is taut and without slack. The method of checking the chain total deflection is shown in **FIG 1:28**. Note that this figure shows the parts on a V8 engine but the method is still the same. If the maximum deflection of the chain (A-B) exceeds .5 inch, the timing chain and sprockets are worn and must be renewed as a set.

Remove the bolt that secures the camshaft sprocket to the camshaft. Grip both sprockets, one in either hand, and slide them off together so that the timing chain is kept in rough alignment. Once both sprockets are free the timing chain can be removed from around them.

Install the parts in the reverse order of disassembly and removal. Make sure that the timing marks on the sprockets are again in alignment after the parts are installed.

It is essential that neither the camshaft or the crankshaft is rotated while the sprockets and timing chain are off. If one is rotated there is a danger that the valves will hit the crowns of the pistons, causing internal damage.

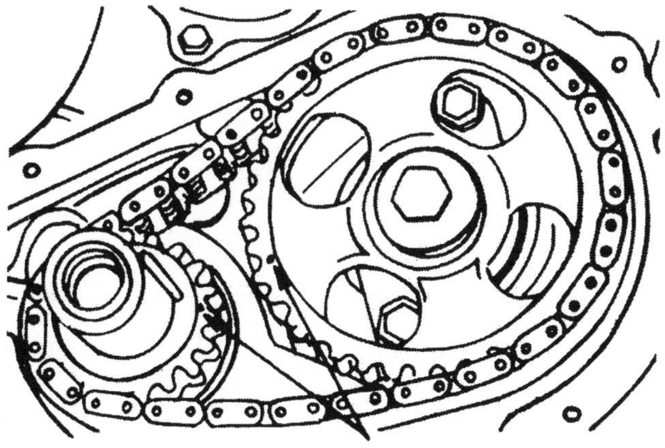

FIG 1:27 The valve timing marks in correct alignment

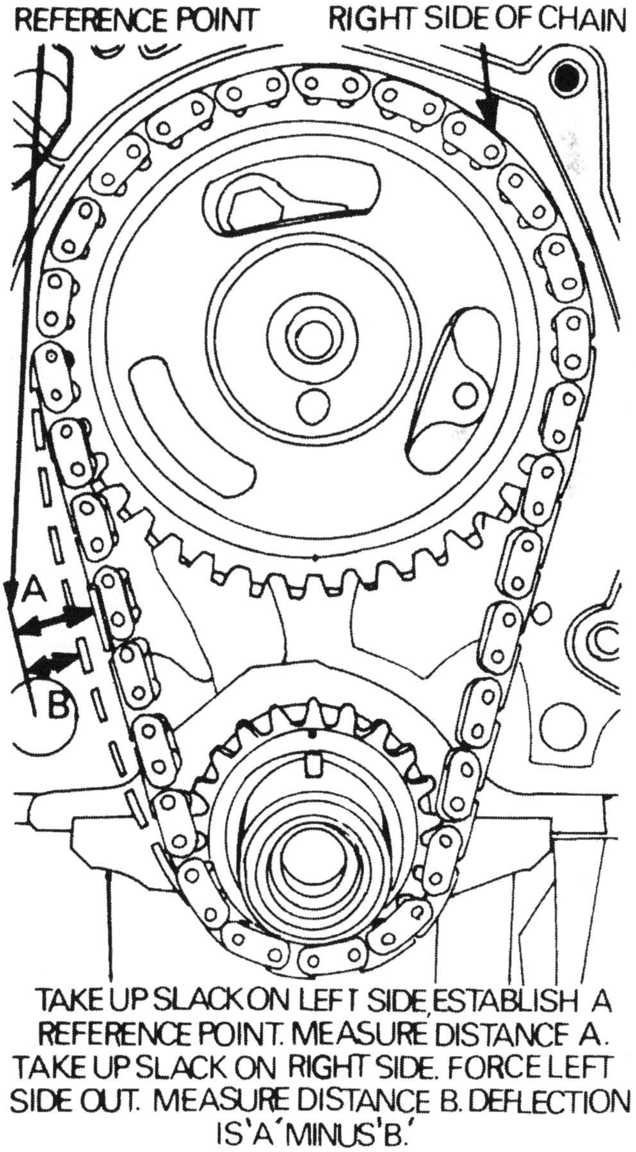

FIG 1:28 Checking the timing chain deflection

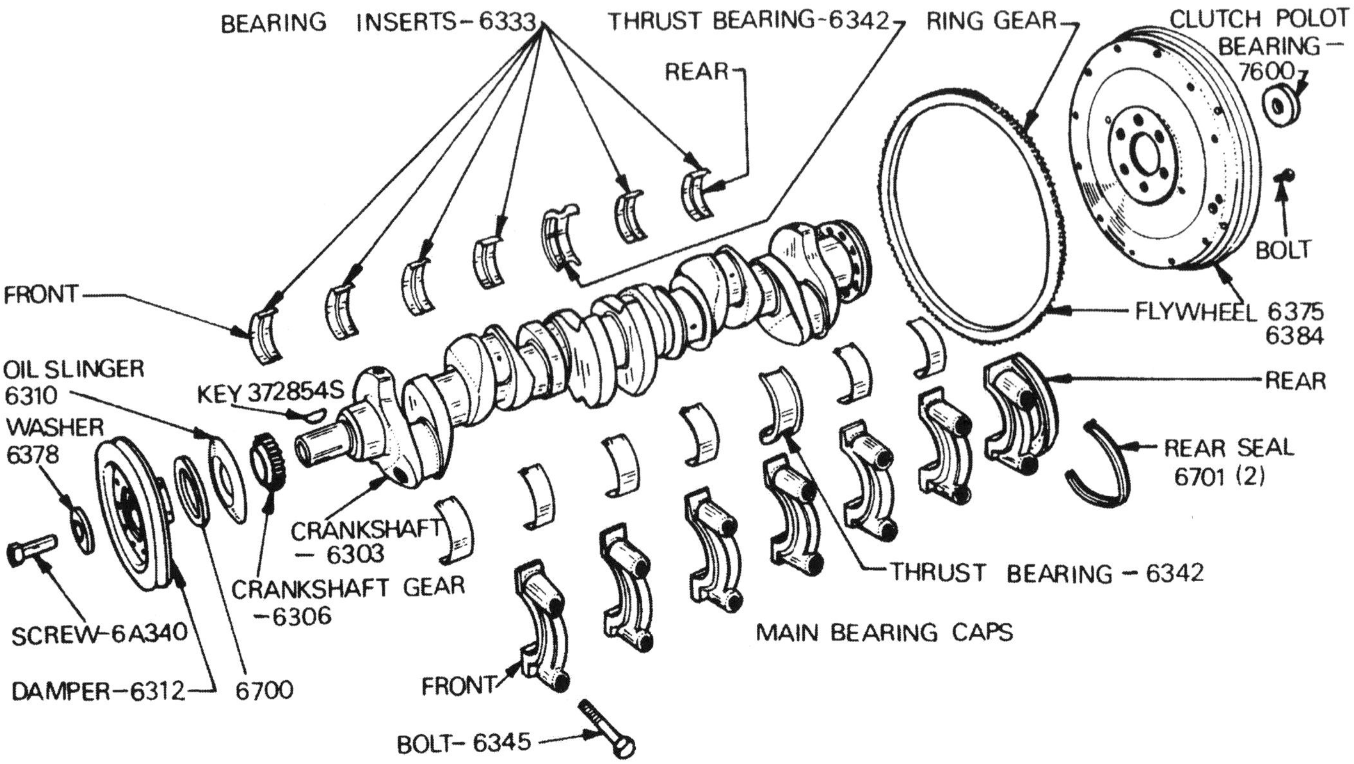

FIG 1 :29 Crankshaft and related parts

The camshaft:

Remove the front cylinder cover and sprockets with timing chain as described previously. Remove the cylinder head as described in **Section 1 :3** and then lift out the valve lifters as shown in **FIG 1 :10**. Remove the oil pan and oil pump as described in **Section 1 :7**. Remove the distributor and the fuel pump.

The camshaft is now ready for removal. Before removing the camshaft, install its sprocket correctly, omitting the timing chain and crankshaft sprocket. Mount a dial gauge onto the cylinder block so that its stylus rests vertically on the end of the sprocket cap bolt. Lever the camshaft firmly in and out so that the end play of the camshaft can be measured. The end play should lie between .001 and .007 inch (wear limit .012 inch). If the end play is excessive, a new thrust plate must be fitted.

Remove the air intake grille, grille support bracket and stone shields from the front of the vehicle to allow the camshaft to be drawn out. Take off the camshaft sprocket and remove the bolts that secure the camshaft thrust plate. Withdraw the camshaft, taking great care not to allow the journals or cams to damage the four bearings.

Install the camshaft in the reverse order of removal. The cam lobes should be lubricated with Lubriplate. It should be noted that slight pitting on the tips of the lobes is normal wear and no action need be taken unless the loss of lift exceeds .005 inch. The remainder of the parts are also installed in the reverse order of disassembly and removal.

Camshaft bearings:

The engine must be removed from the vehicle in order to renew the bearings.

Before fully dismantling the engine, remove the rocker cover and rocker assembly but leave the pushrods in place. Mount a dial gauge on the cylinder head so that it can be used to measure the movement of each pushrod. Slowly crank the engine and note the pushrod lift. If the loss of lift from standard exceeds .005 inch, parts will have to be renewed to cure the fault. The standard lift on all valves should be .232 inch.

The camshaft must be removed and the clutch and flywheel taken off (to gain access to the rear bearing bore plug) before the bearings can be renewed. The work should be left to a service garage as if the special tools are not used there is every likelihood of damaging the new bearings when drawing them back into place.

1 :10 The crankshaft and bearings

All the crankshaft bearings, including the main bearings, can be renewed without removing the engine from the vehicle. If on removing the bearings it is found that the journal or crankpin is excessively worn or scored, the engine must be removed from the vehicle so that the crankshaft can be taken out.

The crankshaft is shown in **FIG 1 :29.**

Bearing inserts:

These are renewable and steel-backed with bearing material lining them. They are fitted exactly as received, apart from cleaning off protective, and do not require boring or scraping to make them fit.

If a crankpin, journal or insert is worn, under no circumstances file the connecting rod or bearing cap in an attempt to reduce the clearance. Similarly never fit shims under the cap to increase the clearance.

Undersize bearing inserts are obtainable to fit a crankshaft that has had its journals or crankpins ground down and these are supplied as a standard fitting for each grade of undersize so that no selective fitting is possible.

Bearings for standard diameter crankpins or journals are supplied in three sizes: standard, —.001 inch and —.002 inch. This allows the inserts to be selectively fitted to give the correct bearing diametrical clearance. The tolerance for clearance on main bearings is .0005 to .0022 inch and the desired setting should lie between .0005 to .0015 inch. The equivalent figures for the connecting rod bearings are .0008 to .0024 allowable and .0010 to .0015 desirable. If .001 or .002 inch undersize bearings have to be used to bring the clearance within tolerances, make sure that all the undersize bearing inserts are fitted on the cylinder block side.

The clearance of a bearing is measured using Plastigage. Remove the cap and make sure that cap, bearing and pin are all scrupulously clean. Lay a strip of Plastigage into the cap, as shown in **FIG 1:30**. Install the cap and tighten its attachments to the correct torque load. **Do not turn the crankshaft while the Plastigage is in place.** Remove the cap and use the gauge to measure the bearing clearance. Measure at the widest and narrowest points so that any taper can be checked.

The oil pan must be removed to gain access to the bearings. Work will be easiest if the vehicle is positioned over a pit. If the car has to be raised to gain access room, make sure that all supports are firmly based and of sufficient strength.

Connecting rod bearings:

Remove the oil pan. Crank the engine to a convenient position for working on the connecting rod bearings. Undo the nuts that secure the bearing caps to the connecting rods. Pull off the bearing caps and remove the inserts from them. Store the bearing caps and inserts in the correct order for installation. The caps and connecting rods are normally marked with numerals to indicate position and order. If these numerals are not present, mark the caps and connecting rods with a code of light dot marks, carefully noting to which side the marks face.

Press the connecting rods up the bores so that they clear the crankpins and remove the bearing inserts from the connecting rods. If the bearing inserts are to be used again, they must be stored in order so that there is no danger of interchanging them on assembly. If the crankshaft is rotated, take care not to trap the connecting rods between the throws and side of the crankcase.

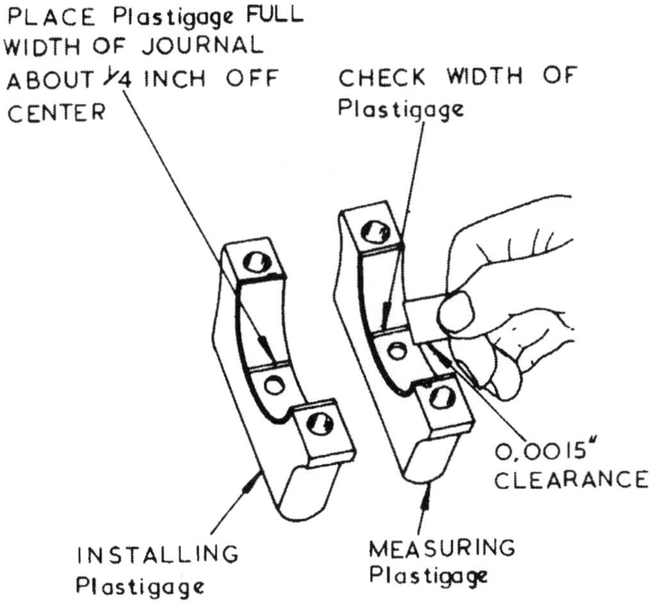

FIG 1:30 Using Plastigage to check the clearance in crankshaft bearings

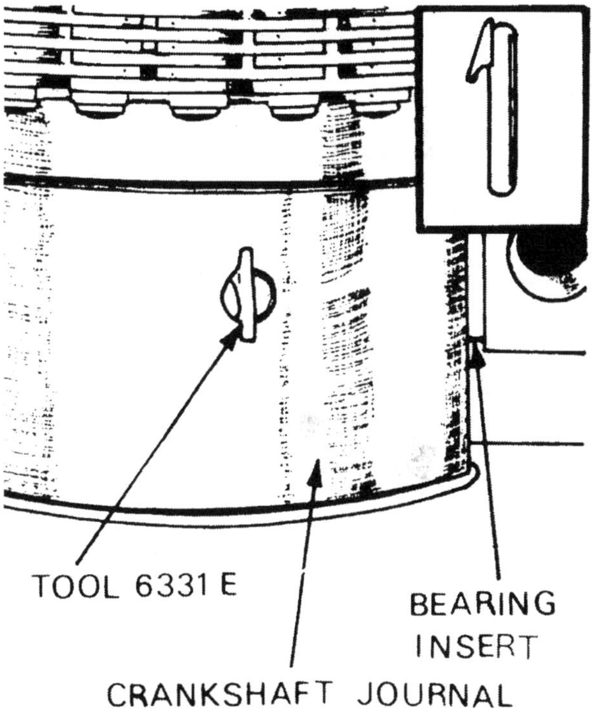

FIG 1:31 Removing the upper main bearing insert with the crankshaft in place

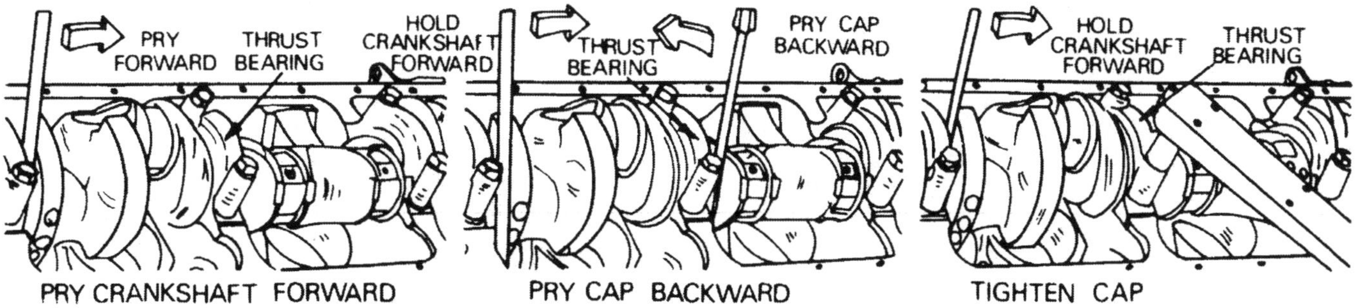

FIG 1:32 The sequence for aligning thrust bearing faces correctly

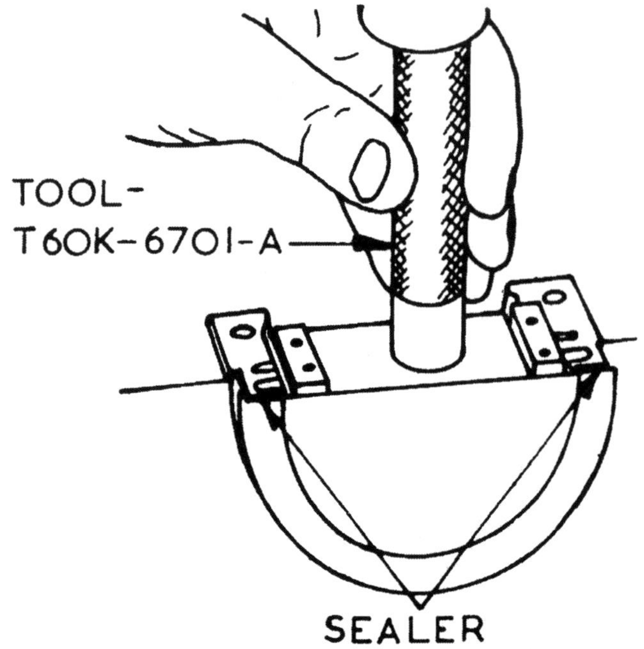

FIG 1:33 Installing a crankshaft rear oil seal

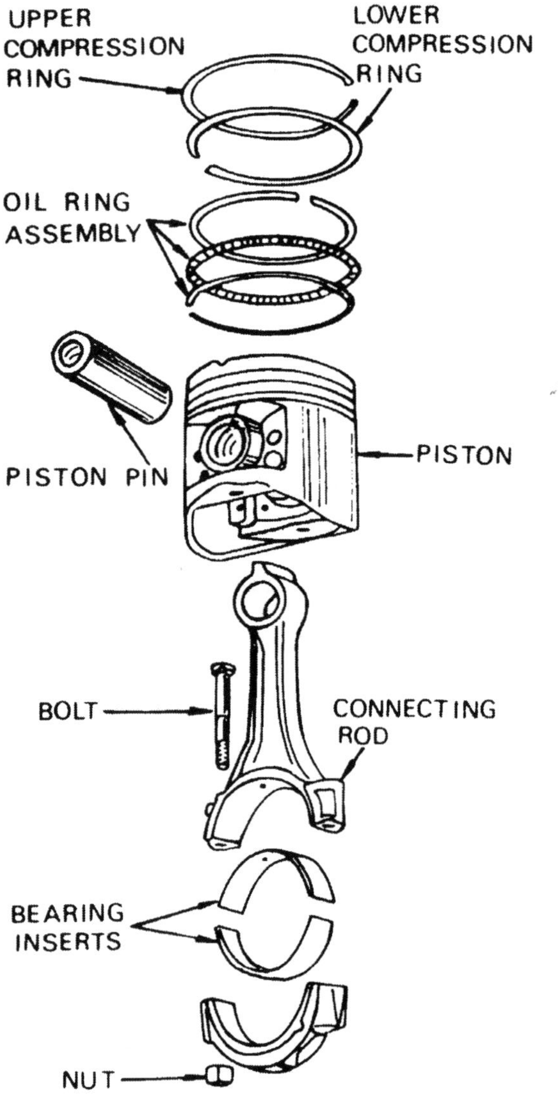

FIG 1:34 Piston, connecting rod and related parts

Examine the bearing inserts and install a new set if any show signs of cracking, pitting, scoring or wear.

Measure each crankpin at several points using a micrometer gauge. If the pin is scored, the maximum out of round exceeds .0004 inch or the taper exceeds .003 inch per inch, the crankshaft must be removed, the crankpins ground down to the next suitable undersize and the crankshaft installed using new undersize bearing inserts.

Thoroughly clean the crankpin and bores for the inserts in the connecting rod and bearing cap. Wipe the inserts clean with a piece of leather or lint-free cloth. Fit the inserts back into position, making sure that their tags locate correctly. Lubricate the crankpins liberally with clean engine oil. Pull the connecting rods down onto the crankpins and fit the bearing caps, making sure that the identification marks are in the correct order and face the correct side of the engine. Tighten the nuts to a torque of 19 to 24 lb ft and check that the crankshaft rotates freely.

Install the oil pump followed by the oil pan. Fill the crankcase to the correct level with fresh engine oil.

Main bearings:

If the engine is fitted to the vehicle, renew one bearing at a time, leaving the crankshaft supported by the remaining bearings. No. 5 (counting from the front) bearing is fitted with special inserts which control the end play of the crankshaft.

Free and remove the bolts securing the bearing cap, noting that the identification number will give the direction which the cap faces, and pull off the cap. Remove the insert from the cap.

Insert the special tool into the oil hole in the crankshaft, as shown in **FIG 1:31**. Carefully rotate the crankshaft so that the tool abuts onto the end of the insert and drives it out from above the crankshaft. Make sure that the tag end of the insert comes out first.

Use a syringe filled with gasoline to clean the area behind the crankshaft. Lubricate the new insert with clean engine oil and drive it back into position using the special tool in the crankshaft oil hole, making sure that its tag locates correctly. Fit the lower insert back into the cap and secure the cap correctly, tightening the bolts to a torque of 60 to 70 lb ft. Check that the crankshaft rotates freely before servicing the next bearing.

When servicing the rear bearing, fit a new oil seal to the bearing cap.

When servicing the bearing which takes the crankshaft thrust, install the cap so that the thrust faces are in line. The method of ensuring this is shown in **FIG 1:32**.

If the crankshaft journals are found to be worn or scored, the crankshaft must be removed and the journals ground down to the next suitable undersize. Install the crankshaft using new undersize bearing inserts.

Crankshaft end play:

Mount a dial indicator onto the crankcase so that the stylus rests vertically on the end of the crankshaft. Lever the crankshaft backwards and forwards so that the end play can be measured on the indicator. The limits of play are .004 to .008 inch but no action need be taken until the wear limit of .012 inch is exceeded. If the play is excessive, new thrust bearing inserts must be fitted to reduce the play. If the play is too small then it is likely that the procedure shown in **FIG 1 :32** has not been correctly followed, though dirt can also be a cause.

Rear oil seal:

The portion in the bearing cap should always be renewed when the cap is removed. The portion in the crankcase can only be renewed after the crankshaft has been removed.

Peel out the old seal and clean the recess. Roughly shape the seal in the fingers and press it into the recess using the thumbs. Start at the centre and work outwards so that the seal protrudes evenly on both sides. Press the seal fully into position with the seal forming tool, as shown in **FIG 1 :33**. Trim the ends of the seal flush with a sharp knife. Apply a little sealer in the area shown on the cap, without allowing sealer onto the seal itself.

Crankshaft removal:

Remove the engine from the vehicle (see **Section 1 :2**). Invert the engine, using old tires to stand it on after removing vulnerable components. Remove the clutch and flywheel (see **Section 1 :8**) as well as the oil pan and oil pump (see **Section 1 :7**). Remove the cylinder front cover and the sprockets with timing chain (see **Section 1 :9**). Disconnect all the connecting rod bearings, and loosely fit the bearing caps back into place after the connecting rods have been pushed into the cylinder bores. Remove all the main bearing caps and carefully lift out the crankshaft.

Thoroughly clean all the parts. Blow through the oilways in the crankshaft with kerosene followed by compressed air to remove any swarf or dirt.

Place the inserts correctly into position so that the tags are located. Lubricate the journals liberally with fresh oil and lay the crankshaft back into place. Install the bearing caps into their correct positions, making sure that they face in their original directions. Secure the caps by torque-loading the bolts. Check the end play and that the crankshaft rotates freely before fitting the connecting rods to the crankpins.

The remainder of the parts are installed in the reverse order of removal and the engine is then fitted back into the vehicle. Leave accessories and vulnerable components until last. Check coolant and lubricant levels before and after starting the engine. Check for leaks with the engine running at a fast idle.

1 :11 The pistons and connecting rods

The parts can be removed from the engine, with the engine installed in the vehicle.

A typical connecting rod and piston assembly is shown in **FIG 1 :34**. When the parts have been installed, the notch on the piston crown must face forwards and the squirt hole in the connecting rod must then face as shown in **FIG 1 :35**.

To remove a piston and connecting rod, remove the cylinder head (see **Section 1 :3**) and the oil pan and oil pump (see **Section 1 :7**). Remove the unworn ridge and ring of carbon from around the top of the bore with garage equipment cutters. If the ridge and carbon is left in place they may damage the piston rings as the piston is removed. Lay rags into the bottom of the bore to catch the swarf and dust as it is cut away. Disconnect

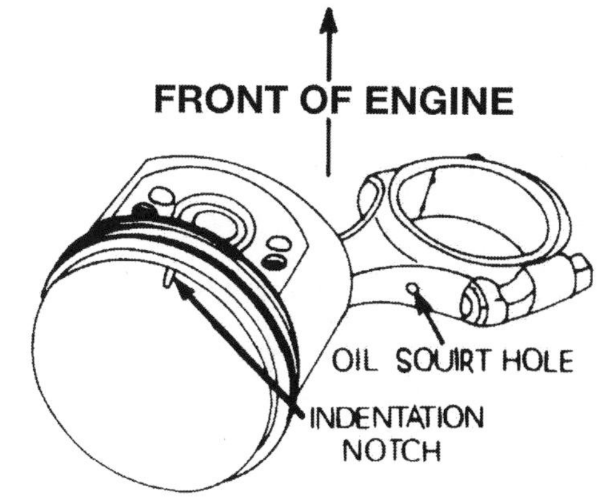

FIG 1 :35 The correct installation of piston and connecting rod

the appropriate connecting rods from the crankshaft, as described in the previous section, and push the connecting rod right up the bore so that the piston can be removed from the top.

Cylinder bores:

These should be examined whenever the cylinder head is removed. A good estimate of the wear can be made by judging the depth of the unworn ridge around the top of the bore. For accurate measurements a special dial gauge should be used and the diameter measured at several points up and down the bore, both in line with

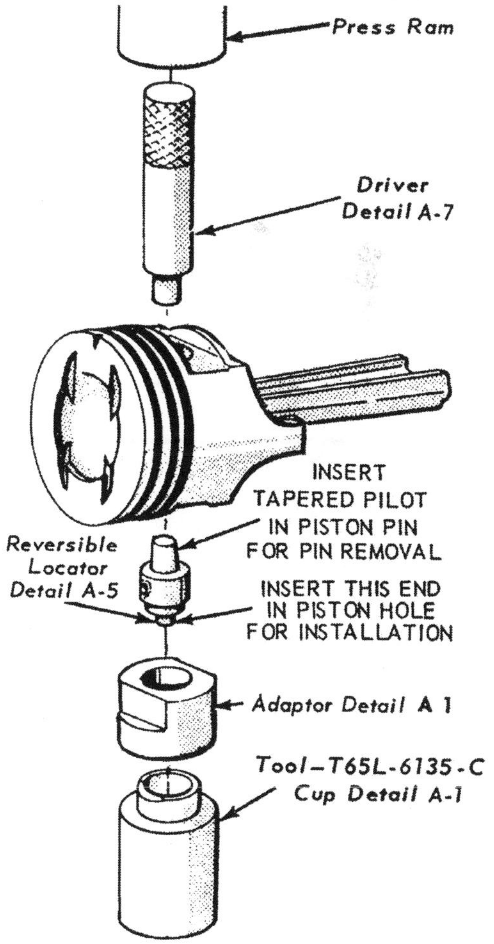

FIG 1 :36 Removing or installing a piston pin

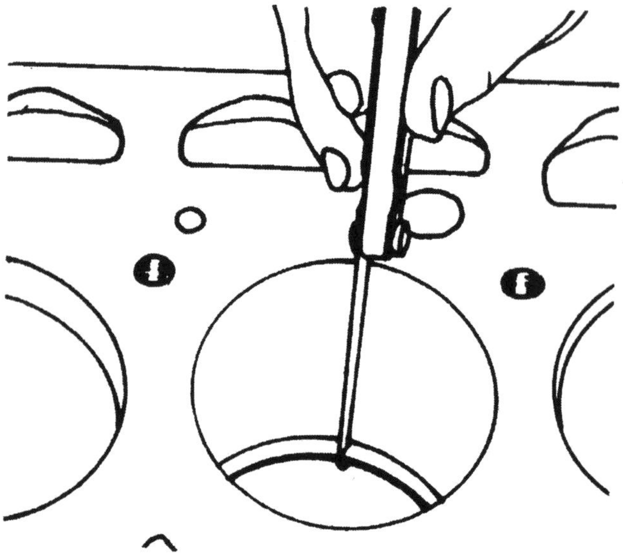

FIG 1:37 Checking piston ring gap

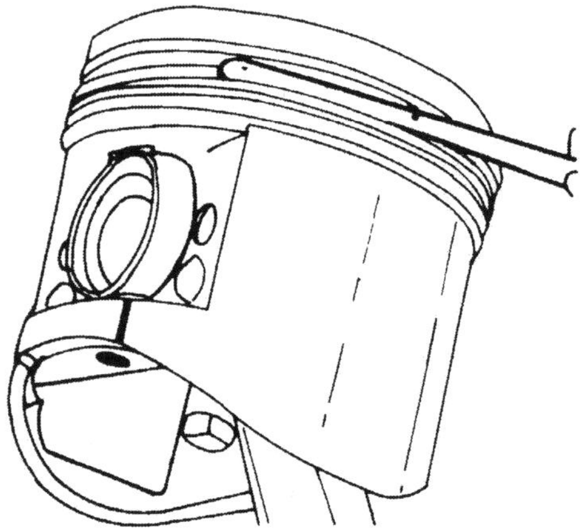

FIG 1:38 Checking piston ring side clearance

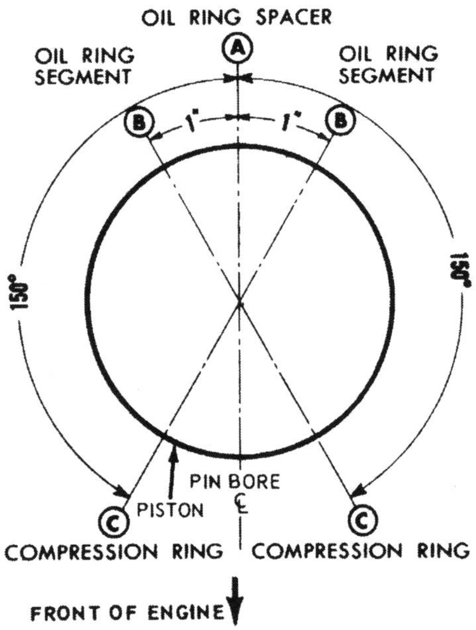

FIG 1:39 Piston ring gap spacing

and at right angles to the piston pin. If the maximum ovality exceeds .005 inch or the taper exceeds .010 inch the bore must be bored out and an oversize piston fitted. Reboring must also be carried out if the bore is scored or damaged. All pistons are of the same weight so if only one bore is damaged it may be rebored, leaving the remainder of the bores untouched.

If the wear is not sufficient to warrant reboring but the high oil consumption requires new rings to cure, the unworn ridge around the top of the bore must be honed out. The old top rings will have worn with the cylinder bore but if a new ring is fitted it will hit the ridge at every stroke and rapidly break up.

Piston pins:

These are an interference fit in the connecting rod and a floating fit in the piston. **The pins should not be removed unecessarily.** The tools required for removing or installing the pin are shown in **FIG 1:36** and a press must be used. If the pistons are removed, and the original pistons used again, it is essential that they are fitted back into their original bores and connecting rods.

All cleaning and servicing operations can be carried out with the pistons fitted to the connecting rods.

Piston rings:

Special ring expanding tools are made for removing or installing the piston rings. The tool grips the ends of the ring and then gently parts them so that the piston can be lifted up and off the piston. If a special tool is not available three steel shims—discarded feeler gauges for instance—can be used instead. Slide a shim under one end of the ring and then work the shim around under the ring, while pressing the ring as it comes free onto the piston land above it. When all the ring is on the land, slide the remaining two shims under the ring so that all three are evenly spaced. Part the ends of the ring gently with the thumbnails and slide the ring up and off the piston, using the shims to protect the piston. Take care not to allow the back of the ring to scrape the piston.

Use a piece of broken ring, or a special tool, to clean the ring groove in the piston. Remove all the carbon but **take care not to remove metal from the piston or the oil consumption will be increased.** Clean out the return holes behind the oil control ring with a piece of blunt-ended wire. If the grooves have a high step on the lower lands it is likely that the clearance will be excessive and a new piston will have to be fitted to cure the fault.

Spring each new ring in turn into the top of a bore and use an inverted piston to press it squarely down the bore to the point of maximum wear. Use feeler gauges, as shown in **FIG 1:37**, to measure the gap between the ends. Try another set of rings but if the gap is too small, file the ends carefully with a smooth file to set the correct gap. The gap for the compression rings should be .010 to .020 inch and the gap for the steel rails of the oil control ring should be .015 to .055 inch.

Install the rings on the piston in the reverse order of removal, taking great care not to force or snap them. Use feeler gauges to check the side clearance as shown in **FIG 1:38**. The oil control ring should be a snug fit in all cases. The side clearance on compression rings should be .002 inch when parts are new but it may be allowed to reach the wear limit of .006 inch before taking action. Check the side clearance using a new piston ring and if the clearance is still outside limits a new piston must be fitted, otherwise oil consumption will be excessive.

Pistons:

The pistons are graded to ensure that the clearance between skirt and cylinder bore can be set with accuracy. Special measuring equipment is required for checking the clearance though a skilled mechanic can check using a set of feeler gauges and spring balance gauge. The selection of pistons is best left to the specialist reboring the block or supplying new pistons. **Make sure that the pistons are fitted to their correct bores and not interchanged.**

When the pistons are out of the engine, the crowns may be lightly polished using worn emerycloth dipped in kerosene as all abrasive particles can be washed off before installation. **Do not use any form of abrasive on the piston skirt.** If there is lacquering on the skirt, indicative of gases blowing past the rings, wash it off with a suitable harsh solvent such as trichlorethylene.

The pistons must be installed so that the notch on the piston crown faces the front of the engine.

Installation:

Check that all the parts are clean and free from particles or dirt. If the piston has been removed from the connecting rod, assemble them as shown in **FIG 1:35**. Fit the rings into position and rotate them so that their gaps are positioned as shown in **FIG 1:39**. Compress the rings into their grooves with a piston ring clamp, after lightly oiling them and the piston skirt. Lower the connecting rod into its correct bore and insert the skirt of the piston. Gently press down on the piston, allowing the clamp to slide off the rings as they enter the bore. **Do not force the rings into place and do not allow them to escape from the clamp before they enter the bore.** The rings are very brittle and will easily snap if mishandled. If the correct ring clamp is not available, a large worm-driven hose clip can be used instead but more care must then be taken.

Install the connecting rod bearings. Refit the oil pump, oil pan and cylinder head in the reverse order of disassembly.

1:12 Thermactor emission control

This type of emission control is usually only used on models fitted with manual-shift transmission. An engine-driven air pump provides a constant flow of air under pressure. The pump is fitted with a relief valve to limit the maximum pressure of the air. Under normal running the air is injected through nozzles directly into the ports by the exhaust valves so that the excess air mixes with the hot exhaust gases to burn any surplus hydrocarbons. A schematic layout of the system is shown in **FIG 1:40**. The check valve is fitted to prevent exhaust gases flowing back through the system if the exhaust pressure rises above the air pressure.

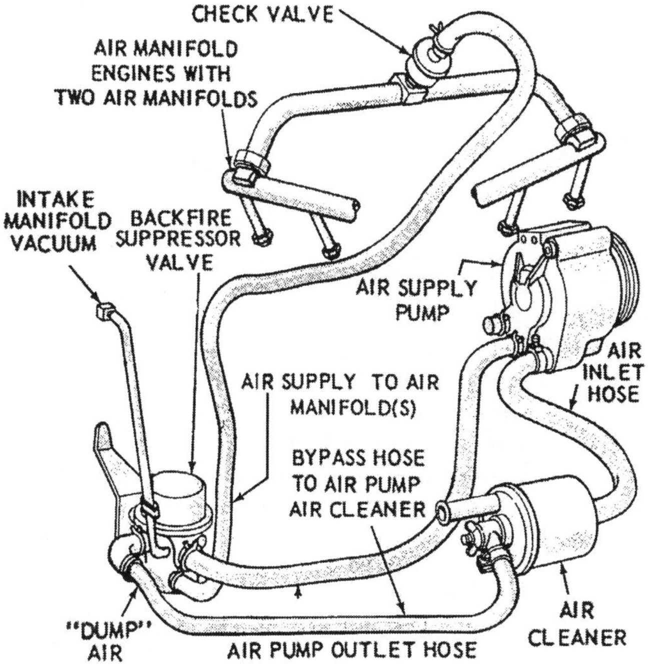

FIG 1:40 Layout of Thermactor system

A dump air valve is fitted to prevent backfiring on deceleration, and a sectioned view of the valve is shown in **FIG 1:41**. Normally the pressure of the spring holds the valve open so that air flows directly to the ports. When the throttle is closed and intake manifold suction is high, the suction operates the valve so that the air no longer passes to the exhaust ports but is dumped through the air cleaner.

The installation of the parts on the engine is shown in **FIG 1:42**.

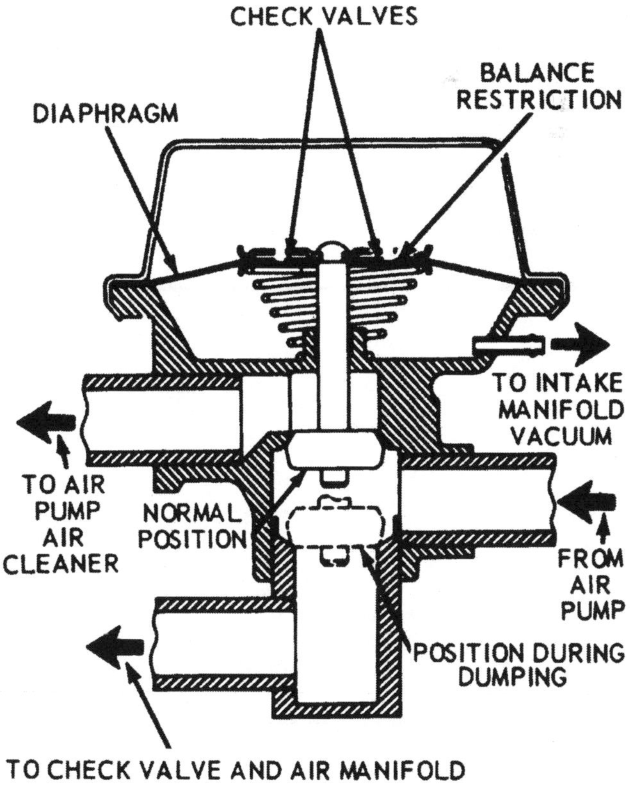

FIG 1:41 Thermactor dump valve

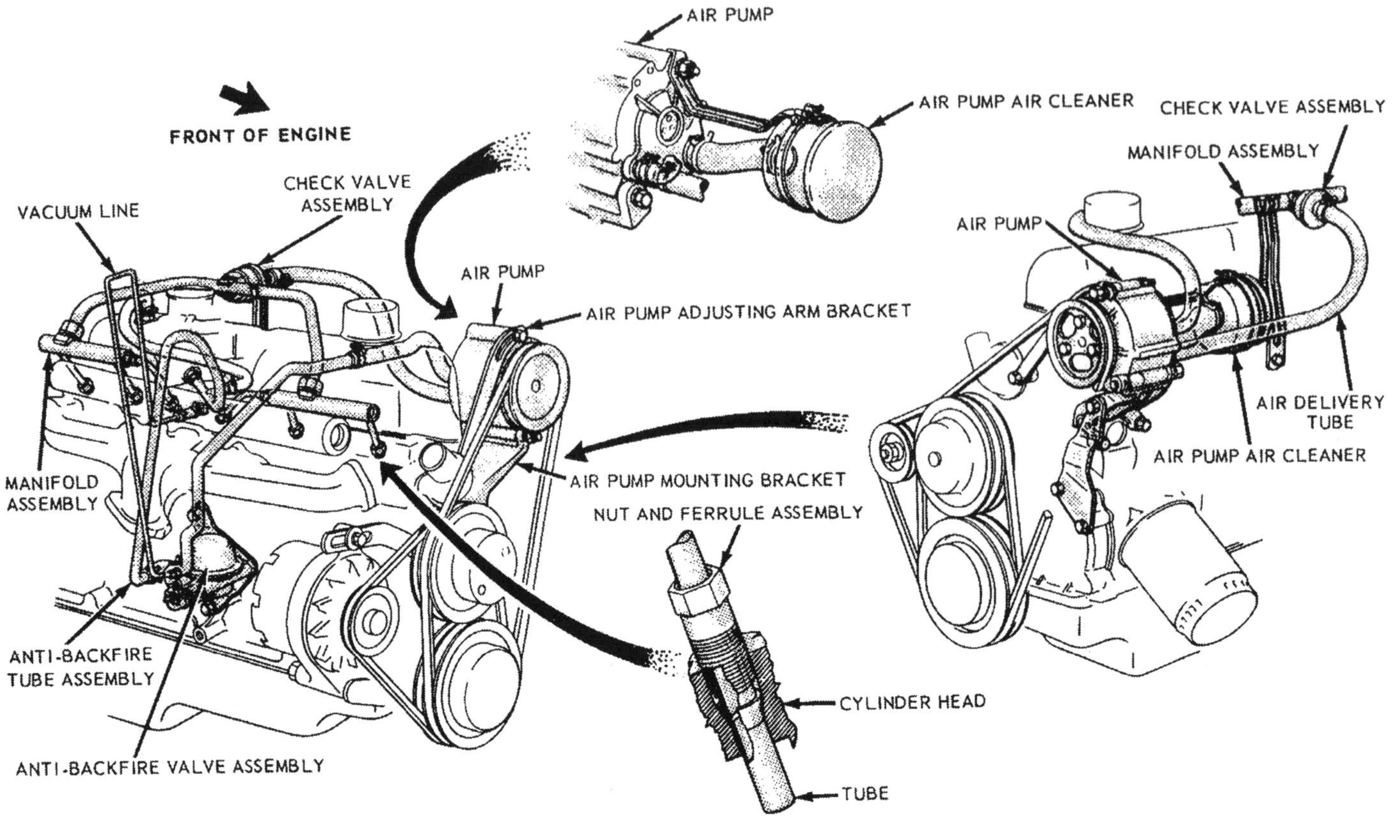

FIG 1:42 Thermactor system installed on the engine

Air cleaner:

The earlier air cleaner components are shown in **FIG 1:43** and the renewal of the filter element is self-explanatory.

On some later models a centrifugal fan filter is attached directly to the pump. To remove the centrifugal type filter, remove the drive pulley from the pump. Pry the outer disc loose and then use a pair of pliers to grip the fan by the vanes and pull it out. If the fan breaks, take care to prevent parts entering the air inlet hole of the air pump. **Do not attempt to remove the metal drive hub.**

The new centrifugal type filter is pulled back into position using the pulley and the bolts to draw it back, making sure that it fits fully into place. The new filter may be tight, giving noisy operation at first, but noise should die away as the filter beds in.

Dump valve test:

Start the engine and run it until it has reached its normal operating temperature. Disconnect the air supply hose from the check valve. At idle the valve should be

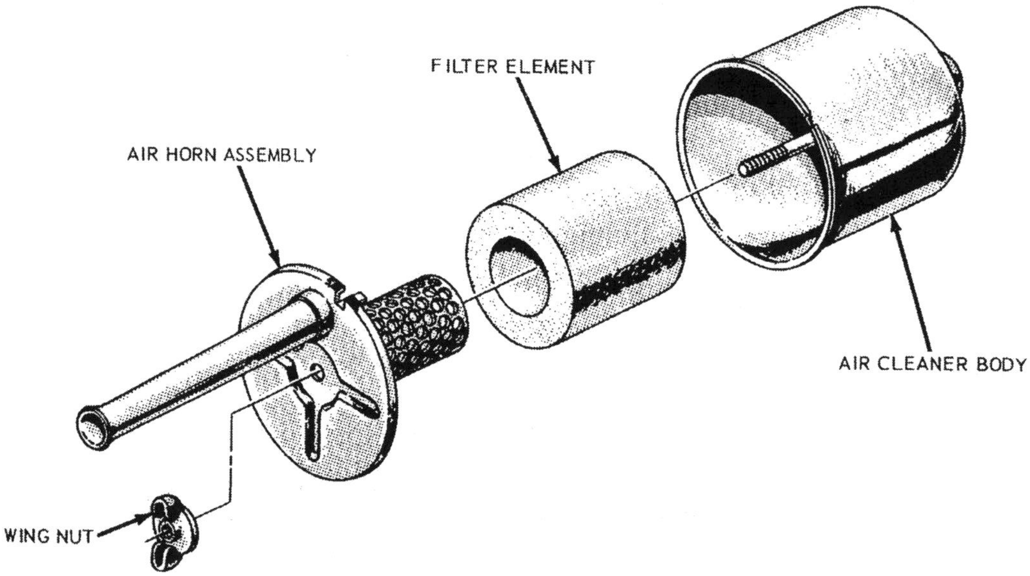

FIG 1:43 The Thermactor air cleaner for the pump inlet

closed with the plate away from the manifold side, though a little flutter may be apparent at idling because of the pulsations in the exhaust. Use a pencil to press the valve open and check that the valve returns immediately when the pencil is removed. Slowly increase the engine speed to 1500 rev/min and check that the valve is tightly closed with no exhaust gas leakage.

If the valve does not return freely after being opened by hand, or if it leaks at the higher engine speed, the valve is defective and must be renewed.

Air leakage:

Air leaks in the system will make it impossible to set a satisfactory curb idle speed. Note that a fault in the crankcase ventilation system will also make it impossible to set curb idle.

Disconnect the vacuum line at the manifold and plug the connector on the manifold. If curb idle can now be satisfactorily set there is a leak and either the hose or the dump valve must be renewed, depending on which component is defective.

Dump valve functional test:

With the engine running, disconnect the hose from the pump at the valve. Snap the throttle open and shut. A sucking noise should be heard from the valve and there should be no backfiring on deceleration. If the engine backfires, the valve is defective and must be renewed.

Reconnect the hose from the pump to the valve and disconnect the dump air hose at the valve. Again snap the throttle open and shut. Compressed air should blow from the valve as the engine decelerates. If air does not blow from the valve it is defective and must be renewed.

1:13 Trouble shooting guide

(a) Engine will not start

1 Water or dirt on spark plug leads, distributor cap or ignition coil
2 Loose wires or defective insulation in ignition low-tension circuit
3 Defective ignition coil
4 Defective ignition contact points
5 Defective ignition condensor
6 Defective automatic choke
7 Faulty or jammed starter motor
8 Spark plug leads incorrectly connected
9 Blocked fuel filter or carburetor jets
10 Leaking valves
11 Sticking valves
12 Valve timing incorrect
13 Ignition timing incorrect

(b) Engine stalls

1 Check 1, 3, 4, 6, 9, 10 and 11 in (a)
2 Spark plugs defective
3 Retarded ignition
4 Mixture too weak
5 Water in fuel
6 Fuel tank vent system blocked
7 Incorrect initial adjustment of valve lifters

(c) Rough engine idle

1 Check 4, 6, 8, 10, 11, 12 and 13 in (a) and 2 and 7 in (b)
2 Incorrectly adjusted carburetor
3 Air leaks at carburetor flange
4 Leaking vacuum fittings
5 Defective crankcase ventilation system
6 Worn piston rings
7 Worn valve stems or guides
8 Weak exhaust valve springs
9 Defective dump valve (Thermactor only)

(d) Engine misfires

1 Check 1, 2, 3, 4, 8, 10, 11, 13 in (a); 2, 3, 4 and 7 in (b)
2 Weak or broken valve springs

(e) Engine overheats

1 Check **Chapter 4**

(f) Compression test pressure low*

1 Check 10, 11 and 12 in (a); 6 and 7 in (c) and 2 in (d)
2 Worn or scored cylinder bores
3 Worn piston rings and grooves
4 Choked air cleaner

*A high test pressure indicates excessive deposits in the combustion chamber

(g) Engine lacks power

1 Check 4, 6, 10, 11, 12 and 13 in (a); 2, 3, 4 and 7 in (b); 6 and 7 in (c) and 2 in (d). Also check (e)
2 Leaking cylinder head gasket
3 Distributor automatic advance and retard not working correctly

(h) Burnt valves or seats

1 Check 10, 11 and 12 in (a) and 7 in (b). Also check (e)
2 Excessive carbon in combustion chamber

(j) Sticking valves

1 Weak valve springs
2 Bent valve stems
3 Scored stem or guide
4 Defective valve lifter
5 Incorrect initial valve lifter adjustment
6 Gum deposits on stem or guide

(k) Excessive cylinder wear

1 Check 6 in (a) and also check (e)
2 Lack of oil
3 Dirty oil
4 Piston rings gummed or broken
5 Badly installed piston rings
6 Connecting rod bent or twisted
7 Piston pin worked loose or incorrectly fitted

(l) Excessive oil consumption

1 Check 6 and 7 in (c); 2 and 3 in (f) and 4 and 5 in (k)
2 Oil return holes in piston blocked
3 Oil level too high in crankcase
4 External oil leaks

(m) Crankshaft bearing failure

1 Check 2, 3 and 6 in (k)
2 Restricted oilways
3 Worn journals or crankpins
4 Loose bearing caps
5 Dirt particles trapped during installation
6 Extremely low oil pressure

(n) Low oil pressure

1 Check 2 and 3 in (k) and also check 2, 3 and 4 in (m)
2 Choked oil filter or pump intake tube strainer
3 Relief valve defective
4 Very worn oil pump
5 Defective gauge or sender unit

(o) Internal water leakage (see Chapter 4)

(p) Poor water circulation (see Chapter 4)

(q) Corrosion (see Chapter 4)

(r) High fuel consumption (see Chapter 2)

CHAPTER 2

THE FUEL SYSTEM

2:1 Description:

The gasoline is stored in a tank at the rear of the vehicle. The fuel is drawn from the tank, along a system of pipes, by the mechanical fuel pump mounted on the engine, and the pump delivers the fuel under low pressure to the inlet of the carburetor fuel bowl. The carburetor controls the flow of air into the engine, and therefore the engine speed, while metering the fuel according to the airflow to provide a correct fuel/air mixture for all conditions.

Single barrel carburetors are fitted to all Mustang six-cylinder engines. 200 engines have a Carter YF 1-V or an Autolite 1100 type, 250 engines a Carter RBS 1-V or an Autolite 1101.

Air to the carburetor passes through an air cleaner which filters out dirt and silences the roar of the carburetor intake. The air cleaner is fitted with a thermostatically controlled intake duct which draws air directly from the area under the hood or from a muff over the exhaust pipe according to the temperature. This ensures a warm air supply to the engine for efficient vapourization of the fuel with no danger of carburetor icing.

A strainer is fitted to the tank gauge sender and inlet unit so that large particles of dirt do not pass into the fuel system. Early models have a cartridge filter attached to the fuel pump but on later models a line filter is fitted between the pump and carburetor. In addition there is a filter screen in the carburetor fuel inlet.

On later models made for use in the State of California, and subsequently on all cars since 1971, a fuel tank evaporative emission control system is fitted. The fuel tank and filler cap are sealed so that no gasoline vapor can escape. Inside the fuel tank itself is a separate expansion container, open at the bottom, which ensures that there is always air space in the tank even when it is brim-full so that fuel is not spilled as it expands with a rise in temperature. The tank is vented through nylon tubes leading first to a separator which returns any liquid fuel to the main tank and then to a carbon-filled cannister in the engine compartment. When the car is parked the carbon granules absorbs fuel vapor. In operation air is drawn through the canister into the engine and the vapor is purged. The components of the system are shown in **FIG 2:1**. No periodic maintenance is required. If liquid gasoline passes into the carbon cannister, the cannister must be renewed.

2:2 Maintenance

This is mainly confined to renewing the filter elements at the specified intervals and checking that the throttle

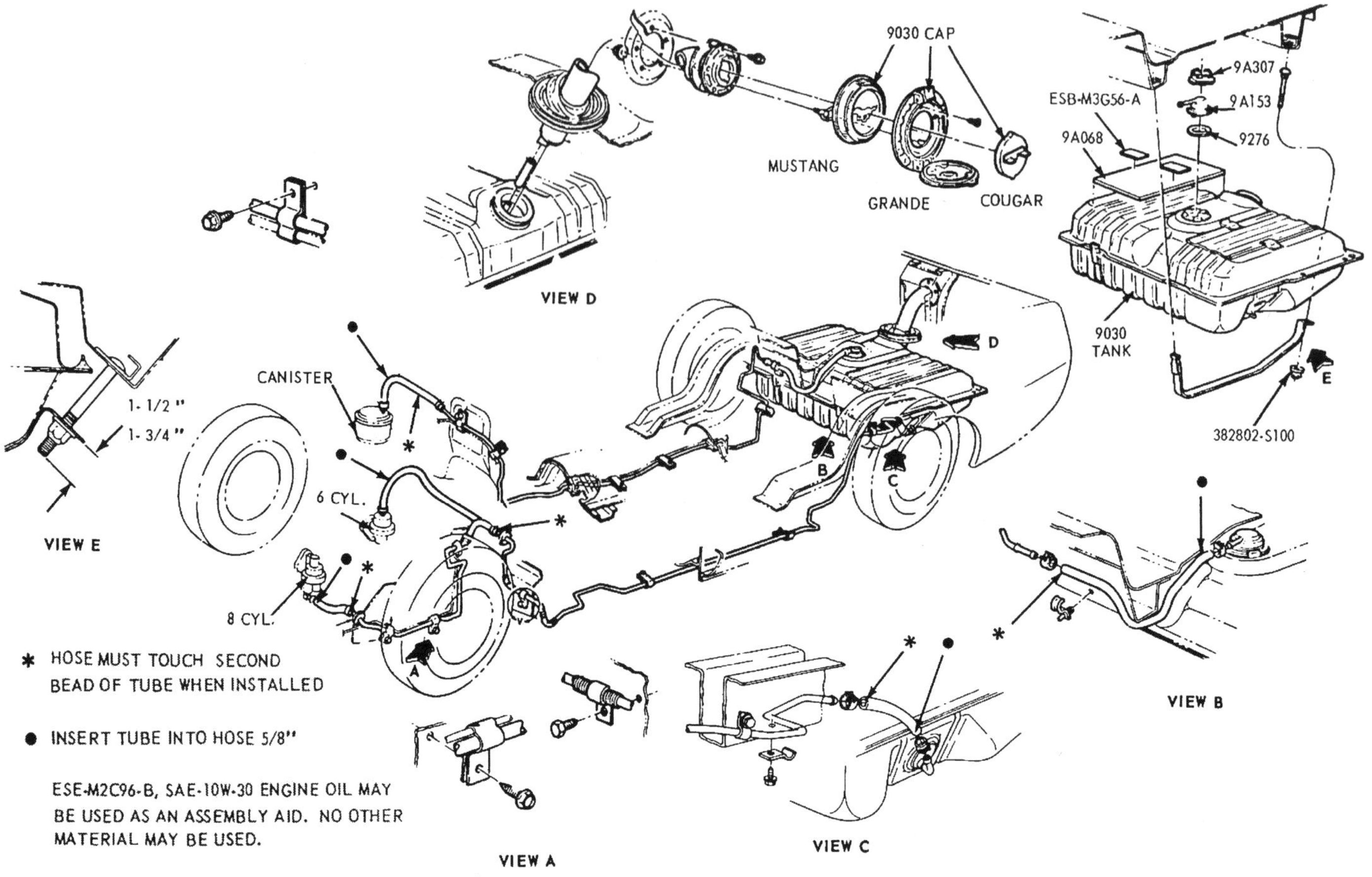

FIG 2:1 Fuel tank, pipe lines, and evaporative emission control system

linkage operates freely and through its full range of movement.

It is advisable to take the vehicle to a service garage equipped with electronic test equipment for regular checks on the fuel system and ignition. This ensures early detection of faults and prevents excessive emissions from the engine.

Modifications to the fuel system or ignition system may be subject to penalties of Federal law.

Fuel filter:

The sealer Carter fuel pump fitted to early models is shown in **FIG 2:2**. Unscrew the filter housing from the pump and discard the old element and gasket. Wash out the housing and its mounting on the pump with cleaning solvent, denatured alchohol or clean gasoline. Fit a new element of the correct type into the housing, making sure that the labeled end is downwards. Lightly coat the new gasket with engine oil and install it on the housing. Hand tighten the housing onto the pump until the gasket just touches the pump and then tighten the housing by a further $\frac{1}{8}$ turn. Start the engine and check for fuel leaks.

On later models a line filter is fitted between the fuel pump and carburetor, screwed into the inlet of the carburetor. The filter is of one-piece construction and cannot be serviced but only renewed. Remove the air cleaner to gain access. Unscrew the filter from the carburetor, disconnect the filter from the fuel line and discard the filter, gasket and retaining clamp. Install the new gasket to the filter and thread the filter back into the carburetor inlet so that it is tight. Fit a new clamp to the hose and slide the hose onto the filter. Tighten the clamp by crimping it. Start the engine and check for fuel leaks before installing the air cleaner.

On all models the fuel filter should be renewed at intervals of 12,000 miles, or earlier if it becomes clogged or choked.

Air cleaner:

The filter element is not cleanable and should normally be renewed at intervals of 12,000 miles. If the vehicle is used in dusty or very dirty conditions, the servicing interval should be reduced.

Disconnect any hoses to the air cleaner. Remove the wing bolt that secures the air cleaner to the carburetor and take off the assembly. Remove the cover and element, discarding the old element. Wipe out the cover and body to remove any dirt or dust.

Install a new gasket onto the carburetor, if required, and install the body onto the carburetor. Place the new element into the body and attach the cover, securing the cover and body with the wing bolt. Reconnect the hoses to the air cleaner.

Crankcase ventilation and vacuum systems:

Check all hoses for blockage or splits as air leaks will alter the idling characteristics of the engine. The regulator valve in the crankcase breathing system must be renewed at regular intervals or whenever it is found to be defective. The hoses can be cleaned using a brush and kerosene, followed by blowing through with compressed air.

2:3 The fuel pump

A Carter permanently-sealed fuel pump is fitted onto the left side of the engine crankcase. The pump is operated from a circular eccentric cam on the engine camshaft.

Only minor repairs can be carried out to the pump and if it is found to be defective a new pump must be installed in its place. The pump is shown in **FIG 2:3**.

Removal:

Disconnect the inlet and outlet fuel lines at the pump. Plug the pipe from the fuel tank to prevent fuel siphoning out. Take out the screws that secure the pump to the engine and withdraw the pump.

Discard the old gasket and remove all traces of sealer from the mounting faces. If the flange of the pump is distorted or damaged (usually caused by overtightening the attachment screws) it can be lapped down on emerycloth spread over a flat surface, after the rocker arm has been removed, or it may be dressed with a fine file.

Install the pump in the reverse order of removal. Use a new gasket lightly smeared with sealer. Be sure that the rocker of the pump is correctly positioned on the cam in the engine and press the pump firmly into position while alternately tightening the attachment screws.

Reconnect the fuel lines and secure them with new clamps, crimping the clamps as required. Start the engine and check for fuel or oil leaks. If the vehicle has stood for some time, a fair amount of cranking may be required to prime the pump and carburetor before the engine will start.

Testing:

A quick field test can be carried out after the fuel line has been disconnected from the fuel filter at the carburetor inlet. Point the hose away from the engine, preferably into a container, and crank the engine on the starter motor. If the fuel pump is operating it will deliver a good spurt of gasoline from the hose at every other revolution of the engine. A worm-driven clip can be used to temporarily connect the fuel line to the filter in place of the correct crimped clamp.

If the pump is off the engine, operate the rocker arm while blocking first the inlet and second the outlet with a finger or thumb. Suction should be felt at the inlet and there should be a pressure build up at the outlet.

For full and accurate tests, a pressure gauge, graduated container, and suitable adaptors will be required. The test equipment is connected as shown in **FIG 2:4**. **The fuel lines must be carefully and tightly connected to prevent spillage of fuel which could lead to fire.** Start the engine and run it until it has reached its normal operating temperature. Reduce the speed to curb-idle and open the hose restrictor momentarily to bleed out any air. Note the pressure reading on the gauge. The pump and filter are satisfactory if the static pressure is 4 to 6 lb/sq in. If the pressure is low, renew the fuel filter and try again. Low pressure can be caused by a blockage in the fuel lines but it is more likely that the pump is defective.

If the static pressure is satisfactory, leave the engine at curb idle and open the restrictor on the hose. Note the time taken to deliver one pint of fuel. If the pump and lines are satisfactory the time taken should be 30 seconds.

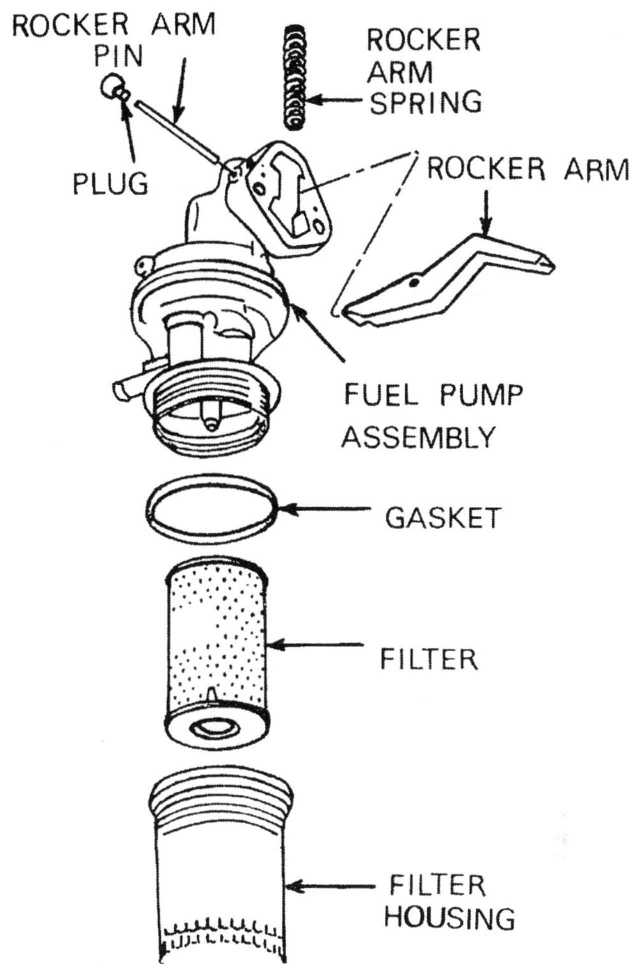

FIG 2:2 Fuel pump with filter fitted to early models

If the time taken is excessive, stop the engine. Disconnect the line from the fuel tank at the pump and plug it to prevent fuel loss. Connect another hose to the pump and dip the free end into a container of clean fuel. Restart the engine, **taking great care not to spill fuel,** and again take the time to deliver one pint of fuel. If the time is now within limits, there is a blockage in the fuel lines on the vehicle. Disconnect the lines and blow through them in a reverse direction with compressed air. If this fails to clear them, new lines must be fitted.

If the static pressure and fuel delivery are low under all conditions, the pump is defective and must be renewed.

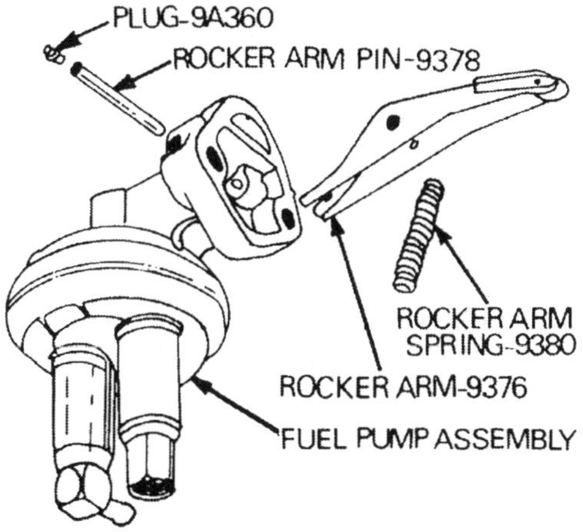

FIG 2:3 Fuel pump—later models

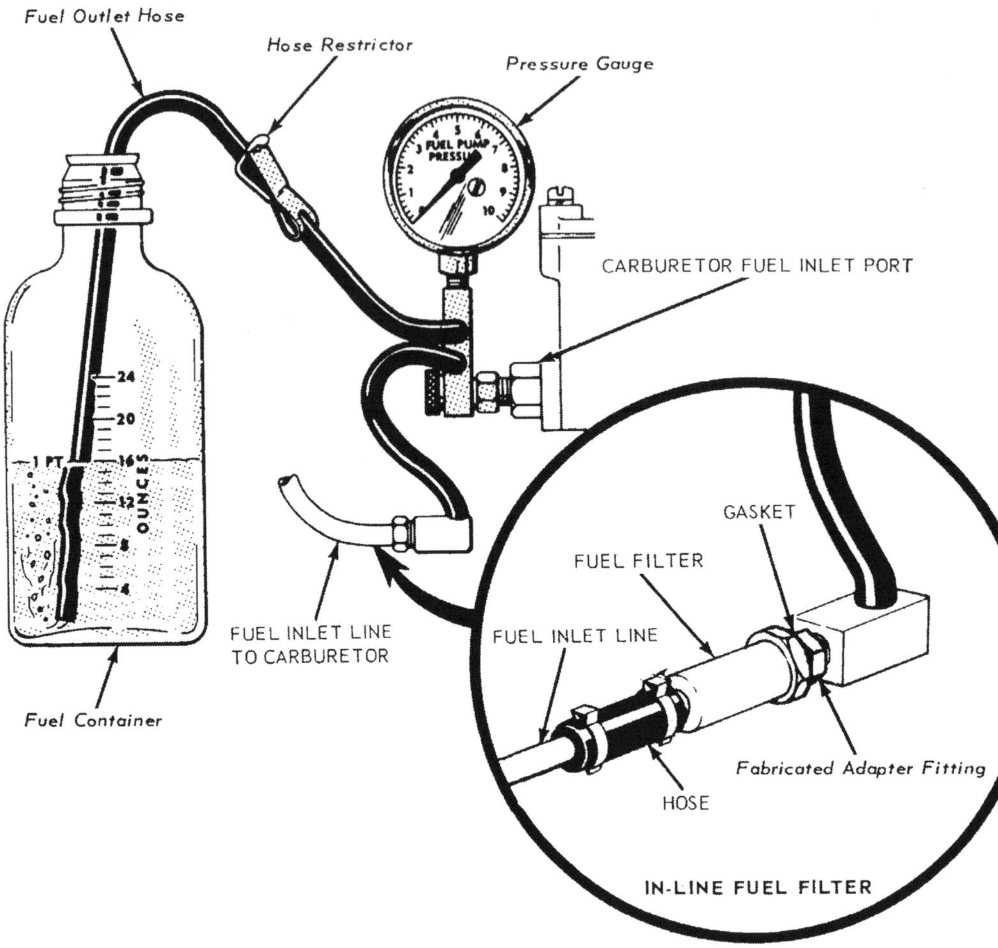

FIG 2:4 Testing the fuel pump output

Servicing:

The only repair that can be carried out is the renewal of the rocker arm. Scrape away the staking that secures the plug and lever off the plug with a screwdriver. Take the pressure of the rocker arm and shake out the pivot pin. The rocker arm and its spring can then be removed. If the pin sticks, tap the body lightly against a firm surface to dislodge it.

Wash the pump with clean gasoline and check it for cracks or damage. Renew the pin, arm and spring if required. Insert the spring back into its bore. Install the rocker arm, making sure that its hook locates into the diaphragm rod. Position the spring onto the locator of the arm and move the arm until the holes align. Insert the pin to hold the rocker arm in place. **Check that the end of the arm is still correctly located in the diaphragm rod.** Press a new plug into position and lock it by staking over the metal.

2:4 The air cleaner duct

A typical duct is shown in **FIG 2:5**. A limited amount of adjustment can be carried out by altering the position of the thermostat.

When the ambient temperature in the engine compartment is below 100°F, the valve should be in the full heat on position. If the valve is not in this position, check the parts for misalignment which could cause the valve to bind.

If no mechanical fault is found, remove the complete duct assembly and immerse it under water. Heat the

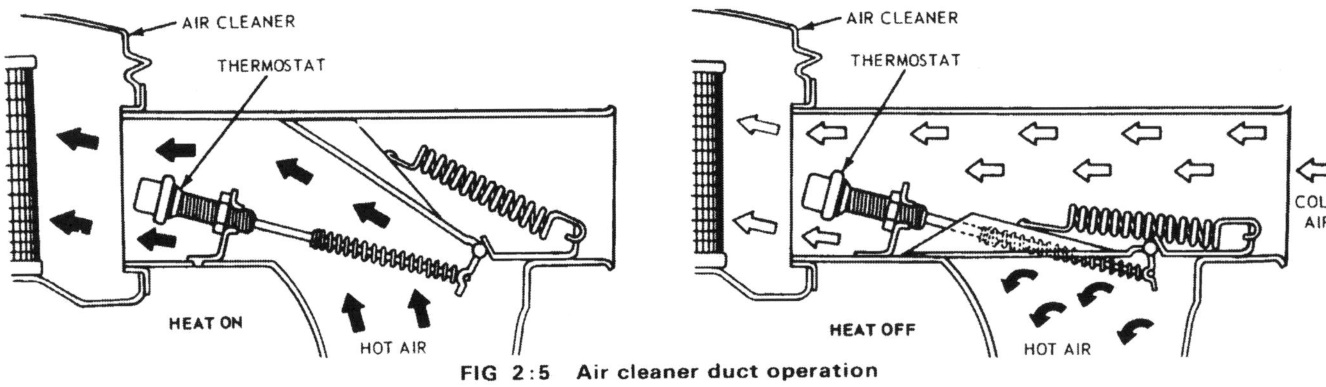

FIG 2:5 Air cleaner duct operation

water until it is at a temperature of 100°F and allow five minutes to soak at this temperature. The valve should be in the full heat-on position. Heat the water further to 135°F and again allow the duct to soak at this temperature for five minutes. The valve should now be in the full heat-off position. If the valve does not move to the correct position and there is no mechanical defect to stop it from moving, a new duct and valve assembly must be fitted.

On some models an additional cold air inlet is fitted to the air cleaner body. The inlet is operated by a vacuum motor so that it is closed when the engine is idling or at low speeds and warmed air is essential to prevent excessive emissions. When the throttle valve is opened the vacuum in the intake manifold is reduced and the valve opens. Check that the valve is closed at idle and opens as the throttle is opened. If the valve sticks in any position, check the vacuum hoses for kinks, splits or blockage and that vacuum is reaching the motor. If the hoses are satisfactory, check the motor on a separate vacuum source and renew it if it fails to operate correctly.

2:5 Carburetor operation:

The four types of carburetor fitted to different years and models vary in details of construction but all operate in a similar way.

The basic part number of all carburetors is 9510 and the variants are distinguished by an identification tag attached to the carburetor which shows a prefix, suffix and design change code (see **FIG 2:6**).

Typical carburetor details are given in Technical Data as a guide. To be absolutely certain of obtaining a correct replacement part refer to your dealer or stockist who has a Master Parts List which identifies the application and interchangeability of components.

Gasoline from the pump enters the fuel bowl of the carburetor. When the fuel reaches the correct level the flow is cut by the action of the float on the needle valve. As the engine uses fuel and the level drops the valve opens again, so that the level is maintained virtually constant.

The flow of air, regulated by the throttle valve, through the venturi of the carburetor barrel generates a drop in pressure. This depression draws fuel into the air stream from the metering jets.

When the engine is at idle, there is insufficient airflow through the carburetor to draw the fuel through the main jet and the maximum suction is at a point beside the throttle valve. Use is made of this to draw fuel from the fuel bowl through the low speed jet and mix it with air so that fuel/air is drawn into the engine, as shown in **FIG 2:7**.

When the throttle valve is partially open, the airflow through the booster venturis is sufficient to draw fuel through the main metering system, as shown in **FIG 2:8**. The metering rod is controlled by two actions. During part throttle operation the vacuum in the manifold pulls down the diaphragm so that the rod is held against the accelerator pump lifter link and, as the link is operated by the throttle linkage, the rod will take up a position dependent on throttle opening. When the throttle is full open, the vacuum in the manifold is insufficient to pull down the diaphragm and the rod moves to its full open position.

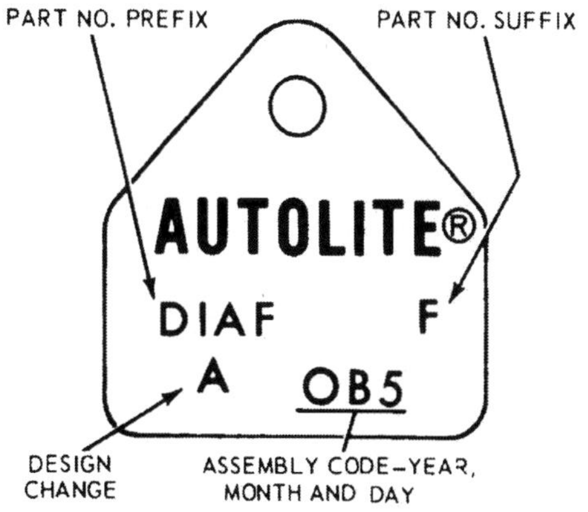

FIG 2:6 Typical carburetor identification tag

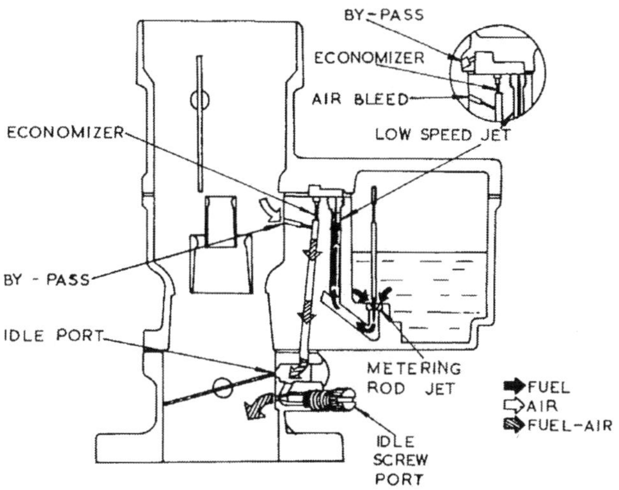

FIG 2:7 Carburetor operation at idle

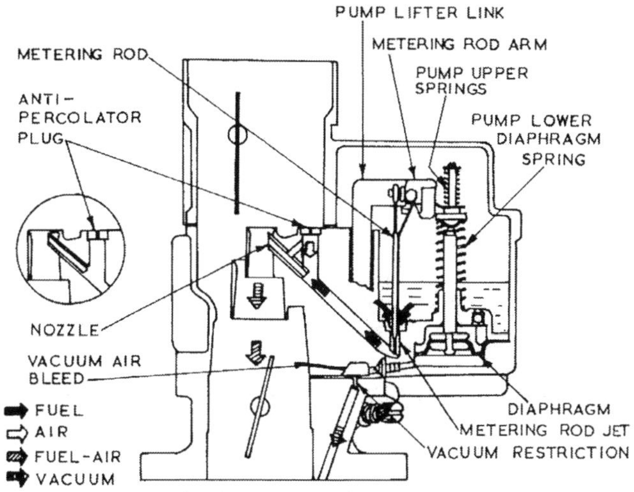

FIG 2:8 Carburetor operation: main system

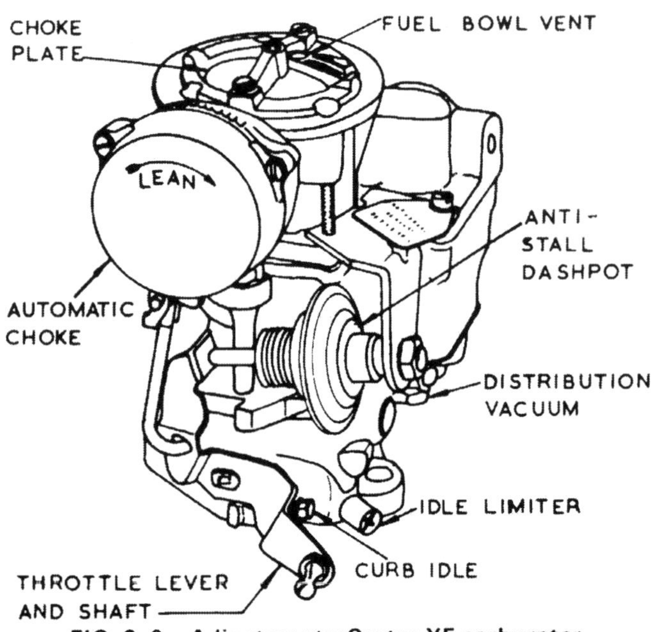

FIG 2:9 Adjustments: Carter YF carburetor

Anti-percolator nozzles and bushes are fitted to the internal passages and these allow vapor to escape and also relieve pressure from the fuel as it expands under heat, so that fuel is not forced into the engine when the engine is stationary. **All the calibrated plugs and bushings are permanent installations, as is the main nozzle, and must not be removed in service.**

When the engine is accelerated the fuel flow will lag behind the airflow, as fuel is heavier, and extra fuel must be injected to keep the mixture at the correct strength. An accelerator pump diaphragm is fitted into the fuel bowl, and the pump is operated by the same linkage as the metering rod. When vacuum is applied to the underside of the diaphragm it is pulled down against the action of the pump spring and fuel is drawn in above the diaphragm, from the fuel bowl through a ball check valve. At part throttle openings, the suction on the diaphragm holds the pump down against the throttle linkage. When the throttle opens on acceleration, the diaphragm is forced upwards by the linkage and the fuel passes through a discharge check needle before being injected directly into the air stream through the carburetor. If the throttle is suddenly opened the pressure of the spring will overcome the diaphragm and the fuel will be injected. An upper spring is fitted to act as a bump stop and to ensure that the fuel is not injected in one gulp but smoothly over a short period of time.

When the engine is cold, a very rich mixture is needed to initiate combustion and then a rich mixture is needed to sustain it. A choke is fitted to give the rich mixture required. The choke consists of a plate valve fitted over the inlet of the carburetor. When the choke valve is closed, the suction throughout the carburetor is increased and excess fuel is drawn through all the jets to enrich the mixture. The choke valve is mounted offset on its shaft so that when the engine starts, the increased air flow partially opens the valve against the pressure of a spring so that the strength of the mixture is partially reduced.

A manually operated choke may be fitted, in which case the choke valve is set by the action of a cable and the control inside the vehicle.

An automatic choke can be fitted. When the engine is cold, the tension of a bi-metallic spring holds the choke valve fully closed, ensuring a very rich mixture as the engine is cranked. As soon as the engine fires, the extra air acts on the offset of the choke valve to open it slightly and partially reduce the mixture strength. A piston is fitted to the housing and this is acted on by the manifold suction. The piston tends to open the choke valve further against the action of the bi-metallic spring. When the engine is running, slots in the piston allow air to be drawn through from the exhaust manifold and into the intake manifold. The heated air from the hot exhaust manifold gradually heats the bi-metallic spring so that it weakens and allows the airflow and piston to open the choke valve further. If the engine is accelerated while it is cold there is less suction on the piston and the spring can then partially close the valve to ensure a richer mixture.

During cold running, a cam on the choke shaft contacts the choke-trip lever and the linkage partially opens the throttle linkage to ensure a fast idle speed. If the engine is accidentally flooded during starting, the linkage is so arranged that the choke valve will be mechanically held open at full throttle and the engine can be dried out by cranking it at full throttle.

2:6 Carburetor adjustment

Performance settings:

Before making adjustments to the carburetor check that the ignition system is in good condition and correctly adjusted (see **Chapter 3**). Similarly check the crankcase ventilation system and Thermactor system if fitted (see **Chapter 1**). Also check all vacuum operated systems for leaks or split hoses. If a system is faulty, air leaks into the inlet manifold will make it difficult or impossible to set the carburetor.

After the performance settings have been made, it is highly recommended that the vehicle is taken to a service garage so that exhaust gas analyzer equipment can be used to check that the emissions from the engine are within legal limits where applicable.

FIGS 2:9, 2:10, 2:11 and **2:12** show the positions of the setting screws on the various carburetors. The plastic limiter caps fitted to home models should not be removed. If a limiter is damaged, a new one must only be installed after the emissions have been properly checked.

If the carburetor adjustments are completely lost, for example after major repair, set them approximately so that the engine can be started and run. On models fitted with limiters turn the idle mixture screw as far anti-clockwise as the limiter allows. On other models, turn the mixture screw fully in to seat lightly but do not overtighten as this will cause damage to the screw seat. If the tapered end of the screw is worn a new screw can be fitted (on non-limiter models only). From the fully closed position undo the screw 1½ turns. Slacken off the curb idle speed screw until the throttle valve is fully shut. Where an anti-stall dashpot or a solenoid throttle positions is fitted it may be necessary to slacken its mountings or unscrew it slightly to allow the throttle to close completely. Turn in the curb idle speed screw until it just touches the stop on the throttle shaft and then turn it in a further 1½ turns.

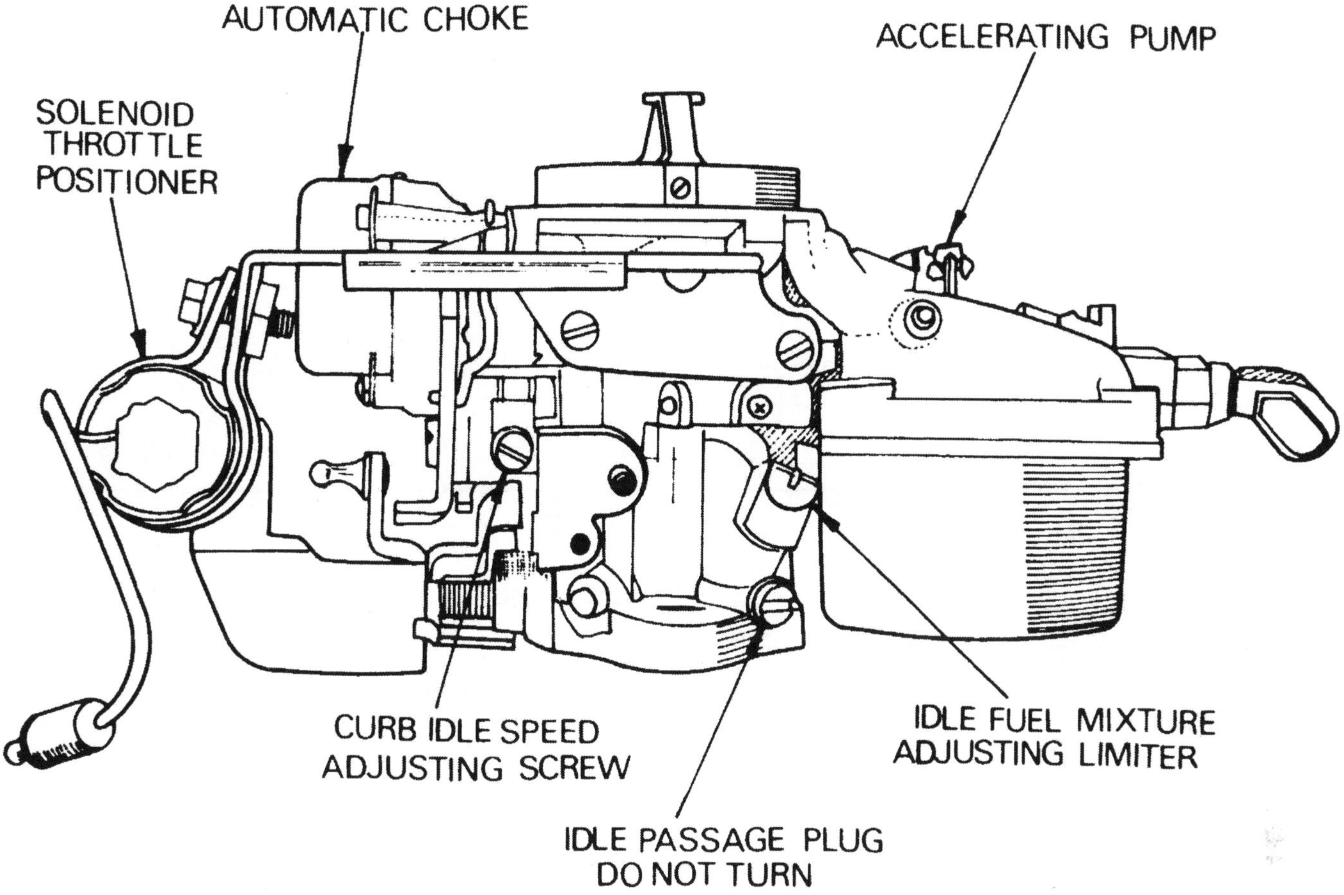

FIG 2:10 Adjustments: Carter RBS carburetor

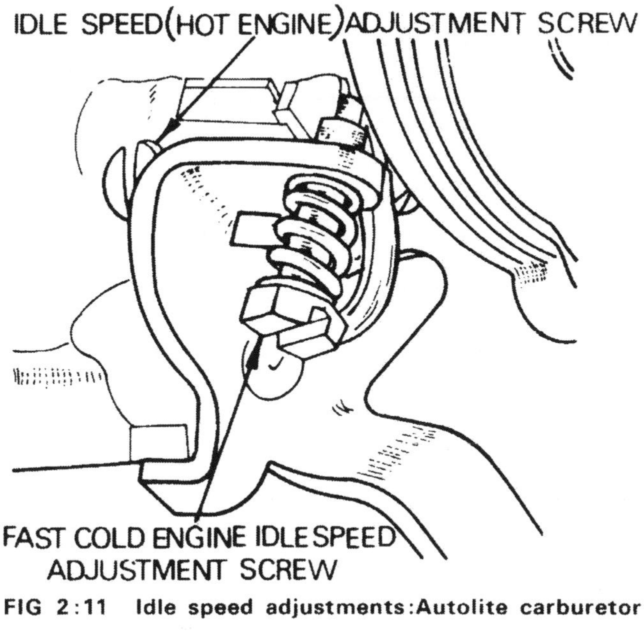

FIG 2:11 Idle speed adjustments:Autolite carburetor

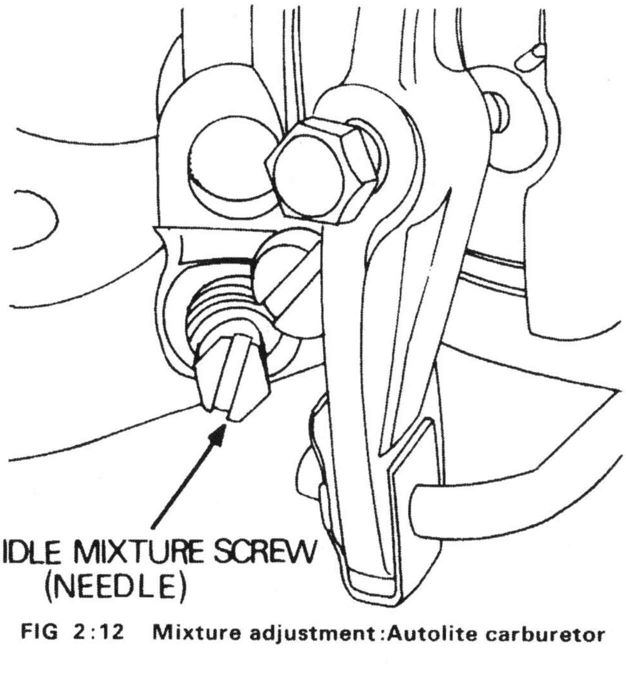

FIG 2:12 Mixture adjustment:Autolite carburetor

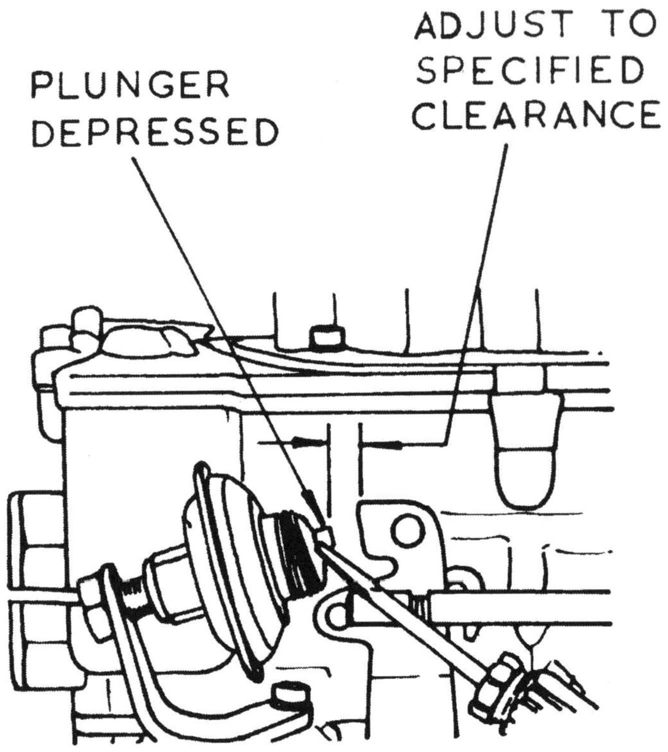

FIG 2:13 Anti-stall dashpot adjustment: Carter YF

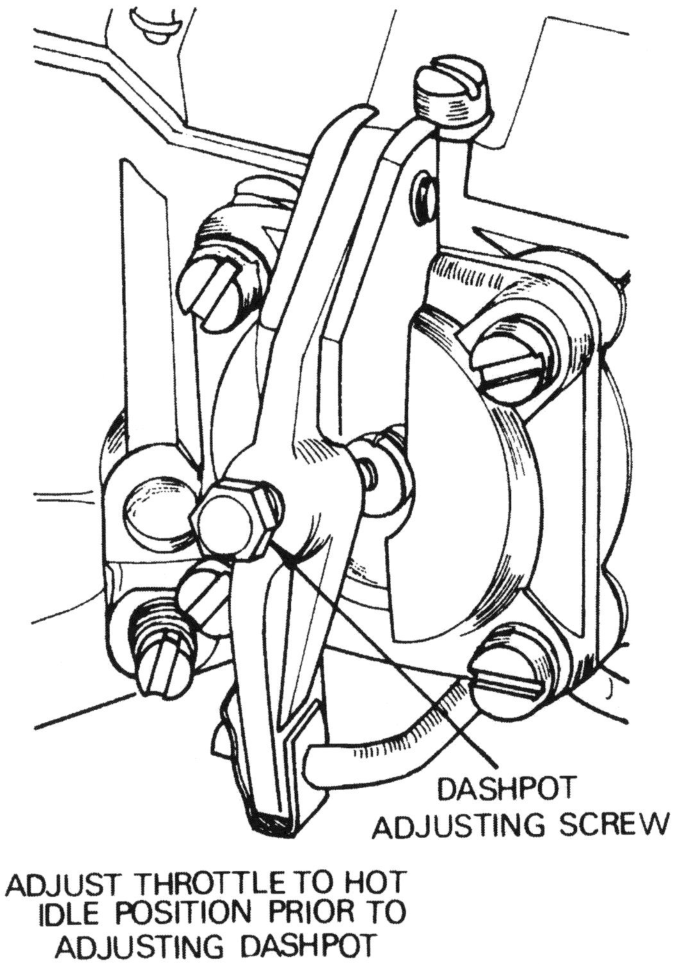

FIG 2:14 Anti-stall dashpot adjustment: Autolite

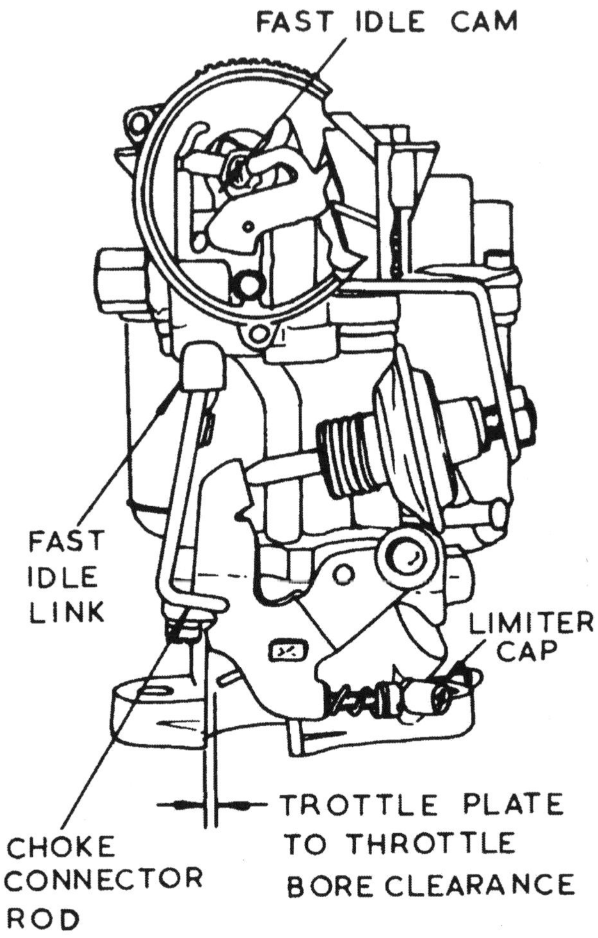

FIG 2:15 Fast idle speed adjustment by throttle clearance

With the rough settings established, or where they have not been disturbed, start the engine and allow it to warm up. Run it for 20 minutes at 1500 rev/min to allow the temperature to stabilize.

Connect an accurate electronic tachometer to the ignition system. On models fitted with manual-shift transmission, select neutral. On models fitted with automatic transmission, chock the rear wheels and firmly apply the park brake, then carry out the adjustments with Drive selected. On all models turn on the headlights to high beam so that the alternator is loaded. If air-conditioning is fitted select Cold on manual-shift transmission models: with automatic transmission the air conditioner must be OFF.

Adjust the curb idle to the specified speed. If the engine has been run for less than 50 miles set the speed about 25 rev/min below specification to allow for the increase as the engine loosens up with use. The correct curb idle speed for models with manual-shift transmission is 750 rev/min while that for models with automatic transmission is 550 rev/min. Where no solenoid is fitted the adjustment is made with the curb idle speed screw. On a carburetor fitted with a solenoid either turn the solenoid assembly in or out of the bracket (Autolite) or turn the solenoid plunger screw in or out (Carter) to obtain the specified speed. Then disconnect the solenoid wire at the bullet connector and adjust the curb idle speed screw to obtain 500 rev/min (with automatic transmission in Neutral). Reconnect the solenoid lead and open the

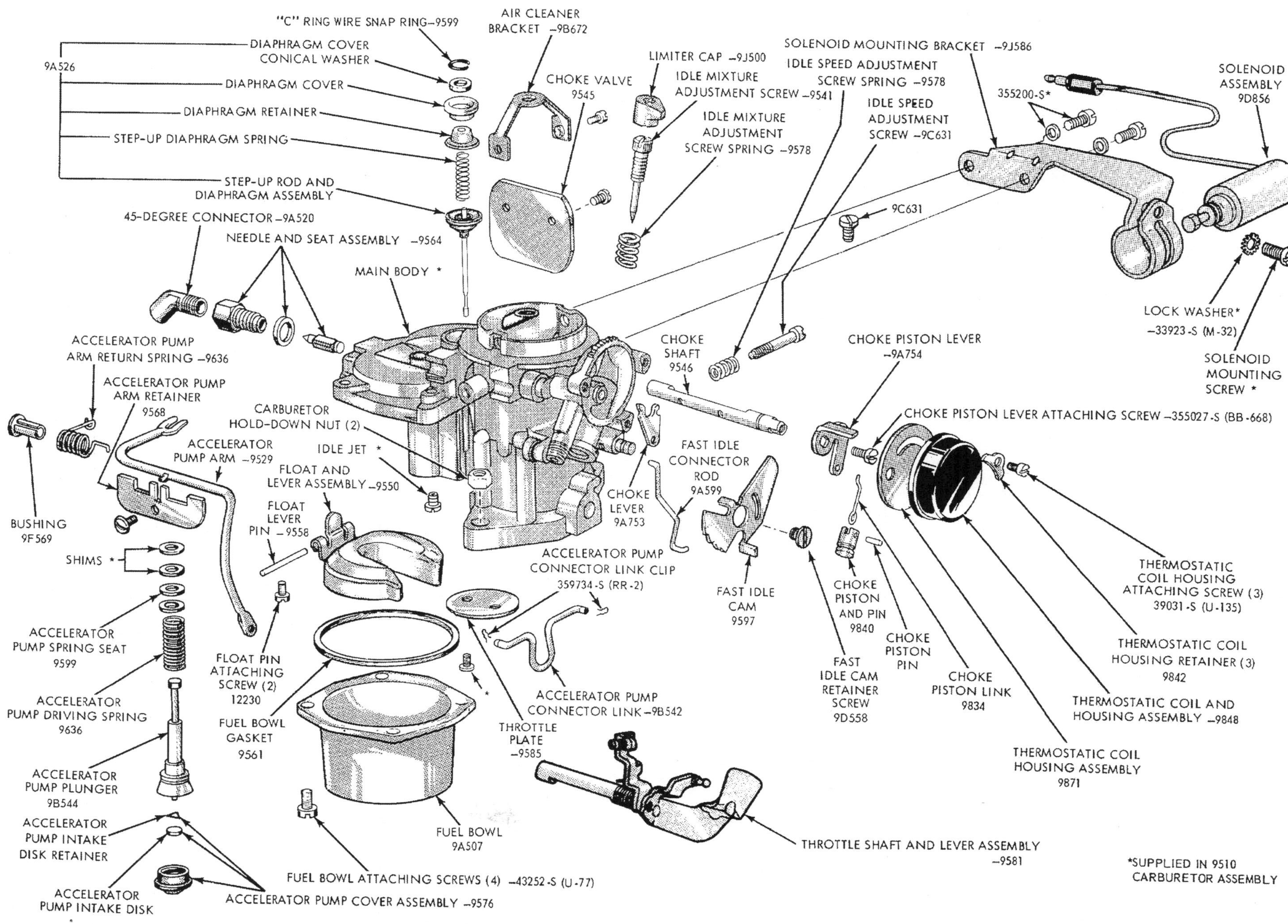

FIG 2:16 Components of Carter RBS carburetor

throttle by hand. The solenoid plunger should follow the throttle lever and remain extended as long as the ignition is switched on.

When the curb idle speed has been set, adjust the mixture screw to give the smoothest idle possible consistent with an even exhaust beat. Slight adjustments may have to be made to the curb idle speed to keep it to specification.

The performance settings should be carried out with the air cleaner installed, or if it has to be removed, checked and if necessary reset after it has been remounted.

Fast idle adjustment:

In addition to restricting the airflow into the carburetor, the choke mechanism raises the idling speed to keep the engine running reliably before it reaches working temperature. A fast idle cam in the linkage progressively lowers the speed towards the normal curb idle rate as the engine warms up and the choke opens. Before adjusting the fast idle speed ensure that the curb idle speed and mixture settings are correct.

Autolite and later Carter carburetors have a fast idle adjuster screw which contacts the edge of the fast idle cam (see **FIG 2:11**). The adjustment must be made with the engine running at normal working temperature and the headlamps switched on.

Remove the air cleaner and connect a tachometer to the ignition. Rotate the fast idle cam by hand until the arrow is aligned with the adjuster screw which should then be resting on the highest step of the cam. Adjust the screw to obtain the specified engine speed (see **Technical Data**).

Earlier Carter carburetors without a fast idle adjuster screw must be removed from the engine should adjustment be necessary. Open the throttle valve and close the choke valve to allow the fast idle cam to rotate to the cold start position. Close the throttle and use a rod or drill of the correct size to check the clearance between the throttle valve and bore as shown in **FIG 2:15**. On manual-shift models the clearance should be .030 inch, and for those with automatic transmission it should be .035 inch. If the clearance is incorrect, adjust it by bending the choke connector rod.

Anti-stall dashpot setting:

The anti-stall dashpot should only be adjusted after the curb idle speed has been set correctly.

FIG 2:13 shows the type of dashpot fitted to the Carter YF and Autolite 1101 carburetors. Hold the throttle linkage closed and use a screwdriver to press in the plunger of the dashpot. Measure the clearance between plunger and throttle lever. If it is not correct at .10 inch (Carter) or .08 inch (Autolite) slacken the dashpot locknut and adjust the position of the dashpot to correct it. Tighten the locknut when the setting is completed.

The integral dashpot fitted to the Autolite 1100 carburetor, shown in **FIG 2:14**, is set differently. Turn the dashpot adjusting screw outward until the end of it is clear of the plunger diaphragm assembly. Turn the screw in again until it just contacts the plunger assembly: then turn it further inward (clockwise) the specified number of turns. The setting varies with the serial number of the carburetor (see **Technical Data**).

2:7 Carburetor servicing

The adjustments covered in the previous section are likely to need regular attention to ensure efficient running and keep engine emissions within limits. This section deals with operations which will have to be carried out if faults occur of if normal adjustment proves difficult or impossible.

The most common cause of faults in the carburetor is dirt or gum blocking the jets or passages. Only a few areas are usually affected by wear: the commonest are the throttle valve spindle bearings and sealing areas in the body. Air leaks resulting from wear at these points will cause faulty idling.

The components of the various carburetors fitted to the range are shown in **FIGS 2:16, 2:17, 2:18** and **2:19**.

When dismantling of the carburetor is called for, it is advisable to have several small containers available so that components can be stored safely and in order as they are removed. Remember that the main casting may contain loose parts such as needles which can be mislaid if the assembly is inverted without care.

Tools:

Special wrenches are made for disassembly of carburetors and these should be used wherever possible. Well fitting tools must be used as the parts are made of soft metals and will be quickly damaged if badly fitting tools are used on them. Where an assembly is held by several screws they should be tightened progressively in rotation to avoid distortion.

Gaskets and seals:

Service kits of new gaskets are available and the appropriate gaskets and seals should always be renewed on assembly.

When installing gaskets, check them carefully. Be sure that all holes are cleanly punched with no shreds of material left to block them and also that they fit correctly.

Removal:

Take off the air cleaner. Disconnect the fuel line. Disconnect other tubes noting their positions for correct reassembly. Free the return spring and disconnect the throttle linkage. Slacken the nuts evenly and remove the carburetor.

Before re-installation, check the carburetor, manifold and spacer for damage to the flanges. High spots or burrs can be smoothed down by careful use of a fine file. Excessive distortion (usually caused by overtightening the nuts) can be lapped down using fine emery cloth, or grinding paste, spread on plate glass.

Cleaning:

Special carburetor cleaning fluids are available but they must be used with caution following the makers' instructions, as some are caustic and can cause injury. Denatured alcohol or clean gasoline may be used on all parts of the carburetor and harsher solvents such as trichlorethylene may be used provided that they only contact metal parts.

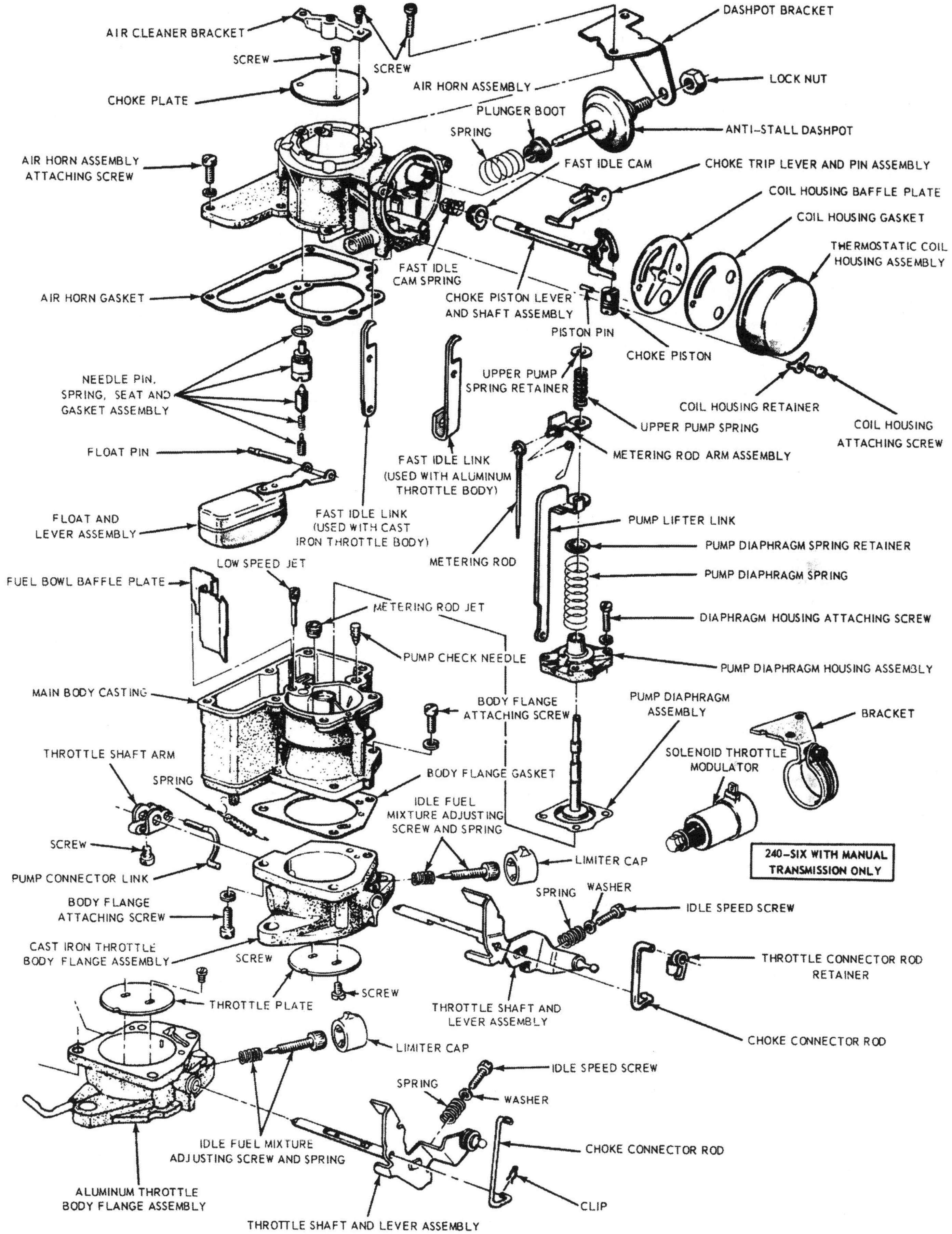

FIG 2:17 Components of Carter YF carburetor

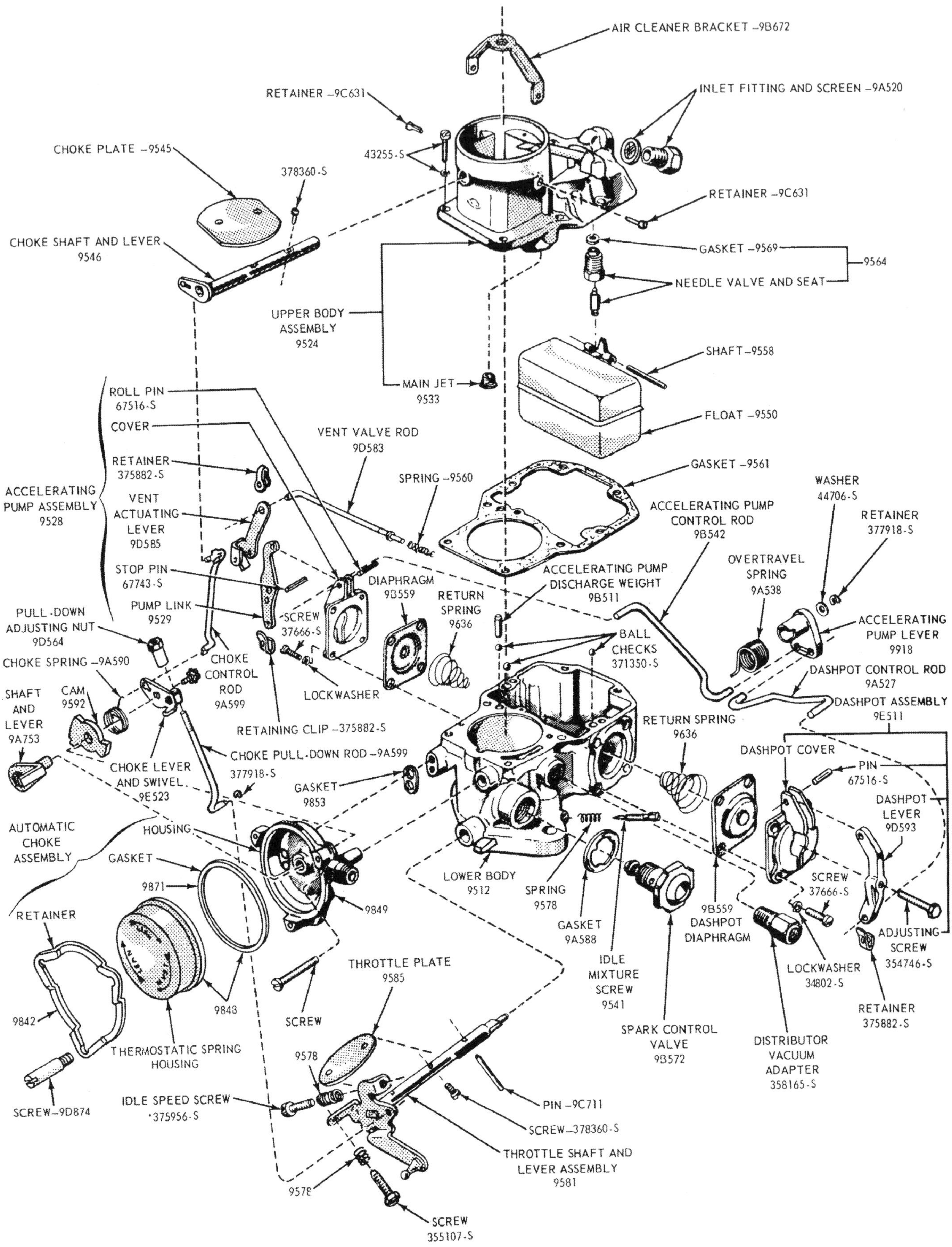

FIG 2:18 Components of Autolite 1100 carburetor

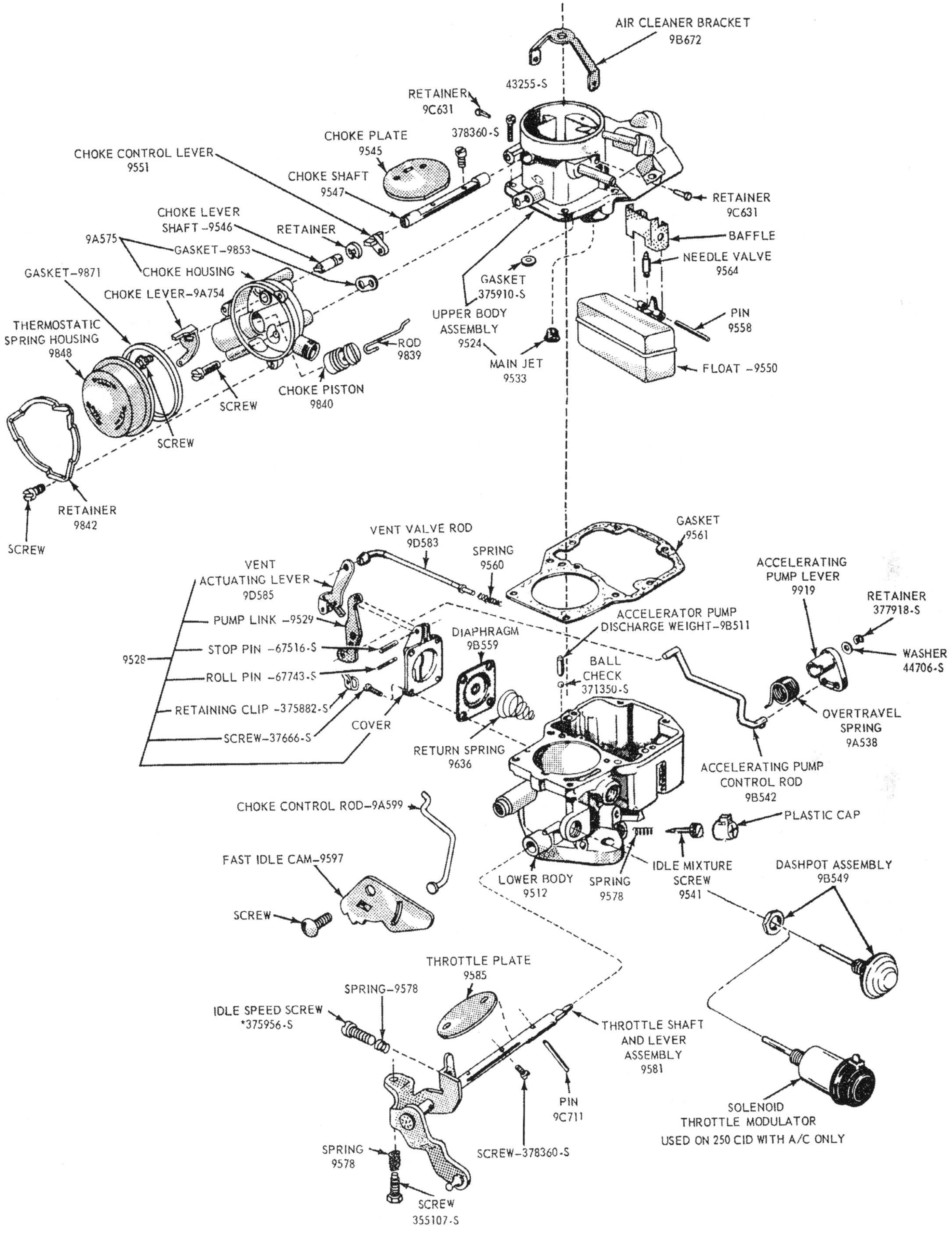

FIG 2:19 Components of Autolite 1101 carburetor

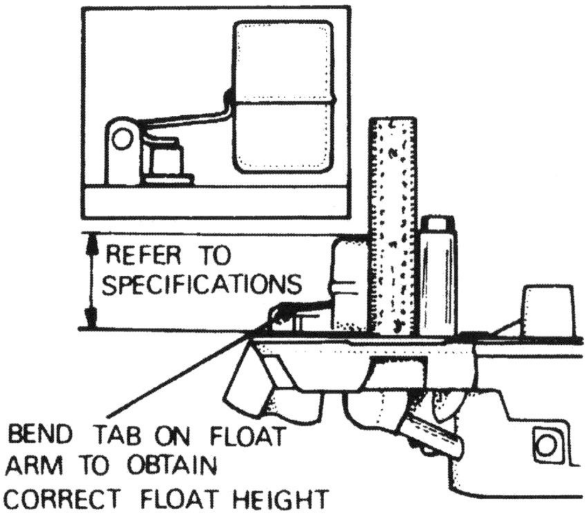

FIG 2:20 Setting fuel level: Autolite

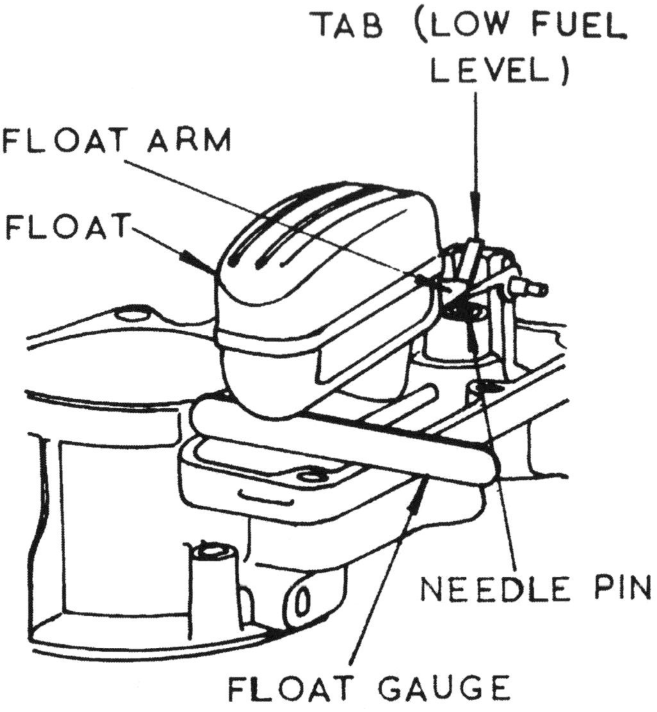

FIG 2:22 Setting fuel level: Carter YF

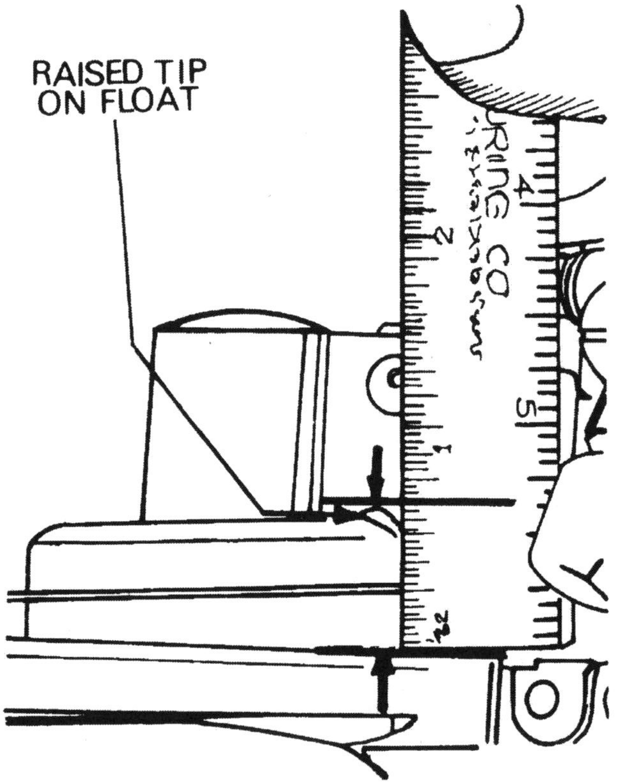

FIG 2:21 Setting fuel level: Carter RBS

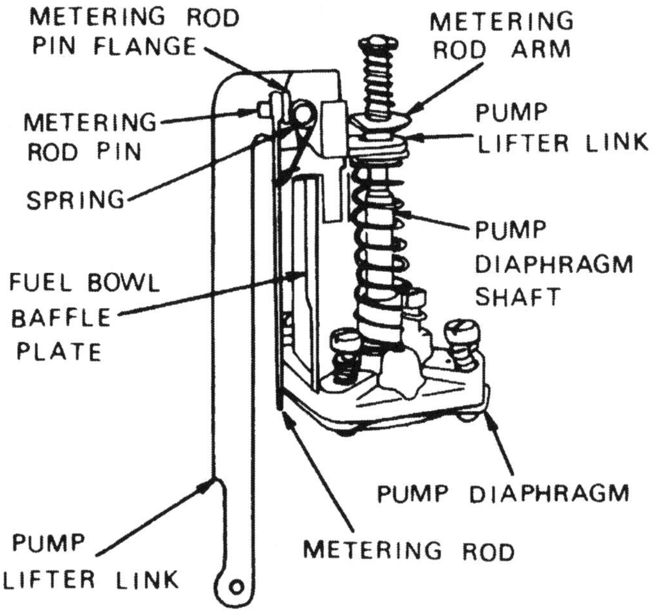

FIG 2:23 Carter YF accelerator pump assembly

Use a small stiff brush to remove deposits. Jets and passages must not be poked through with wire, bristle or any form of tool as they will become worn by this treatment. Flush them with cleaning solvent and then blow through with compressed air against the direction of normal flow. It is advisable to clean and replace each jet in turn to avoid confusing one with another and fitting them in the wrong position.

Throttle and choke valves:

Before removing the valves, lightly mark them with letters or a code on valve and casting so that they can be put back in their original positions.

Remove the linkage from the shaft and then file off the staking from the screws holding the valves. Take out the screws, slide the valves out of their slots and remove the shafts.

To cure wear, try installing a new shaft only, but if there is still excessive play the body must be renewed. On Carter models the separate throttle housing can be renewed, but care must be taken in ordering the correct replacement as some types have light alloy housings while others have cast-iron housings. On Autolite models a complete carburetor will be required.

Install the shaft and slide the plate back into position, noting that the notch if present must align with the idle

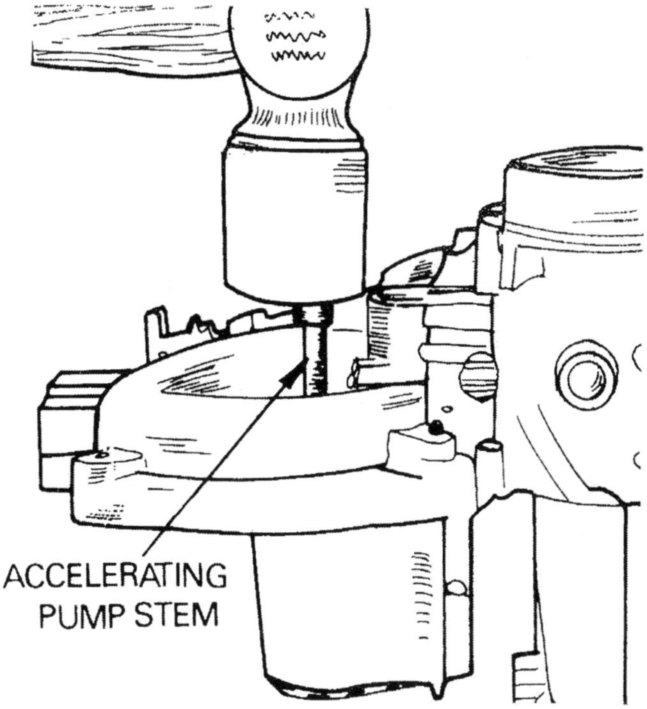

FIG 2:24 Removing Carter RBS accelerator pump assembly

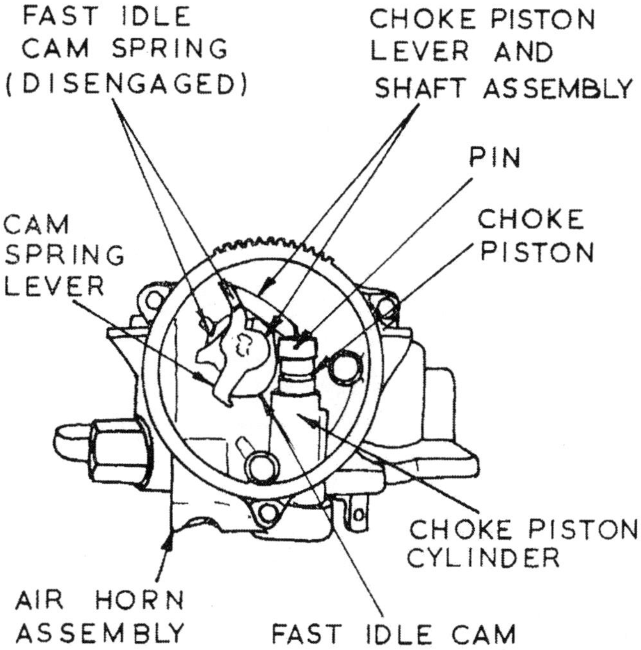

FIG 2:25 Automatic choke assembly with spring housing removed

ports. Tighten the screws just enough to hold the plate. Operate the shaft and gently move the plate until the valve opens and closes freely with no binding. Tighten the screws fully and check for free movement again. Lock the screws by staking the ends. When staking the screws, support the shaft on a block of wood or metal to avoid damage or bending.

Flooding or fuel starvation:

In the absence of any leaking gasket or blockage, the most likely cause of flooding or fuel starvation is a defective float needle valve assembly.

In the case of the Carter RBS carburetor, detach the fuel bowl. With the other types of carburetor, remove the air horn assembly which incorporates a fuel bowl cover.

Remove the float lever pin (which may be a push fit or located by a set screw), detach the float and lever, take out the needle valve and check the end of the needle and the seating. If worn or damaged, these should be replaced. The needle must move freely. Check the float for leaks or damage and renew if necessary. (A leaking float should only be repaired in emergency.) Reinstall the needle valve and float.

Fuel level setting:

An incorrect fuel level in the carburetor bowl will affect the mixture strength throughout the speed range and so can impair performance and economy as well as affecting engine emissions. Very high or low levels will also cause flooding or starvation.

To set the float level, first invert the air horn or body assembly to which the float is hinged so that the float is resting against the needle valve. On an Autolite carburetor, measure the vertical distance from the air horn sealing face to the top of the float (see **FIG 2:20**): this should be $1\frac{3}{32}$ inch. On the Carter type RBS, measure the vertical distance from the bowl mounting flange to the raised tips on the outer ends of the float, which should be $\frac{9}{16}$ inch at both sides of the float (see **FIG 2:21**). With the Carter YF carburetor a gauge strip cut to $\frac{7}{32}$ or $\frac{3}{8}$ inch width (see **Technical Data**) is used as shown in **FIG 2:22**. Holding the assembly at eye level check that the float arm is just resting on the needle valve when the float is supported by the gauge.

In each case bend the float lever if necessary to obtain the required setting, keeping the float square with the hinge pin in the lateral plane. When bending the lever do not put any load on the needle valve or the seating will be damaged. Where a tab is fitted to limit the downward movement of the float take care that this is not bent.

Poor acceleration or hesitation:

The accelerator pump should inject a good spurt of fuel when the throttle linkage is moved smartly through its full range: poor acceleration or hesitant pick-up will result from faulty operation.

Remove the check valves and weight for the accelerator pump, after taking out the screw that holds the weight in place, and check that they are operating correctly and not blocked, gummed or damaged. Disconnect the linkage and remove the accelerator pump diaphragm by taking out the attachment screws. On the Carter YF carburetor the pump and metering needle are removed as a sub-assembly (shown in **FIG 2:23**). The plunger-type pump on the Carter RBS is removed as shown in **FIG 2:24**. First remove the snap ring from the shaft, then use a light hammer on the top of the shaft to tap the assembly out of the lower end of the casting.

Check the parts, renewing them as necessary, and reassemble in the reverse order of dismantling. On the Carter RBS type fit a new pump cover to provide a good seal, using a $\frac{5}{8}$ inch socket to drift it gently into place. Apply pressure only round the edges, not on the center of the cover.

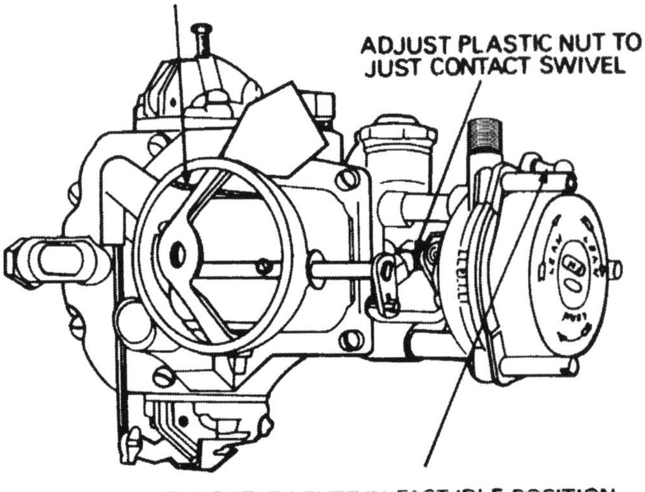

FIG 2:26 Autolite 1100 choke linkage adjustment

Automatic choke:

The parts of a typical automatic choke mechanism, with the thermostatic coil spring and housing removed, are shown in **FIG 2:25**. If the spring or its housing is defective, the two must be renewed as an assembly.

Note the relative positions of the index marks on the spring housing and choke body. Remove the housing retainer springs and take off the retainer, housing assembly, gasket and baffle plate.

Swing the choke lever and check that the piston is free to move in its bore. If there is any binding, the piston must be removed and cleaned thoroughly but without the use of abrasives. If the choke piston lever and fast idle cam assembly has to be dismantled, it is most important that the relative positions of the parts are noted for correct reassembly. To remove the choke valve spindle, refer to the information on throttle and choke valves already given.

The parts are installed in the reverse order of dismantling. Be sure that the eye of the thermostatic spring locates on the tang of the lever. Turn the housing to align the correct index marks before finally tightening the screws which secure it.

2:8 Carburetor static settings

These are set on manufacture and do not normally require altering. They may need checking after dismantling the carburetor or if faulty operation occurs.

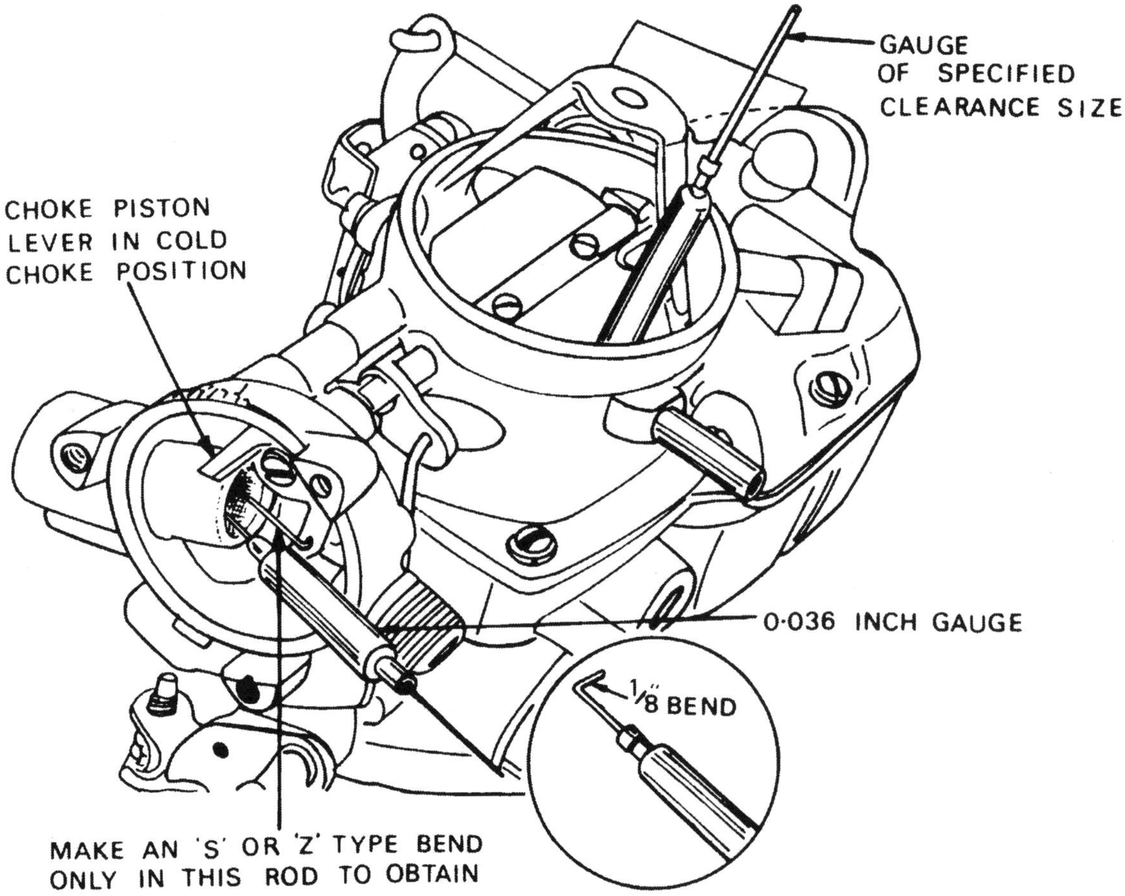

FIG 2:27 Automatic choke adjustment: RBS and 1101 types

Autolite 1100 choke adjustment:

Be sure the fast idle speed is set to specification: then position the fast idle screw on the highest step of the cam (cold position). Insert a gauge or drill of the correct thickness (see **Technical Data**) between the lower edge of the choke valve plate and the bore of the body, as shown in **FIG 2:26**. With the valve plate held against the gauge, adjust the plastic nut on the choke pull down rod until it contacts the swivel on the choke lever assembly.

Autolite 1101 and Carter RBS choke setting:

Remove the choke spring housing, and refer to **FIG 2:27**. Gauge thicknesses are given in **Technical Data**. Bend the wire gauge as shown and insert it between the lower edge of the choke piston slot and the upper edge of the righthand slot in the housing. Turn the choke lever counterclockwise and keep light pressure on it to hold the gauge in place. Open the throttle enough to ensure that the fast idle screw is not contacting the cam. Check the clearance between the choke valve plate and bore with a suitable gauge or drill. If necessary make a careful Z-bend in the link between the choke-piston and lever to adjust the choke plate clearance. Be sure that the bend in the link will not foul in operation.

Install the spring housing as detailed in **Section 2:7** and check the performance adjustment.

Accelerator pump setting:

On the Carter RBS carburetor, the stroke of the pump can be checked as shown in **FIG 2:28**. Measure the height of the pump shaft from the flat surface of the main casting with the throttle both open and closed. The stroke, which is the difference between the two measurements, should be .40 inch. To adjust it, use a screwdriver to open or close the U-bend in the connector link.

Carter YF metering needle:

This is adjusted as shown in **FIG 2:29**. Back out the curb idle adjuster screw until the throttle valve plate is fully closed in the bore. Press down on the diaphragm shaft until the metering rod arm contacts the lifter link (see **FIG 2:23**) and then turn the adjusting screw until the metering rod just bottoms in the casting. Release the diaphragm shaft and turn the adjusting screw one more turn clockwise for the correct setting.

2:9 Trouble shooting guide

(a) Leakage or insufficient fuel delivery

1 Air vent to fuel tank blocked
2 Fuel lines partially blocked
3 Fuel filter partially blocked
4 Defective fuel pump
5 Defective float in fuel bowl of carburetor
6 Defective carburetor needle valve assembly
7 Loose hose connections
8 Filter screen under carburetor needle valve partially blocked
9 Loose carburetor parts or defective gaskets and seal

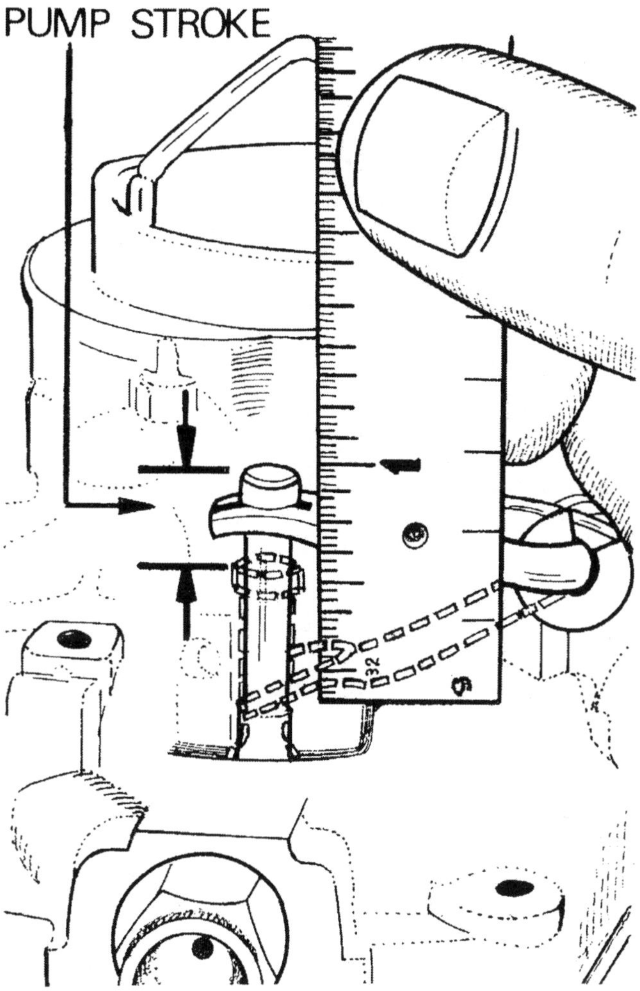

FIG 2:28 Measuring the pump stroke: Carter RBS

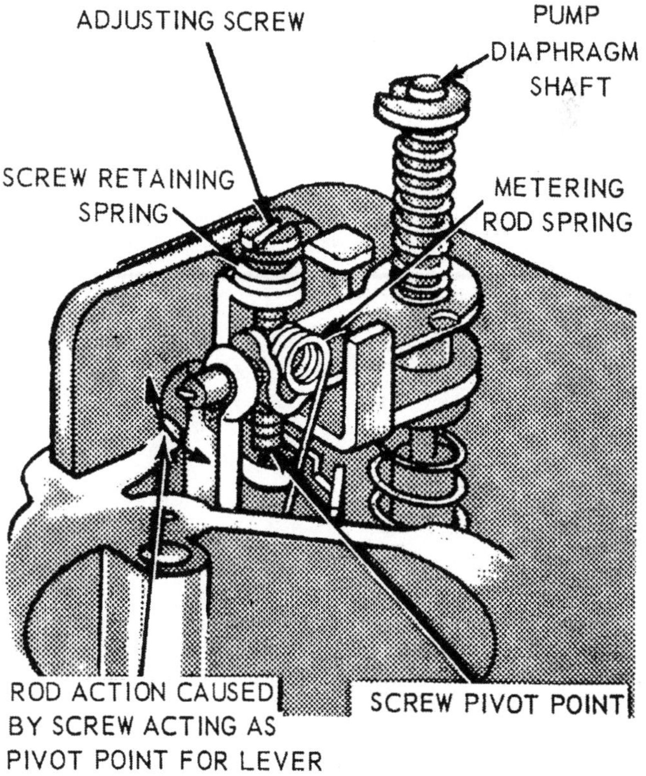

FIG 2:29 Adjusting the metering rod: Carter YF

(b) High fuel consumption

1 Carburetor incorrectly adjusted
2 Dirty air cleaner
3 Automatic choke defective
4 Fuel leakage
5 Fuel level too high in carburetor
6 Loose or worn jets
7 Incorrect jets fitted
8 Brakes binding
9 Tires underinflated
10 Engine running too hot
11 Vehicle regularly overloaded

(c) Idling speed too high or difficult to set

1 Defective air cleaner
2 Defective Thermactor system
3 Air leaks from a vacuum operated system
4 Air leaks around carburetor flange
5 Carburetor incorrectly set
6 Throttle linkage sticking
7 Defective automatic choke
8 Worn carburetor throttle valve plate or shaft

(d) No fuel delivery

1 Carburetor needle valve jammed closed
2 Vent in fuel tank blocked
3 Fuel filter, strainer, screen or lines blocked
4 Defective fuel pump
5 Bad air leak on suction side of pump
6 No fuel in tank

CHAPTER 3

THE IGNITION SYSTEM

3:1 Description

A schematic layout of the ignition system is shown in **FIG 3:1**. The illustration shows a V8 distributor cap, but the six cylinder system is otherwise similar. The distributor shaft, driven by the engine, carries a rotor arm at the upper end and a cam which operates a low tension contact breaker. The body of the distributor is set so that the contact breaker points are just opening and the rotor arm is pointing towards the correct electrode in the distributor cap as a piston reaches the firing point in its cylinder.

When the contact breaker points are closed, current flows through the primary windings of the ignition coil inducing a magnetic field around them. When the contact breaker opens the current is cut off. A capacitor (condenser) in the distributor controls arcing across the points and ensures that the cut off is sharp, so the magnetic field around the primary windings collapses rapidly. The collapse induces a very high voltage in the many turns of the secondary windings in the coil. This high tension current is led to the centre of the distributor cap and from there by a carbon brush to the rotor arm. It jumps the small gap between the rotor arm and the appropriate electrode and is taken by high tension lead to the spark plug. The resulting spark across the electrodes of the plug ignites the fuel/air mixture in the cylinder. Both high and low tension circuits are completed on the earth return principle through the structure of the engine.

The ignition coil is designed to function at a lower voltage than that of the battery. When the engine is running it is fed through a resistance wire which provides this reduced voltage. When the starter motor is operated the coil is fed directly from a terminal on the starter relay with no intervening resistance. This compensates for the voltage drop caused by the heavy current drain of the starter motor so that the coil operates at full efficiency for easy starting.

The optimum point of ignition for efficient combustion will vary with the speed and load on the engine, so automatic controls are provided to advance or retard the ignition as appropriate. On most engines in the range both centrifugal and vacuum operation is employed, but some earlier engines without emission controls have only the latter.

In the centrifugal system weights are fitted between the distributor shaft and contact breaker cam, articulated so that their outward movement advances the cam in relation to the shaft. The weights are held in by calibrated

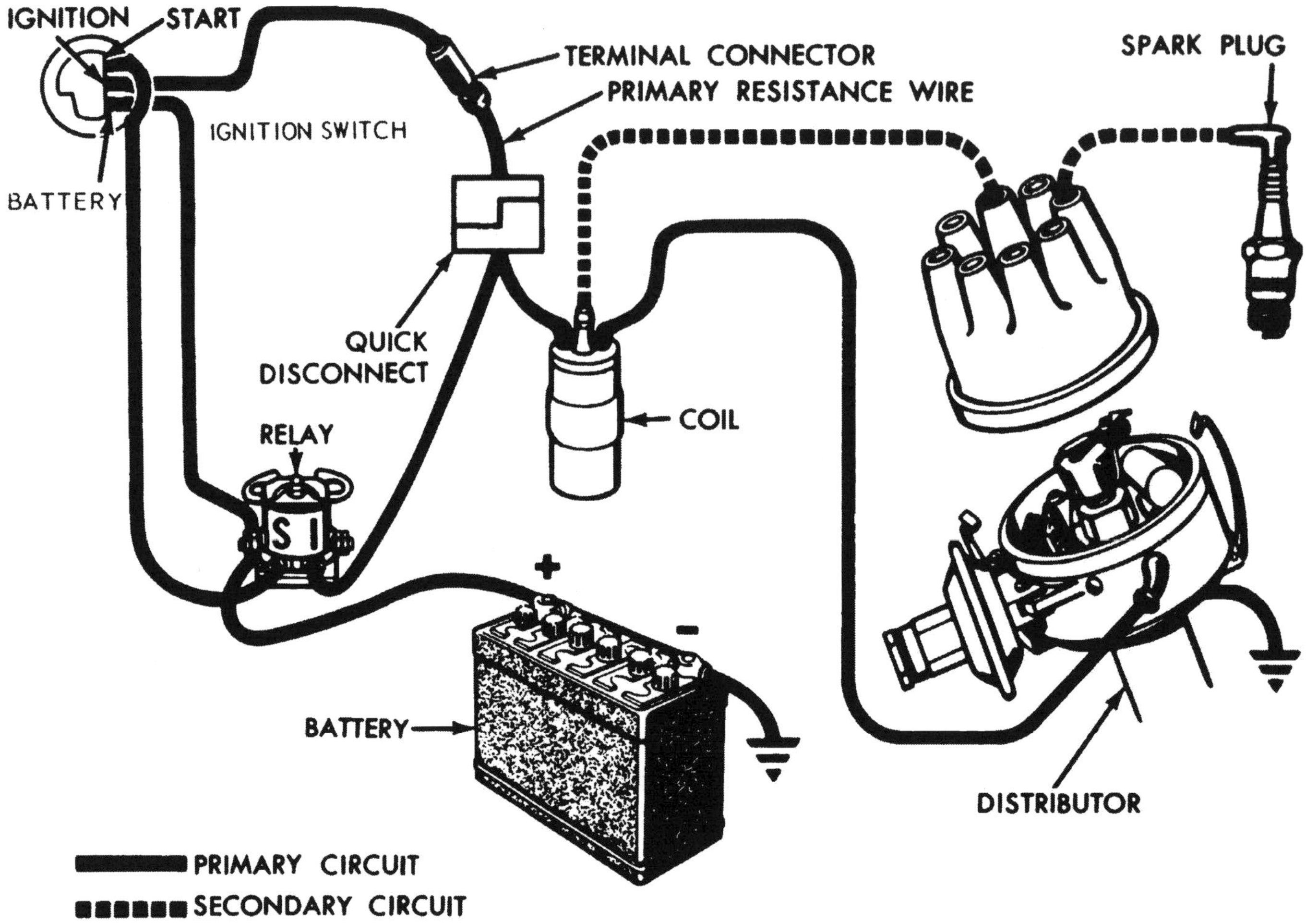

FIG 3:1 Typical ignition system

springs and move outward progressively under the action of centrifugal force as the speed of rotation rises.

The vacuum unit utilizes the depression generated in the intake system of the engine, which is proportional to load and throttle opening. The suction displaces a diaphragm which rotates the breaker base plate about the cam to alter the timing. Earlier models have a single diaphragm unit linked by small bore tube to the carburetor.

Later models have a dual diaphragm unit linked both to the carburetor and the intake manifold: this system is more fully described in **Section 3:8**.

3:2 Maintenance

Regularly wipe the HT leads, ignition coil and distributor cap clean with a soft dry cloth. A build up of dirt or moisture will cause arcing and reduced efficiency in the system.

The distributor points should be checked and adjusted at intervals of 12,000 miles and at the same time the distributor should be lightly lubricated. Clean the spark plugs at 6000 miles and renew them at 12,000 miles.

Lubrication:

Free the clips and remove the distributor cap from the distributor, leaving it attached to the HT leads. Firmly but squarely pull the rotor arm off from the cam. Pour a few drops of oil onto the felt pad in the top of the cam. With a clean finger smear a little high temperature lubricant around the cam face. Pour a little SAE.10W oil into the oil cup on the distributor body. Wipe away all surplus lubricant and make sure that none goes onto the actual contact points. Wipe the rotor arm and distributor cap all over with a clean soft cloth and install them on the distributor, making sure that the lug on the rotor arm fits into the slot in the cam.

If the distributor cap and rotor arm are very dirty, clean them with a stiff brush and denatured alcohol as a solvent. Be sure to clean out all crevices in the distributor cap.

3:3 Breaker points

The installation of the breaker plate is shown in **FIG 3:2**.

Cleaning:

For maximum efficiency the points should have a frosted, slate grey color. Oil or grease on the points will blacken them and also show up as a smudgy black line on the breaker plate under them. Cleaning is carried out by brushing the points with a small stiff brush using chloroform or denatured alcohol as a solvent. Renew the points if they are badly blackened.

Points which are blue in color or badly burnt should be renewed. A defective capacitor, defective, or incorrectly connected radio suppressor, incorrect alignment of the points, or the voltage regulator in the charging system set too high, will cause the points to burn or blacken.

In normal use metal will be transferred from one point to the other, causing a build-up of metal on one and removing it to form a pit on the other. No action need be taken until the build-up reaches the same dimension as the correct points gap. Once the build-up has reached this size, renew the contact breaker set. **Only in an emergency should the points be cleaned by use of a smooth file or by grinding.**

Renewing the breaker points:

Free the clips and lift off the distributor cap. Pull the rotor arm squarely and firmly off the cam. Disconnect the wires to the breaker baseplate and disconnect the link for the vacuum unit. Remove the screws and take out the breaker assembly complete with the capacitor. Take out the screws and remove the contact breaker set from the baseplate.

Install the new parts in the reverse order of removal, after washing them with denatured alcohol to remove any protective. Crank the engine until the points are fully closed and check their alignment. Misalignment, particularly on vented points, will cause rapid wear, pitting and overheating. The correct and incorrect alignments of the points are shown in **FIG 3:3**. If the points are not correctly aligned, bring them into line by carefully bending the bracket of the fixed contact point. Do not attempt to bend the moving contact arm. Once the points are installed and aligned, set the gap correctly.

Spring tension:

The tension of the moving contact spring should be checked using a suitable spring gauge. Apply the load so that it is at right angles to the contact faces and pull until the contacts just part, noting the reading. A limited variation in the tension can be made by slackening the nut that secures the spring and moving it as shown in **FIG 3:4**. If the tension cannot be brought to specifications a new contact breaker set must be fitted.

Adjusting gap:

New contact points can be set using feeler gauges, as the surfaces are not yet pitted. On used contacts the feeler gauge will bridge the pit to give a false reading. The method of setting the points with feeler gauges is shown in **FIG 3:5**.

Used points should be set using a dwell meter or oscilloscope and therefore garage equipment is required.

The gap should be .027 inch and the dwell angle 35 to 40 deg. at idle speed.

3:4 Ignition faults

A persistent misfire in the engine can be caused by defective ignition. The quickest way of determining faults is to connect the vehicle to electronic test equipment and the oscilloscope will then quickly pinpoint the fault.

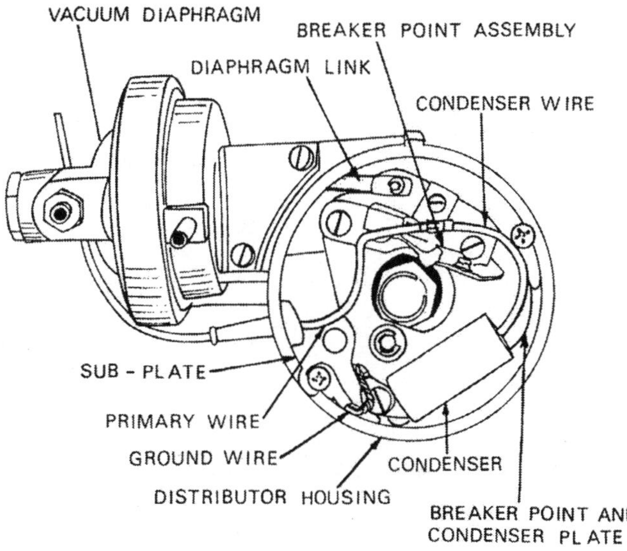

FIG 3:2 Breaker plate installation

If test equipment is not available, check the fuel system for faults which can be the cause of misfiring (see **Chapter 2**). Remove the distributor cap and check the contact breaker points for damage or incorrect gap as well as making sure that they are clean.

When likely faults have been checked for, start the engine and set it running at a fast idle speed. Either block the accelerator linkage open or adjust the curb idle screw on the carburetor.

Use rags, dry gloves or special insulated tongs to disconnect each plug lead from its plug in turn while listening carefully to the note of the engine. **When disconnecting the plug leads, twist and pull directly on the moulded cap and not on the lead itself.** If a cylinder is not firing, the note of the engine will not change but if a cylinder is firing there will be a drop in speed. Listen carefully as the other cylinders can easily mask one cylinder not firing.

When the defective cylinder(s) has been found, stop the engine. Insert an old spark plug or length of metal rod into the moulded cap of the suspect lead and restart

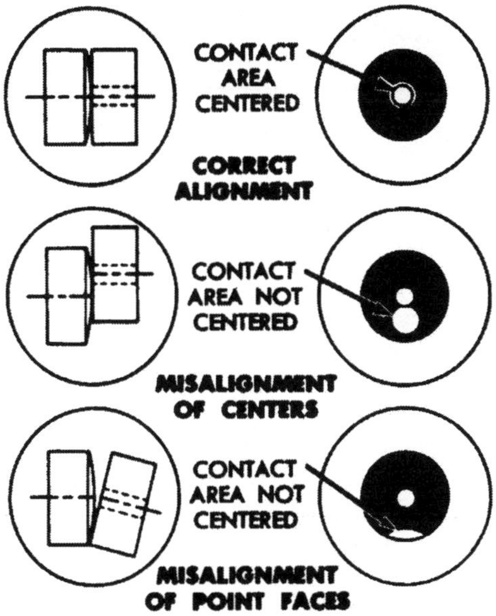

FIG 3:3 Breaker point alignment

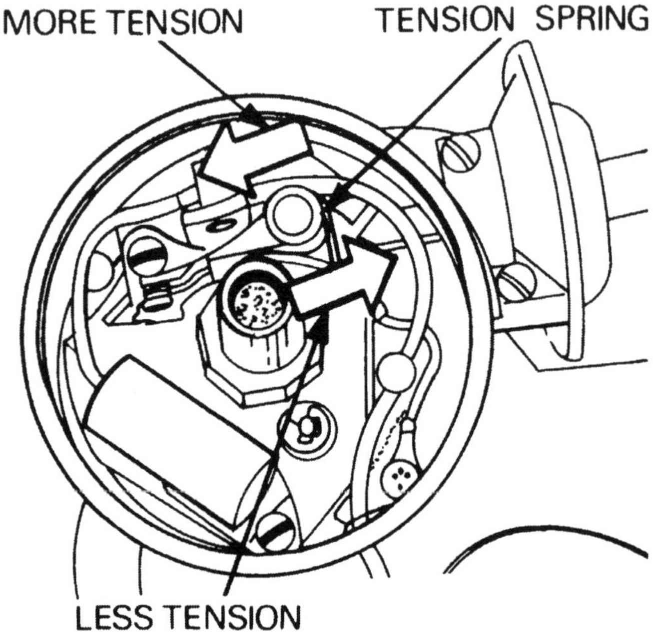

FIG 3:4 Adjusting contact breaker spring tension (Typical)

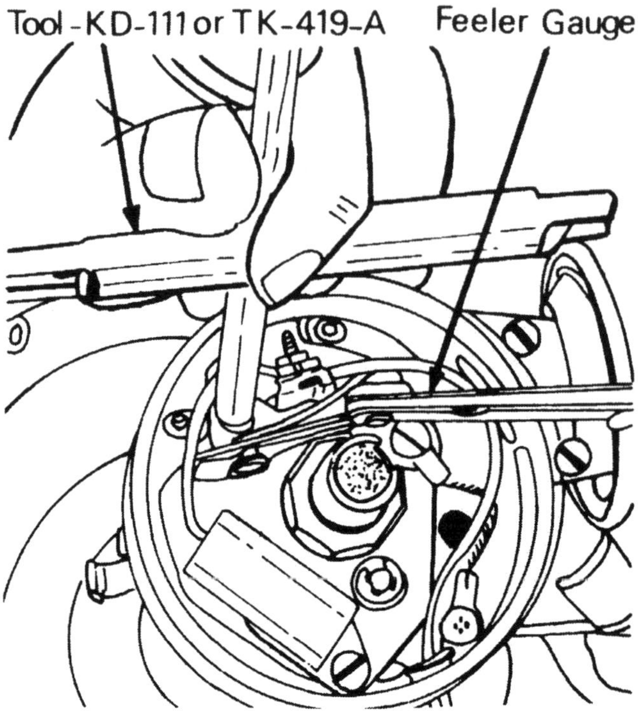

FIG 3:5 Adjusting contact breaker point gap (Typical)

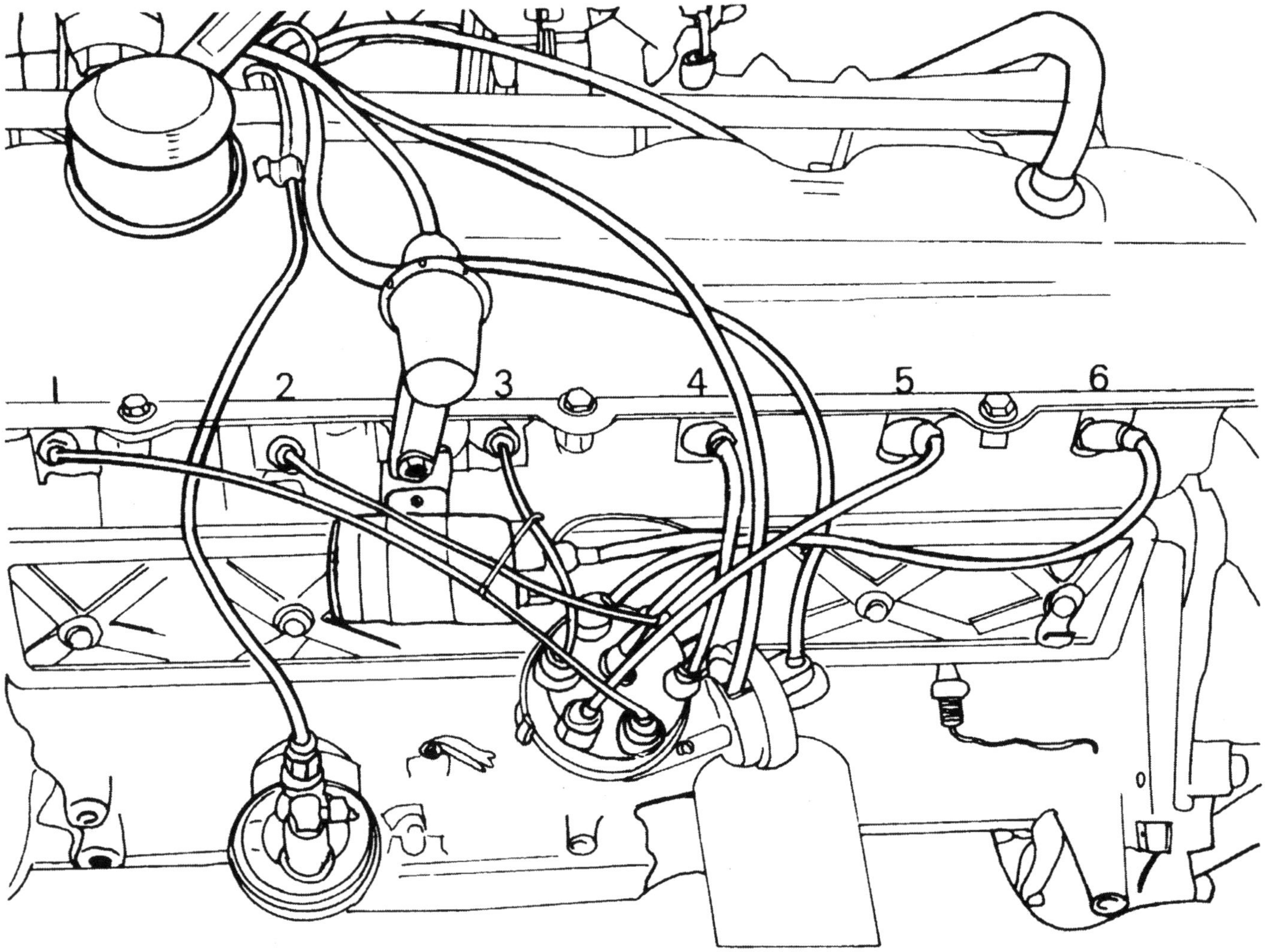

FIG 3:6 Layout of ignition leads

the engine. Hold the rod or spark plug approximately $\frac{3}{16}$ inch away from a convenient grounded metal part of the vehcile. **Do not allow the lead near the fuel system because of the fire risk.** If the lead and ignition are satisfactory there will be a regular blue spark. If it is found that the sparks are weak on all cylinders, disconnect the HT lead from the central electrode of the distributor cap and repeat the test using the lead while cranking the engine on the starter motor. The sparks will be faster than for a single cylinder but they should still be fat, healthy and blue.

If the sparks are weak, irregular or thin and reddish in color, check the HT lead and renew it if it appears burnt, cracked or perished. Repeat the test to see whether there is an improvement.

If the sparks are still weak, remove the distributor cap and rotor arm. Thoroughly clean both of them, removing the HT leads if necessary, using a small stiff brush and chloroform or denatured alcohol as a solvent. Check the rotor arm for damage and renew it if defective. Check the distributor cap carefully for hairline cracks or thin black lines between electrodes or from an electrode to the edge of the cap. The thin black lines are 'tracking' where the high voltage has burnt a path. In either case renew the distributor cap as no repairs are possible.

If the sparks are good, then the spark plug is at fault. Remove the spark plug so that it can be cleaned and tested or fit a new spark plug into its place. If the sparks from the central HT lead are poor and the lead is in good condition then the fault lies with the ignition coil.

Testing the primary circuit:

Crank the engine until the contact points are closed, with the distributor cap removed. Switch on the ignition and flick the points apart with a fingernail. If there is current flowing there will be a small low voltage spark across the contacts.

A more accurate test is to disconnect the primary wire from between the terminal on the ignition coil and the low-tension terminal on the distributor. Reconnect it using a low-wattage 12-volt test lamp in series. Remove the distributor cap so that the contacts can be observed. Slowly crank the engine over with the ignition switched on. If the system is satisfactory the lamp will come on as the points close and then go out again as they open. It is advisable to remove all the spark plugs so that the engine can be cranked over by hand, rotating the crankshaft damper.

If the lamp fails to come on at all, trace back through the wiring until the fault is found. Use a voltmeter or test lamp to find the live points in the circuit and then work forwards until the break or fault is found. Note that the resistance wire will lower the voltage, so do not expect full battery voltage beyond the resistance wire. Stuck open contacts can also be a cause.

If the lamp comes on and does not go out at all there is a shortcircuit in the distributor. Recheck the contacts to make sure that they are opening, and that they are correctly installed. Check the wires in the distributor for faulty connection or frayed insulation. If the points and wires are satisfactory, repeat the test with the condenser disconnected. If the lamp now operates correctly there is a shortcircuit in the condenser.

Resistance wire:

This can be checked using a suitable resistance meter, and the resistance should be 1.3 to 1.4 ohms. The wire can also be checked using a voltmeter. Disconnect the leads from the ignition coil and connect the DIST terminal (normally connected to the distributor) to a good ground using a jumper lead. Undo the quick disconnect connector and join the voltmeter between the end of the resistance wire and the other terminal on the ignition coil. Switch on the ignition and if the voltage reading lies between 4.5 and 6.6 volts the resistance wire is satisfactory. Renew the wire if the voltage is outside specifications.

Condenser:

Condenser failure is not a common fault. If the paper insulation in the unit fails the spark usually erodes away the foil in the area so that the condenser is self-healing. A shortcircuited condenser is easily detected as described earlier.

Open circuit is more difficult to detect without special equipment though it may be suspected if starting is difficult and the contacts are excessively 'blued' or burnt. The best check is to fit a new condenser and see whether this cures the fault.

3:5 Removing and installing the distributor

Disconnect the primary wire at the distributor body. Disconnect the vacuum hoses from the vacuum unit, labelling them if the colors have faded. Remove the distributor cap, leaving it attached to the HT leads.

Make two aligning marks, one on the distributor body and the other on the engine. Make a mark on the top edge of the distributor case by the brass end of the rotor arm. Provided that the engine is not turned and the distributor installed so that all the marks are again in alignment, the ignition timing will not be lost. Remove the distributor clamp hold down bolt and withdraw the distributor from the engine.

Install the distributor in the reverse order of removal. When pressing the distributor back into place, allow for the rotation of the gears as they mesh, to ensure that the rotor arm is still pointing at the mark when the distributor is fully in place. Be sure that the distributor shaft engages fully and correctly with the oil pump intermediate shaft. If the crankshaft has been rotated since the distributor was removed it will be necessary to set the ignition timing from scratch.

If the HT leads have been removed, install them in the correct order into the distributor cap. When removing or installing the leads pull and twist directly on the moulded cap itself, as pulling on the wire can damage it. The correct installation of the leads is shown in **FIG 3:6**. The ends of the leads must be clean and a little silicone grease may be used as waterproofing.

3:6 Setting the ignition timing

If the timing has been completely lost, leave the distributor out until the engine has been set. Turn the engine until it is at TDC after the compression stroke on No. 1 (front cylinder). The timing marks are shown in **FIG 3:7**.

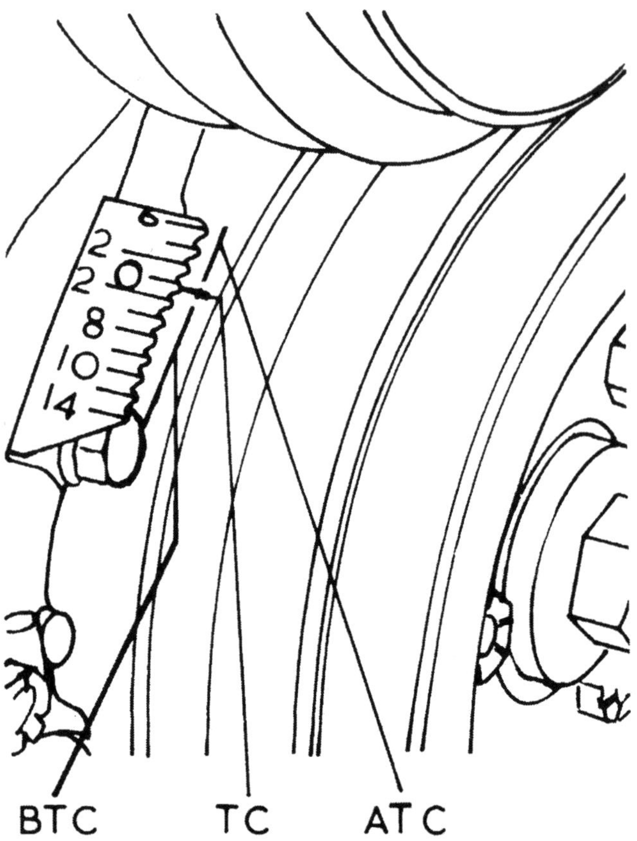

FIG 3:7 Ignition timing marks

The compression stroke of the engine can be found after the spark plug has been removed. Crank the engine slowly with a thumb blocking the spark plug hole and there will be a pressure rise felt in the cylinder as the piston moves up on the compression stroke. The position can also be found if the rocker cover is off the engine, as both valves on No. 1 cylinder will be fully closed when the timing marks are at the correct TDC position.

Insert the distributor back into the engine. The distributor casing must be installed so that the vacuum unit is pointing in the original direction, otherwise it may be impossible to rotate the case to set the timing accurately. Use the original engine and case aligning marks if the old distributor is being reinstalled. The distributor should be installed so that when it is fully in position the rotor arm is pointing to the position of the electrode in the cap connect to No. 1 spark plug. Leave the distributor clamp just tight enough to prevent the distributor from being moved accidentally.

Crank the engine forward by two turns. Just before the end of the second revolution, turn the engine very slowly and stop turning when the correct timing marks are in line. If the marks are overshot, turn the engine a further two revolutions or else turn it back far enough to ensure that all backlash will be taken up when the engine is again rotated forwards to the correct timing position. The correct timing for all models is given in **Technical Data**.

Disconnect the primary wire from between the distributor and DIST terminal on the ignition coil and reconnect it with a low-wattage test lamp in series. Switch on the ignition. Apply light backward pressure on the rotor arm to take up all backlash. Rotate the distributor body backward or forward slightly until the point is found where the test lamp just goes out, indicating that the contacts have just opened. Tighten the distributor clamp in this position. Crank the engine forward for another two turns, slowing the rate of turning at the end of the second revolution, and stop turning the instant that the test lamp goes out (noting that it will have come on and gone out during the revolutions). If the timing is correct and the engine was stopped immediately the lamp went out, the timing marks will again be in exact alignment.

The timing may also be checked with a stroboscpic light when the engine is running. Connect the light as instructed by the maker of the instrument. Mark the timing points on the crankshaft damper and the edge of the pointer with thin white lines so that they will be more easily seen. Start the engine and run it until it has reached its normal operating temperature. Disconnect the vacuum lines from the distributor, and plug them to prevent the air from altering the idling. Slacken the distributor clamp slightly so that the body can be rotated. Start the engine and shine the light onto the timing marks. **Take great care not to wear loose clothing and avoid the moving drive belts, keeping tools and hands well away from them.** With the engine at idle, rotate the distributor body until the timing mark and pointer are apparently exactly in line. Tighten the clamp that secures the distributor. Remove the stroboscopic light and reconnect the vacuum lines.

3:7 The distributor

The components of the distributor are shown in **FIG 3:8**. A press and special tools will be required to disassemble and service the distributor, so the best solution for the owner is to fit a new tested distributor if the old one is excessively worn.

The end play of the distributor shaft can be checked with a dial gauge after the cap and rotor arm have been removed, without removing the distributor from the engine.

If the distributor is dismantled, it is essential to mark the weights, springs, and drive gear so that they are installed into their original positions.

3:8 The dual-diaphragm distributor

On this type of distributor the vacuum unit contains two diaphragms. The outer diaphragm is operated by the suction in the carburetor while the inner diaphragm is operated by the suction in the intake manifold. Typical vacuum systems are shown in **FIGS 3:9** and **3:10**. The actual layout and color scheme of the lines will vary from model to model. When disconnecting a system, it is most advisable to label the lines to ensure correct reconnection. The operation of the dual-diaphragm vacuum unit is shown in **FIG 3:11**. The unit can be adjusted by altering the number of shims under the stop bolt at the end of the unit.

Testing:

Start the engine and run it until it has reached its normal operating temperature, with the choke valve fully open on the carburetor. Disconnect the vacuum lines from the vacuum unit, plugging the line that was connected to the inner diaphragm. Increase the idle speed by

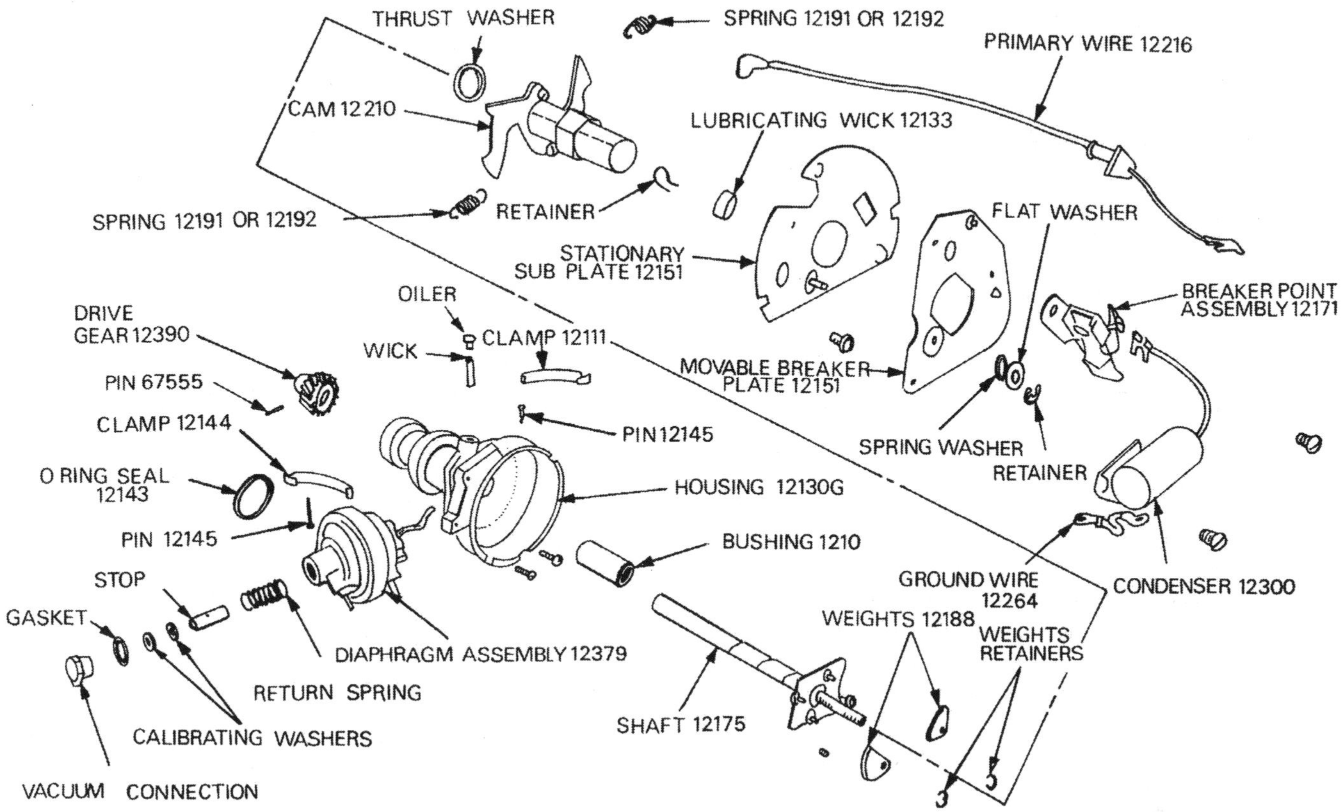

FIG 3:8 Distributor components

setting on the first step of the fast-idle cam. Check the ignition timing with a stroboscopic light (see previous section) noting the setting. Connect the vacuum line from the carburetor to the outer diaphragm and the ignition should advance immediately, as shown by the stroboscopic light. Adjustment can be carried out by varying the shims under the stop bolt.

Return the engine to normal idle speed and note the timing position. Un-plug the other line and connect it to the inner diaphragm. The ignition should immediately retard. No adjustment or repair is possible in this case and if there is no retard the vacuum unit must be renewed.

3:9 Temperature sensing valve

A distributor vacuum valve is fitted into the system and this unit senses the temperature of the coolant. The unit contains a valve which connects the ports of the unit in such a way that when the coolant temperature becomes excessive the ignition timing is altered to speed up the curb idle of the engine, so that during prolonged periods of idling the engine will be speeded up if it starts to overheat.

To check the valve, connect a tachometer to the ignition. Check that all vacuum hoses are correctly routed and in good condition. Start the engine and run it

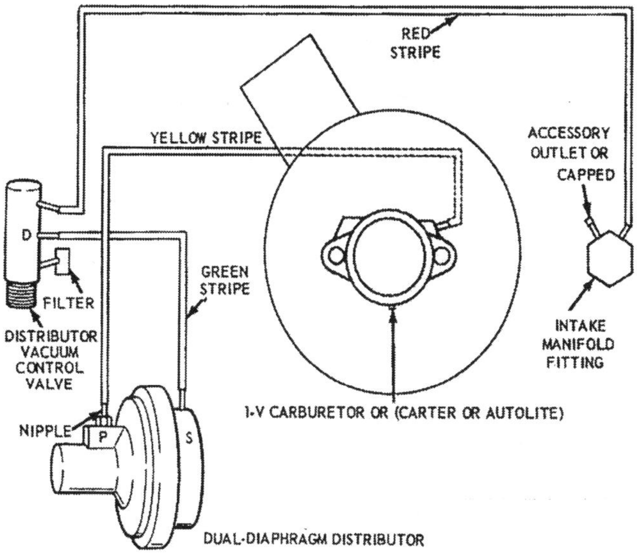

FIG 3:9 Schematic vacuum systems with temperature sensing valve only

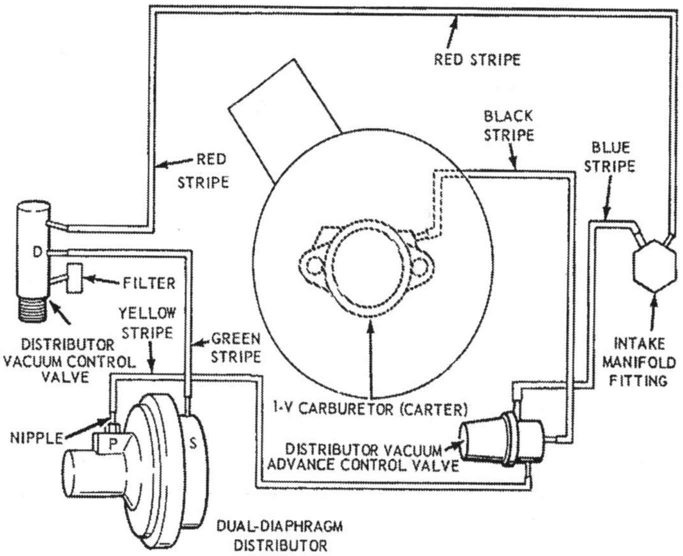

FIG 3:10 Schematic vacuum system with temperature sensing valve and vacuum advance control valve

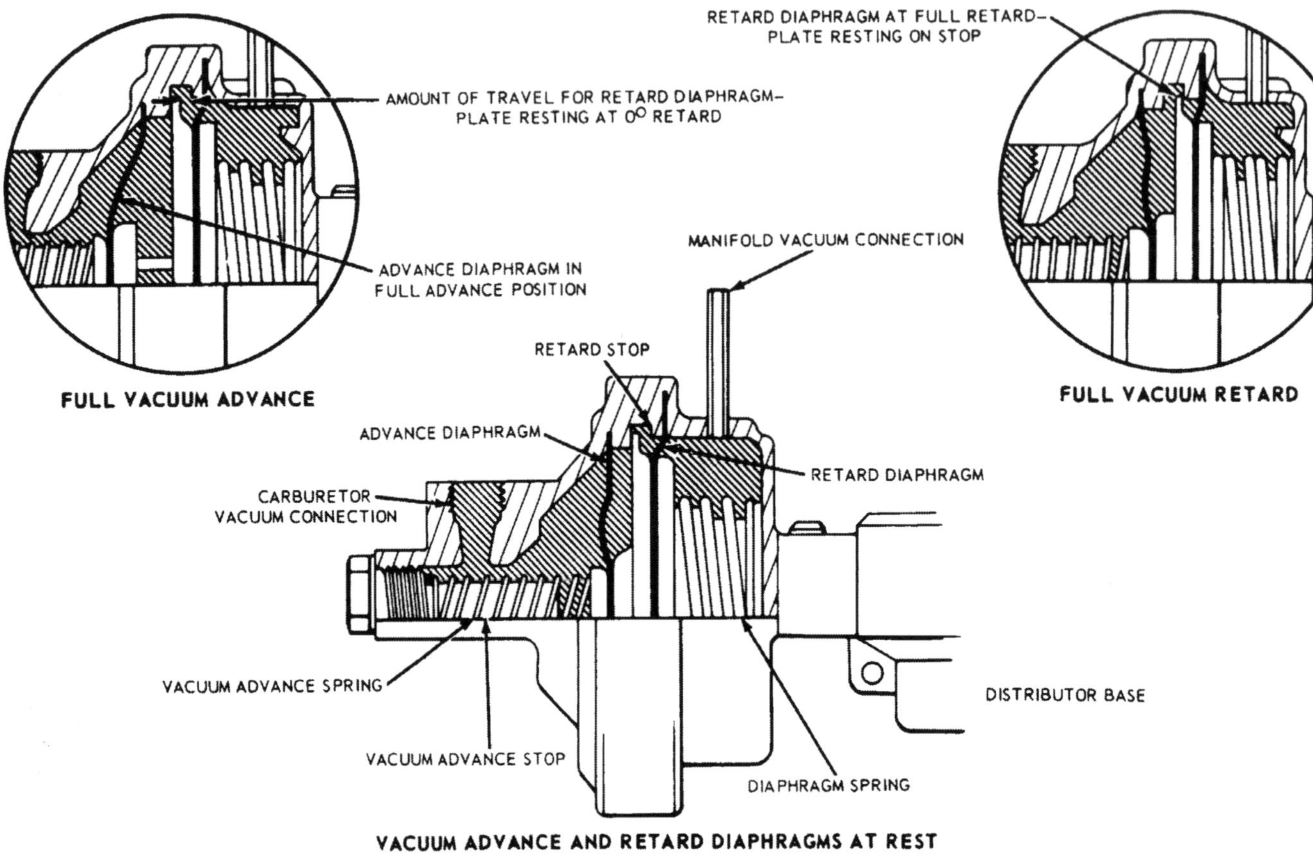

FIG 3:11 Dual diaphragm vacuum unit

until it has reached its normal operating temperature, with the choke plate valve fully open. **At this stage make sure that the engine does not overheat.** Note the idle speed with the transmission in neutral and the throttle valve fully closed to the curb idle stop. Disconnect the vacuum hose that connects the valve to the intake manifold at the manifold. Plug the hose. If the valve is satisfactory the idle speed will not alter. If the idle speed drops by 100 rev/min or more, the valve is defective and must be renewed. Reconnect the vacuum hose.

Cover the radiator and allow the engine to run until the overheat warning light comes on, or the needle on the gauge moves into the overheat sector. Before carrying out this part of the test it is essential that the cooling system is checked and the level in the system also checked (see **Chapter 4**). If the valve is satisfactory the idle speed will increase by 100 rev/min when the engine overheats. If the idle speed does not increase, remove and discard the old valve—fitting a new one in its place.

Uncover the radiator and allow the engine to run at a fast idle to cool before switching off.

3:10 The vacuum advance control valve

On some models a distributor vacuum advance control valve is also fitted. During deceleration this valve switches the manifold suction to the outer diaphragm of the distributor vacuum unit, so that the extra advance of the ignition will prevent excessive emissions.

The valve with its cover removed is shown in **FIG 3:12**. Make sure that the engine idle speed is correctly set (see **Chapter 2**). Turn off the headlights and note the engine speed. Remove the plastic cover from the valve and turn the adjusting screw in a counterclockwise direction, without exerting excessive inward pressure. After 5 to 6 turns the idle speed should suddenly increase to approximately 1000 rev/min. **Do not unscrew the adjusting screw further than this point otherwise the compressed spring in the valve will be released.** If, after the sixth turn, the idle speed does not increase, press in on the adjusting screw and release it to trigger the engine to 1000 rev/min. Turn the adjusting screw clockwise until the idle speed returns to the original level and then make one further turn clockwise. Increase the engine speed to 2000 rev/min, using the throttle, hold this speed for approximately 5 seconds and release the throttle. If the engine takes longer than 4 seconds to return to the idle speed, free the anti-stall dashpot and repeat the test from speed. If, with the dashpot released, the engine still takes longer than 3 seconds to return from 2000 rev/min to idle, turn the adjusting screw a further quarter-turn clockwise and repeat the test. Up to one further full turn may be made from the original setting, taking it in quarter-turn increments, to bring the time taken to return from 2000 rev/min within specifications. If the valve requires more than one full turn clockwise from the original setting, the valve is defective and a new one must be fitted.

3:11 The spark plugs

The spark plugs should be cleaned and tested at intervals at 6000 miles and renewed at intervals of 12,000 miles.

Removal:

Slacken the spark plug using a well fitting box wrench or plug wrench. Blow away all loose dust and dirt from around the base of the plug with an airline or tire pump. Unscrew the spark plug by hand, unless it is too stiff. If the spark plug is stiff to remove, use the wrench to take it out as far as possible without exerting heavy pressure and then wind the plug in and out to this point a few times before attempting to unthread it further. In extreme cases, wrap a piece of rag around the base, soak the rag with kerosene or penetrating oil and leave it overnight to free the threads.

As the spark plugs are removed they should be stored in order for examination of their firing ends.

Examination:

The color and condition of the deposits on the firing ends of the plugs will give a good guide to the conditions inside the combustion chambers.

After examination, throw away any plugs that have cracked insulators or badly burnt electrodes, without bothering to clean or test them.

A light powdery deposit ranging in color from brown to greyish tan, coupled with light wear on the electrodes, indicates that the conditions are normal. Much city- or constant-speed driving will leave the deposits white or yellowish. Cleaning and testing the spark plugs is all that is required.

If the deposits are wet and black they are caused by excess oil leaking into the combustion chamber. The only cure is an engine overhaul but fitting a hotter running grade of plug may help to alleviate the problem.

Black but dry and fluffy deposits are caused by incomplete combustion. This may be caused by excessive idling—particularly if all the spark plugs are similarly affected—but it may also be caused by defective ignition or spark plug, or running with too rich a mixture.

Overheated spark plugs have a white blistered look about the central electrode. If lead-based fuels are used there may be glints of metal on the central electrode. The electrodes will also be excessively burnt away. Some possible causes are; engine overheating, very weak mixture, incorrectly set ignition timing, incorrect grade of plug or running at high-speeds with the vehicle overloaded.

Cleaning, adjusting and testing:

Wash oily plugs in gasoline before cleaning them further. Have the plugs cleaned on an abrasive-blasting machine and then pressure tested after attention to the electrodes. Use a steel-wire brush to clean the threads and spark plug in an emergency.

Trim the electrodes square with a fine file. Adjust to the correct gap of .032 to .036 inch by bending the side electrode only. **Never bend the central electrode as this will crack the insulator.**

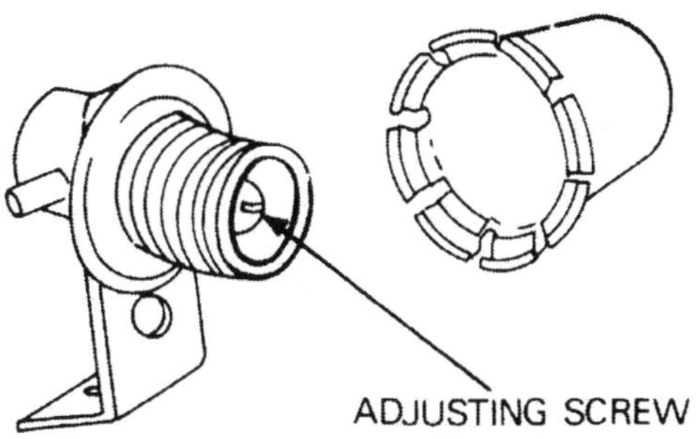

FIG 3:12 **Vacuum advance control valve**

Check that the plugs spark under pressure. While still applying air pressure to the plug, drip a little thin engine oil around the joint of the insulator and body. If air bubbles are seen in the oil, reject the plug.

Clean the external portion of the insulator with denatured alcohol, unleaded gasoline or similar suitable solvent.

Installing:

If threads in the cylinder head are dirty, clean them using a well-greased tap. Failing a tap, use an old spark plug with crosscuts down the thread.

Graphite grease only may be used on the threads as any other grease will bake hard and lock the plug to the head.

Thread the plugs in by hand until they bottom. Tighten them to a torque of 15 to 20 lb ft. If a torque wrench is not available, tighten the plug by at most a further half-turn from the handtight position. **Overtightening is likely to cause damage and defects.**

3:12 Trouble shooting guide

(a) Engine will not fire

1 Battery discharged or terminals dirty
2 Breaker contact points dirty or incorrectly set
3 Distributor cap dirty, cracked or tracking
4 Rotor arm not making contact with carbon button in cap
5 Distributor rotor arm cracked, dirty or omitted
6 Defective ignition coil
7 Defective HT lead between coil and distributor
8 Broken or stuck contact breaker spring
9 Contact points stuck open
10 Internal shortcircuit in distributor

(b) Engine misfires

1 Check 2, 3 and 6 in (a)
2 Weak contact breaker spring
3 Defective HT lead
4 Spark plug(s) loose
5 Spark plug insulation cracked or dirty
6 Spark plug gap incorrectly set
7 Ignition timing too far advanced
8 HT leads installed in wrong order

Inches	Decimals	Milli-metres	Inches to Millimetres		Millimetres to Inches	
			Inches	mm	mm	Inches
1/64	.015625	.3969	.001	.0254	.01	.00039
1/32	.03125	.7937	.002	.0508	.02	.00079
3/64	.046875	1.1906	.003	.0762	.03	.00118
1/16	.0625	1.5875	.004	.1016	.04	.00157
5/64	.078125	1.9844	.005	.1270	.05	.00197
3/32	.09375	2.3812	.006	.1524	.06	.00236
7/64	.109375	2.7781	.007	.1778	.07	.00276
1/8	.125	3.1750	.008	.2032	.08	.00315
9/64	.140625	3.5719	.009	.2286	.09	.00354
5/32	.15625	3.9687	.01	.254	.1	.00394
11/64	.171875	4.3656	.02	.508	.2	.00787
3/16	.1875	4.7625	.03	.762	.3	.01181
13/64	.203125	5·1594	.04	1.016	.4	.01575
7/32	.21875	5.5562	.05	1.270	.5	.01969
15/64	.234375	5.9531	.06	1.524	.6	.02362
1/4	.25	6.3500	.07	1.778	.7	.02756
17/64	.265625	6.7469	.08	2.032	.8	.03150
9/32	.28125	7.1437	.09	2.286	.9	.03543
19/64	.296875	7.5406	.1	2.54	1	.03937
5/16	.3125	7.9375	.2	5.08	2	.07874
21/64	.328125	8.3344	.3	7.62	3	.11811
11/32	.34375	8.7312	.4	10.16	4	.15748
23/64	.359375	9.1281	.5	12.70	5	.19685
3/8	.375	9.5250	.6	15.24	6	.23622
25/64	.390625	9.9219	.7	17.78	7	.27559
13/32	.40625	10.3187	.8	20.32	8	.31496
27/64	.421875	10.7156	.9	22.86	9	.35433
7/16	.4375	11.1125	1	25.4	10	.39370
29/64	.453125	11.5094	2	50.8	11	.43307
15/32	.46875	11.9062	3	76.2	12	.47244
31/64	.484375	12.3031	4	101.6	13	.51181
1/2	.5	12.7000	5	127.0	14	.55118
33/64	.515625	13.0969	6	152.4	15	.59055
17/32	.53125	13.4937	7	177.8	16	.62992
35/64	.546875	13.8906	8	203.2	17	.66929
9/16	.5625	14.2875	9	228.6	18	.70866
37/64	.578125	14.6844	10	254.0	19	.74803
19/32	.59375	15.0812	11	279.4	20	.78740
39/64	.609375	15.4781	12	304.8	21	.82677
5/8	.625	15.8750	13	330.2	22	.86614
41/64	.640625	16.2719	14	355.6	23	.90551
21/32	.65625	16.6687	15	381.0	24	.94488
43/64	.671875	17.0656	16	406.4	25	.98425
11/16	.6875	17.4625	17	431.8	26	1.02362
45/64	.703125	17.8594	18	457.2	27	1.06299
23/32	.71875	18.2562	19	482.6	28	1.10236
47/64	.734375	18.6531	20	508.0	29	1.14173
3/4	.75	19.0500	21	533.4	30	1.18110
49/64	.765625	19.4469	22	558.8	31	1.22047
25/32	.78125	19.8437	23	584.2	32	1.25984
51/64	.796875	20.2406	24	609.6	33	1.29921
13/16	.8125	20.6375	25	635.0	34	1.33858
53/64	.828125	21.0344	26	660.4	35	1.37795
27/32	.84375	21.4312	27	685.8	36	1.41732
55/64	.859375	21.8281	28	711.2	37	1.4567
7/8	.875	22.2250	29	736.6	38	1.4961
57/64	.890625	22.6219	30	762.0	39	1.5354
29/32	.90625	23.0187	31	787.4	40	1.5748
59/64	.921875	23.4156	32	812.8	41	1.6142
15/16	.9375	23.8125	33	838.2	42	1.6535
61/64	.953125	24.2094	34	863.6	43	1.6929
31/32	.96875	24.6062	35	889.0	44	1.7323
63/64	.984375	25.0031	36	914.4	45	1.7717

UNITS	Pints to Litres	Gallons to Litres	Litres to Pints	Litres to Gallons	Miles to Kilometres	Kilometres to Miles	Lbs. per sq. In. to Kg. per sq. Cm.	Kg. per sq. Cm. to Lbs. per sq. In.
1	.57	4.55	1.76	.22	1.61	.62	.07	14.22
2	1.14	9.09	3.52	.44	3.22	1.24	.14	28.50
3	1.70	13.64	5.28	.66	4.83	1.86	.21	42.67
4	2.27	18.18	7.04	.88	6.44	2.49	.28	56.89
5	2.84	22.73	8.80	1.10	8.05	3.11	.35	71.12
6	3.41	27.28	10.56	1.32	9.66	3.73	.42	85.34
7	3.98	31.82	12.32	1.54	11.27	4.35	.49	99.56
8	4.55	36.37	14.08	1.76	12.88	4.97	.56	113.79
9		40.91	15.84	1.98	14.48	5.59	.63	128.00
10		45.46	17.60	2.20	16.09	6.21	.70	142.23
20				4.40	32.19	12.43	1.41	284.47
30				6.60	48.28	18.64	2.11	426.70
40				8.80	64.37	24.85		
50					80.47	31.07		
60					96.56	37.28		
70					112.65	43.50		
80					128.75	49.71		
90					144.84	55.92		
100					160.93	62.14		

UNITS	Lb ft to kgm	Kgm to lb ft	UNITS	Lb ft to kgm	Kgm to lb ft
1	.138	7.233	7	.967	50.631
2	.276	14.466	8	1.106	57.864
3	.414	21.699	9	1.244	65.097
4	.553	28.932	10	1.382	72.330
5	.691	36.165	20	2.765	144.660
6	.829	43.398	30	4.147	216.990

CHAPTER 4

THE COOLING SYSTEM

4:1 Description

A sectional view of the engine showing the coolant flow is shown in **FIG 4:1**. The coolant circulates around the outside of the cylinder bores and through the passages in the cylinder heads, taking away surplus heat. Hot water normally tends to rise, because of expansion, and the heated coolant therefore passes out through the elbow outlet on the engine and from this elbow through a hose to the header tank of the radiator. The coolant passing through the radiator core is cooled by the passage of air over the radiator fins and then returns to the engine. To ensure a rapid and positive flow of coolant at all times, the thermo-siphon action of the coolant is assisted by a centrifugal engine-driven water pump. The pump draws the coolant from the radiator and passes it into the engine.

A cooling fan is fitted to the water pump hub to ensure that there is an adequate flow of air through the radiator core when the vehicle is stationary or travelling slowly. The fan is designed to operate only when required. On some models a special temperature-sensitive clutch is fitted to the hub so that the fan is only rotating at full speed when the engine is hot and in the low speed range. On other models the fan is fitted with special blades which flex and reduce the airflow and load at high engine speeds.

On vehicles fitted with automatic transmission, a heat exchanger is fitted into the cooler radiator tank to cool the transmission fluid. On downflow radiators the heat exchanger is fitted in the lower tank but on crossflow radiators the unit is fitted in the lefthand chamber.

A thermostat valve is fitted to the elbow outlet of the engine. The valve ensures that the engine temperature stays constant and also gives a shorter warm-up period. When the coolant is cold the valve stays shut and the coolant is bypassed directly back into the engine. The amount of coolant to be heated is therefore reduced and there are no heat losses from the radiator, so the engine warms up rapidly. When the coolant heats, the valve opens and allows the coolant to pass through the radiator where it is cooled normally.

The filler cap to the radiator contains two concentric valves. The outer valve seals against the neck of the filler and only opens when the pressure of the system exceeds specification. Surplus coolant or air are then vented overboard. Pressurizing the system raises the boiling point of the coolant so that the engine can be run hotter without danger of boiling in localized hot spots. **If the filler cap is removed while the engine is overheated the pressure will force scalding water over the operator's hand and the subsequent drop in**

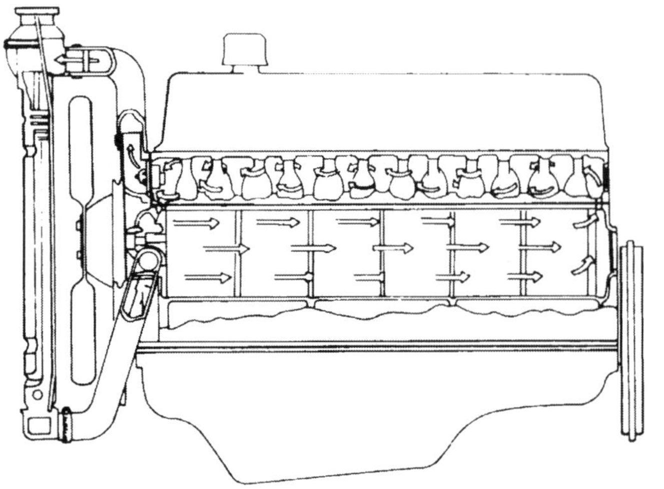

FIG 4:1 Coolant flow through the cooling system

pressure will allow the coolant to boil, forcing out even more scalding coolant. The cap has two positions, one of which is a lock that holds the cap in place but allows excess pressure to escape if the engine is warm. The inner valve of the cap opens as the system cools and allows air to enter and prevent a vacuum from forming in the cooling system.

A mixture of antifreeze and water should always be used. Even in hot climates the mixture should protect the system down to a temperature of 0°F. The antifreeze in these cases is not to protect against frost, but it does contain corrosion inhibitors as well as raising the boiling point of the water. In temperate climates a 50 per cent mixture of water and antifreeze should be used and this strength will protect the engine down to temperatures of —35°F and raise the boiling point to 250°F.

4:2 Maintenance

Regular maintenance is confined to checking that all the drive belts are in good condition and that their tension is correct, as well as checking the coolant level in the radiator. When checking the level make sure that the coolant is clean and top up using a mixture of antifreeze and water. If water alone is used for topping up, the antifreeze in the system will be diluted.

The life of the antifreeze is two years. At yearly intervals have the strength of the mixture tested and add extra antifreeze as required. At the end of two years, drain the system and discard the coolant. Flush the system and refill with fresh coolant.

Coolant level:

The level should always be checked with the engine cold. On downflow radiators, top up until the level is one inch below the bottom of the ring in the filler neck. On crossflow radiators the level is marked on the outside of the header tank.

Draining:

The radiator is fitted with a drain cock underneath it. If the system is being drained for any reason other than renewing the coolant, attach a hose to the cock and drain the coolant into a clean container so that it can be poured back into the system after the work has been carried out.

When completely draining the cooling system, remove the block drain plug.

Flushing:

The use of Rotunda Cooling System Cleanser will remove scale and dirt if the instructions on the tin are followed.

The system can be partially cleaned by inserting a hosepipe into the filler neck, with all the drains open, and running water through until it comes out clean from the drains.

If the system is so clogged that Rotunda cleanser will not remove all the dirt and scale have the system pressure-flushed at a service garage. For pressure-flushing, the thermostat must be removed from its housing.

In some cases it will help to remove the radiator, invert it and flush through in a reverse direction so that water and gravity combined remove scale trapped in the tubes.

Hoses:

Pinch the hoses lightly between the fingers and check them for perishing, hardening, softening or cracking. Renew any hoses that are defective. Slacken off the hose clamps and pull the hoses free with a twisting motion.

Fit the hose clamp onto the hose so that it is at least $\frac{1}{8}$ inch from the end of the hose. Lubricate the inside of the hose with water, soft soap or an approved water-resistant sealer and slide the hose back into position. Make sure that the hose clamp is well past the bead on the pipe or connector and position it centrally on the portion of hose on the pipe. Tighten the hose clamps sufficiently to prevent leaks without cutting into the hoses.

Filling:

Make sure that all drains are closed and hoses connected. Fill the cooling system through the radiator filler cap until it is full, pouring it in slowly to allow air to escape. Disconnect the heater outlet hose at the water pump and wait until coolant flows out through it, with the heater set to hot, and reconnect the hose. Fill the engine to the correct level. Start the engine and check for coolant leaks. Leave the engine running at 2000 rev/min for 20 minutes and then switch it off and allow it to cool. When the system is cool, top up to the correct level. As much as a quart of coolant may be required for topping up to allow for the air that was trapped in the system. If the heater fails to operate on hot, disconnect the heater outlet hose and reconnect it when coolant flows out. A tin of Rotunda Rust Inhibitor should be added when filling with fresh coolant.

4:3 The radiator

Removal:

Drain the cooling system as described in the previous section. On models fitted with automatic transmission, disconnect the oil cooler lines at the radiator. Plug the hoses and ports on the oil cooler to prevent dirt from entering. Remove the radiator bolts and supports (if fitted) while supporting the radiator and then lift out the radiator.

Install the radiator in the reverse order of removal. Fill the cooling system. On models fitted with automatic transmission, check the fluid level in the transmission (see **Chapter 7**).

Cleaning:

The internal flushing of the radiator has been dealt with in the previous section.

Periodically use an airline or hosepipe to blow through the radiator fins, from the back, so as to remove accumulations of mud and dead insects. In extreme cases, remove the radiator so that it can be soaked and scrubbed in a hot weak detergent solution.

Repairs:

If the radiator or transmission oil cooler are defective, the repairs should be left to a specialist firm.

4:4 Drive belts

Belts are used to transmit the drive to the water pump and alternator, as well as to the Thermactor air pump, air conditioning and power steering pumps if these are fitted.

All the belts are independently adjustable. If a belt is glazed or slack it will slip and drive will not be efficiently transmitted to the accessory, while if the belt is too tight the bearings of the accessory may be damaged. The best method of checking the tension is to use the special gauge T63L-8620-A and this is the method to use wherever possible. A rough guide is that the tension is correct if the belt deflects by approximately $\frac{1}{4}$ inch from the straight line when moderate thumb pressure is applied at the center of a run. The settings using the gauge are given in Technical Data. Note that a belt is considered used when it has run for 10 minutes or more.

Water pump and alternator:

The alternator is pivoted about two bolts and secured by a third bolt to a slotted link. Slacken all three bolts and use a wooden lever behind the drive end bracket of the alternator to pull the alternator away from the engine until the correct belt tension is reached. Hold the alternator with the lever and tighten all three bolts. This also sets the belt tension for the air-conditioner, if one is fitted.

Thermactor:

Slacken the Thermactor pump attachments and pull the pump away from the water pump until the correct belt tension is reached. Hold the pump and tighten the attachments.

Power steering:

The pump must not be levered by the casing of the oil container otherwise the container may distort and leak. A boss is cast in with the pump and it is designed so that a wrench can be fitted over it. Slacken the pump attachments and use a wrench on the boss to move the pump so that the correct belt tension is set. Hold the pump and tighten its attachments.

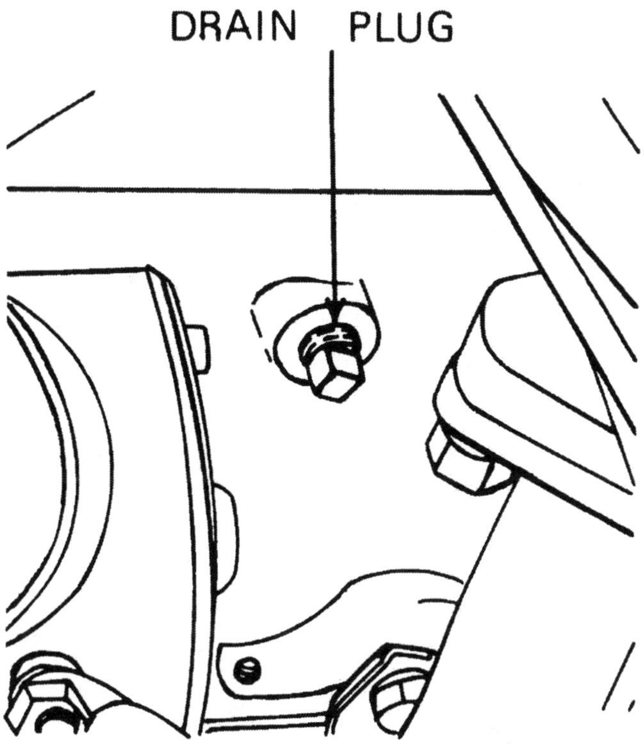

FIG 4:2 Typical engine drain plug

Belt removal:

Slacken off the belt as far as possible. Start removing the belt by rolling it forwards off the pulley and work in the direction of normal rotation. If the belt is too tight to remove fully, crank the engine to free it from the pulley. Once the belt is off one pulley it is easy to remove it from the vehicle.

Install the belt in the reverse order of removal, after checking it for defects or glaze on the sides, and set the correct tension. If two belts are fitted to drive the water pump and alternator, they must be renewed as a pair and not individually.

4:5 The cooling fan

On some models a Ford Flex-fan is fitted. The fan blades are flexible so that they alter pitch when under heavy load. When the engine is running fast, and the vehicle therefore moving fast, the blades reduce pitch and the fan takes less power from the engine.

On other models a clutch is fitted between the fan blades and water pump pulley. The unit operates by utilizing the drag of a special silicone-based fluid between the inner and outer members. When drive is required, centrifugal force and the pumping action in the unit forces the fluid outwards between the inner and outer member so that drag drives the outer member. Immediately the engine is started there will be maximum drive. A bi-metallic spring, either flat as shown in **FIG 4:3** or spiral as shown in **FIG 4:4**, is fitted to the clutch unit. The spring senses the temperature of the air coming from the radiator and adjusts the flow of fluid in the clutch accordingly. During the warm-up cycle, when the air temperature from the radiator is less than 150°F to 180°F, the spring moves the plunger to allow all the fluid to return to the central chamber and stay there. With no fluid in the outer chamber there is very little drag and

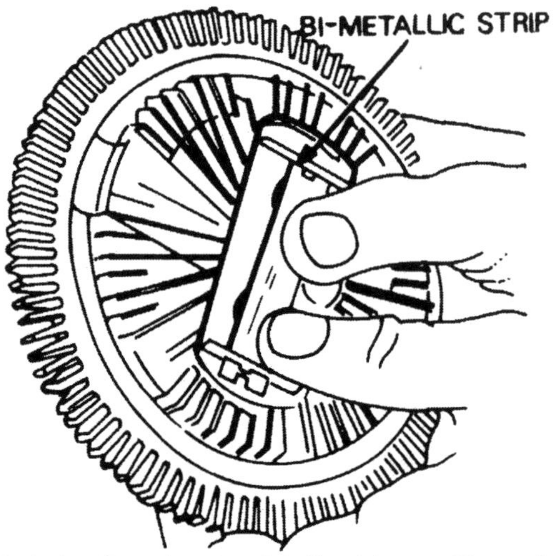

FIG 4:3 Fan clutch with flat bi-metallic spring

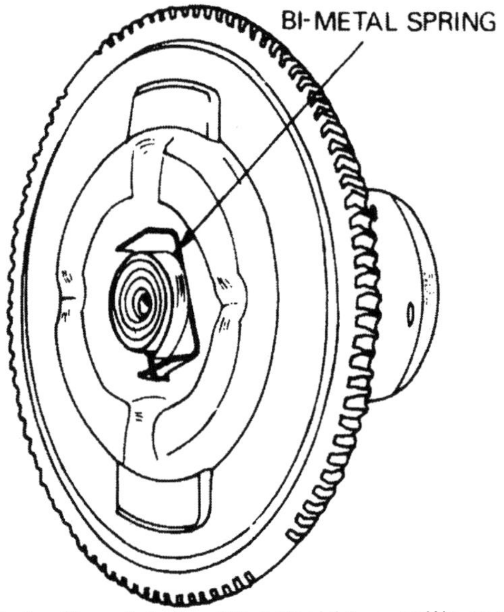

FIG 4:4 Fan clutch with spiral bi-metallic spring

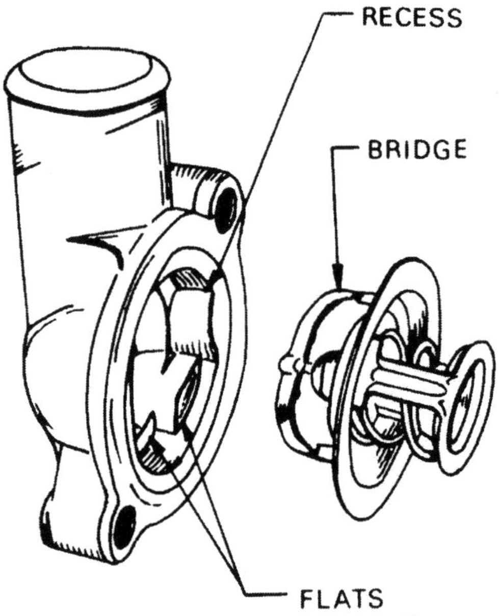

FIG 4:5 Thermostat and outlet elbow

the fan free-wheels, running at a comparatively slow speed. When the air temperature from the radiator rises, the spring senses this and moves the plunger so that fluid is allowed to return to the outer chamber. The fluid causes drag and the fan is then driven at engine speed. The spring will modulate the speed of the fan between the high and low settings according to the road conditions and temperature of the engine. At high speed, the clutch action will limit the speed of the fan irrespective of the engine temperature.

To check the clutch, start the engine and allow it to run for a few minutes. Stop the engine and the fan should turn freely. Restart the engine and allow it to run until it has reached its normal operating temperature, covering the radiator to speed the process. Remove the blocking from the radiator and allow the engine to run for a minute so that the bi-metallic spring can soak to the temperature of the air from the radiator. Stop the engine. Protect the hands with rag and try to turn the fan blades. The blades should only turn with considerable pressure. If the fan turns easily the clutch is defective.

4:6 The thermostat

The attachment of the thermostat valve is shown in **FIG 4:5**, and the thermostat is installed and turned clockwise to lock it in the elbow.

Drain the cooling system and take out the two bolts that secure the elbow in place, noting that the upper radiator hose is sufficiently flexible to leave it attached to the elbow. Pull off the elbow and remove the thermostat and gasket from it.

Before installing the thermostat, remove all traces of old gasket and sealing compound from the faces. Smear the faces of a new gasket with sealer and install it on the engine. Correctly fit the thermostat to the elbow and secure the elbow to the engine. Fill the cooling system and check for leaks.

Testing:

Suspend the thermostat in a container of water and gradually heat the water while stirring it with a thermometer. Note the temperature at which the valve starts to open and, less important, the temperature at which it is fully open. If the valve does not start to open within a few degrees of the correct temperature (see **Technical Data**) or the valve sticks in one position, the thermostat is defective and must be renewed.

Hold the thermostat up against a strong light and if there are gaps between valve and housing, the thermostat is again defective. A defective thermostat cannot be repaired and must be renewed.

4:7 The water pump

The pump has a bleed hole in the casting for the pump to allow coolant that has leaked past the seals to escape. This hole is not for lubrication purposes.

If the pump leaks or is in any way defective it cannot be repaired and a new pump must be installed in its place.

Removal:

Drain the cooling system. Remove all the drive belts and take off the cooling fan with the pump hub and pulley assembly. Take out the bolts securing the pump and remove the pump, after disconnecting the radiator and heater hoses. On some models a bypass hose is fitted and this too must be disconnected from the pump. Discard the old gasket.

Clean the mounting faces free of old gasket or sealer. Lightly smear a new gasket with waterproof sealer and lay it back into position. Attach the pump, using the bolts, and make sure that the bypass hose (if fitted) connects to the pump. Install the fan, pulley and hoses in the reverse order of disassembly and reconnect the hoses. Set the correct belt tensions and fill the cooling system. Check for leaks both before and after starting the engine.

4:8 Trouble shooting guide

(a) Internal coolant leakage

1 Loose cylinder head bolts
2 Cracked cylinder head
3 Cracked cylinder block
4 Defective cylinder head gasket
5 Warped cylinder head

(b) Corrosion

1 Impurities in the coolant
2 Lack of inhibitor or antifreeze
3 Infrequent draining and flushing

(c) Poor circulation

1 Perished or collapsed coolant hoses
2 Defective thermostat
3 Low coolant level
4 Loose, slipping or broken drive belt
5 Engine water passages restricted
6 Radiator tubes or fins blocked

(d) Overheating

1 Check (c)
2 Defective cooling fan clutch
3 Sludge in oil pan or crankcase
4 Too low oil level in oil pan
5 Tight engine
6 Choked exhaust
7 Binding brakes
8 Slipping clutch (manual transmission only)
9 Incorrect valve timing
10 Retarded ignition
11 Mixture too weak
12 Automatic transmission overheating

Inches	Decimals	Millimetres	Inches to Millimetres: Inches	Inches to Millimetres: mm	Millimetres to Inches: mm	Millimetres to Inches: Inches
1/64	.015625	.3969	.001	.0254	.01	.00039
1/32	.03125	.7937	.002	.0508	.02	.00079
3/64	.046875	1.1906	.003	.0762	.03	.00118
1/16	.0625	1.5875	.004	.1016	.04	.00157
5/64	.078125	1.9844	.005	.1270	.05	.00197
3/32	.09375	2.3812	.006	.1524	.06	.00236
7/64	.109375	2.7781	.007	.1778	.07	.00276
1/8	.125	3.1750	.008	.2032	.08	.00315
9/64	.140625	3.5719	.009	.2286	.09	.00354
5/32	.15625	3.9687	.01	.254	.1	.00394
11/64	.171875	4.3656	.02	.508	.2	.00787
3/16	.1875	4.7625	.03	.762	.3	.01181
13/64	.203125	5·1594	.04	1.016	.4	.01575
7/32	.21875	5.5562	.05	1.270	.5	.01969
15/64	.234375	5.9531	.06	1.524	.6	.02362
1/4	.25	6.3500	.07	1.778	.7	.02756
17/64	.265625	6.7469	.08	2.032	.8	.03150
9/32	.28125	7.1437	.09	2.286	.9	.03543
19/64	.296875	7.5406	.1	2.54	1	.03937
5/16	.3125	7.9375	.2	5.08	2	.07874
21/64	.328125	8.3344	.3	7.62	3	.11811
11/32	.34375	8.7312	.4	10.16	4	.15748
23/64	.359375	9.1281	.5	12.70	5	.19685
3/8	.375	9.5250	.6	15.24	6	.23622
25/64	.390625	9.9219	.7	17.78	7	.27559
13/32	.40625	10.3187	.8	20.32	8	.31496
27/64	.421875	10.7156	.9	22.86	9	.35433
7/16	.4375	11.1125	1	25.4	10	.39370
29/64	.453125	11.5094	2	50.8	11	.43307
15/32	.46875	11.9062	3	76.2	12	.47244
31/64	.484375	12.3031	4	101.6	13	.51181
1/2	.5	12.7000	5	127.0	14	.55118
33/64	.515625	13.0969	6	152.4	15	.59055
17/32	.53125	13.4937	7	177.8	16	.62992
35/64	.546875	13.8906	8	203.2	17	.66929
9/16	.5625	14.2875	9	228.6	18	.70866
37/64	.578125	14.6844	10	254.0	19	.74803
19/32	.59375	15.0812	11	279.4	20	.78740
39/64	.609375	15.4781	12	304.8	21	.82677
5/8	.625	15.8750	13	330.2	22	.86614
41/64	.640625	16.2719	14	355.6	23	.90551
21/32	.65625	16.6687	15	381.0	24	.94488
43/64	.671875	17.0656	16	406.4	25	.98425
11/16	.6875	17.4625	17	431.8	26	1.02362
45/64	.703125	17.8594	18	457.2	27	1.06299
23/32	.71875	18.2562	19	482.6	28	1.10236
47/64	.734375	18.6531	20	508.0	29	1.14173
3/4	.75	19.0500	21	533.4	30	1.18110
49/64	.765625	19.4469	22	558.8	31	1.22047
25/32	.78125	19.8437	23	584.2	32	1.25984
51/64	.796875	20.2406	24	609.6	33	1.29921
13/16	.8125	20.6375	25	635.0	34	1.33858
53/64	.828125	21.0344	26	660.4	35	1.37795
27/32	.84375	21.4312	27	685.8	36	1.41732
55/64	.859375	21.8281	28	711.2	37	1.4567
7/8	.875	22.2250	29	736.6	38	1.4961
57/64	.890625	22.6219	30	762.0	39	1.5354
29/32	.90625	23.0187	31	787.4	40	1.5748
59/64	.921875	23.4156	32	812.8	41	1.6142
15/16	.9375	23.8125	33	838.2	42	1.6535
61/64	.953125	24.2094	34	863.6	43	1.6929
31/32	.96875	24.6062	35	889.0	44	1.7323
63/64	.984375	25.0031	36	914.4	45	1.7717

UNITS	Pints to Litres	Gallons to Litres	Litres to Pints	Litres to Gallons	Miles to Kilometres	Kilometres to Miles	Lbs. per sq. In. to Kg. per sq. Cm.	Kg. per sq. Cm. to Lbs. per sq. In.
1	.57	4.55	1.76	.22	1.61	.62	.07	14.22
2	1.14	9.09	3.52	.44	3.22	1.24	.14	28.50
3	1.70	13.64	5.28	.66	4.83	1.86	.21	42.67
4	2.27	18.18	7.04	.88	6.44	2.49	.28	56.89
5	2.84	22.73	8.80	1.10	8.05	3.11	.35	71.12
6	3.41	27.28	10.56	1.32	9.66	3.73	.42	85.34
7	3.98	31.82	12.32	1.54	11.27	4.35	.49	99.56
8	4.55	36.37	14.08	1.76	12.88	4.97	.56	113.79
9		40.91	15.84	1.98	14.48	5.59	.63	128.00
10		45.46	17.60	2.20	16.09	6.21	.70	142.23
20				4.40	32.19	12.43	1.41	284.47
30				6.60	48.28	18.64	2.11	426.70
40				8.80	64.37	24.85		
50					80.47	31.07		
60					96.56	37.28		
70					112.65	43.50		
80					128.75	49.71		
90					144.84	55.92		
100					160.93	62.14		

UNITS	Lb ft to kgm	Kgm to lb ft	UNITS	Lb ft to kgm	Kgm to lb ft
1	.138	7.233	7	.967	50.631
2	.276	14.466	8	1.106	57.864
3	.414	21.699	9	1.244	65.097
4	.553	28.932	10	1.382	72.330
5	.691	36.165	20	2.765	144.660
6	.829	43.398	30	4.147	216.990

CHAPTER 5

THE CLUTCH

5:1 Description

A mechanically operated clutch is fitted to all models covered by this manual. The linkage for the 1969 model is shown in **FIG 5:1**, later models are shown in **FIG 5:2** and this is fairly typical for the range.

A sectioned view of the clutch in the flywheel housing is shown in **FIG 5:3**. The mechanical linkage operates the clutch release lever. The clutch cover assembly is bolted to the flywheel and therefore rotates with the engine. The clutch disc has an internally splined hub which fits onto the splines of the transmission input shaft so that they rotate together.

The clutch cover assembly contains springs and a pressure plate, all of which rotate with the cover. When the clutch is engaged, the springs press the pressure plate forward so that the clutch disc is gripped by its friction linings between the pressure faces of the pressure plate and flywheel. The clutch disc therefore revolves with the flywheel and cover assembly, and drive is transmitted to the transmission.

When the clutch pedal is pressed, the linkage moves the release lever so that it presses the release bearing forwards. The release bearing first contacts and then presses on the levers in the clutch cover. When actuated, these levers draw the pressure plate back against the springs, releasing the clutch disc so that it and the transmission are free to revolve independently or even come to a stop with the engine running.

As the clutch pedal is released, there is a gradual increase in pressure on the clutch disc. At first the disc slips, taking up drive smoothly, but when full pressure is reached it is revolving at the same speed as the flywheel.

The clutch and brake pedal components and parts on a 1965 model are shown in **FIG 5:4**.

5:2 Clutch adjustments

As the linings on the clutch disc wear, the clutch pedal free travel will alter. **It is essential that there is always clearance between the clutch release bearing and the levers in the clutch cover when the clutch is disengaged.** If there is no clearance the release bearing will wear rapidly and in extreme cases the clutch will slip at all times. If the clearance is excessive the clutch may not be released fully when the pedal reaches the end of its travel, causing difficulty in engaging gear.

The method of checking is shown in **FIG 5:5**. On earlier models the total clutch pedal movement should be checked. If the total travel is not at the correct specification of $6\frac{1}{2}$ inch, set the bump stop for the pedal to bring

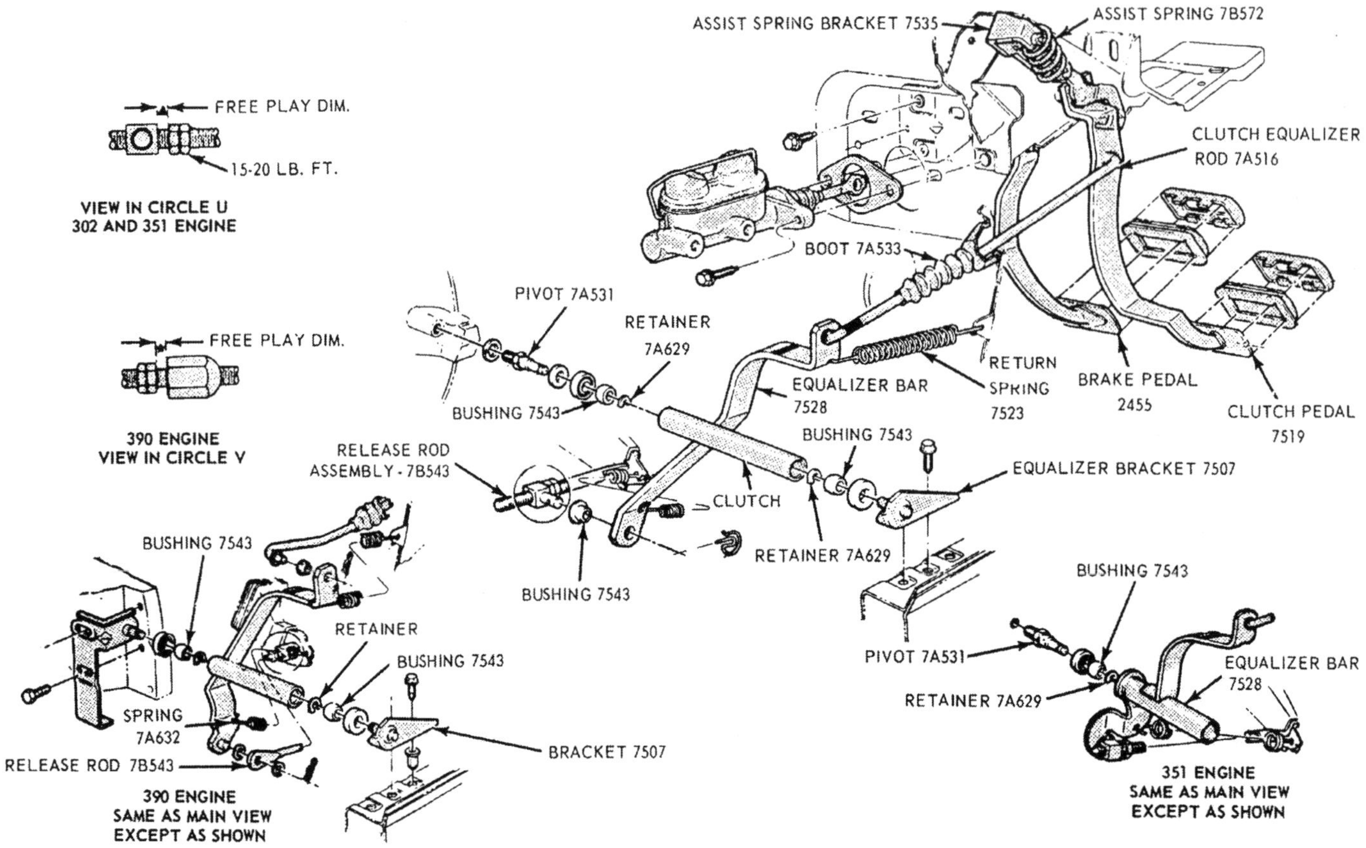

FIG 5:1 1969 model clutch linkage

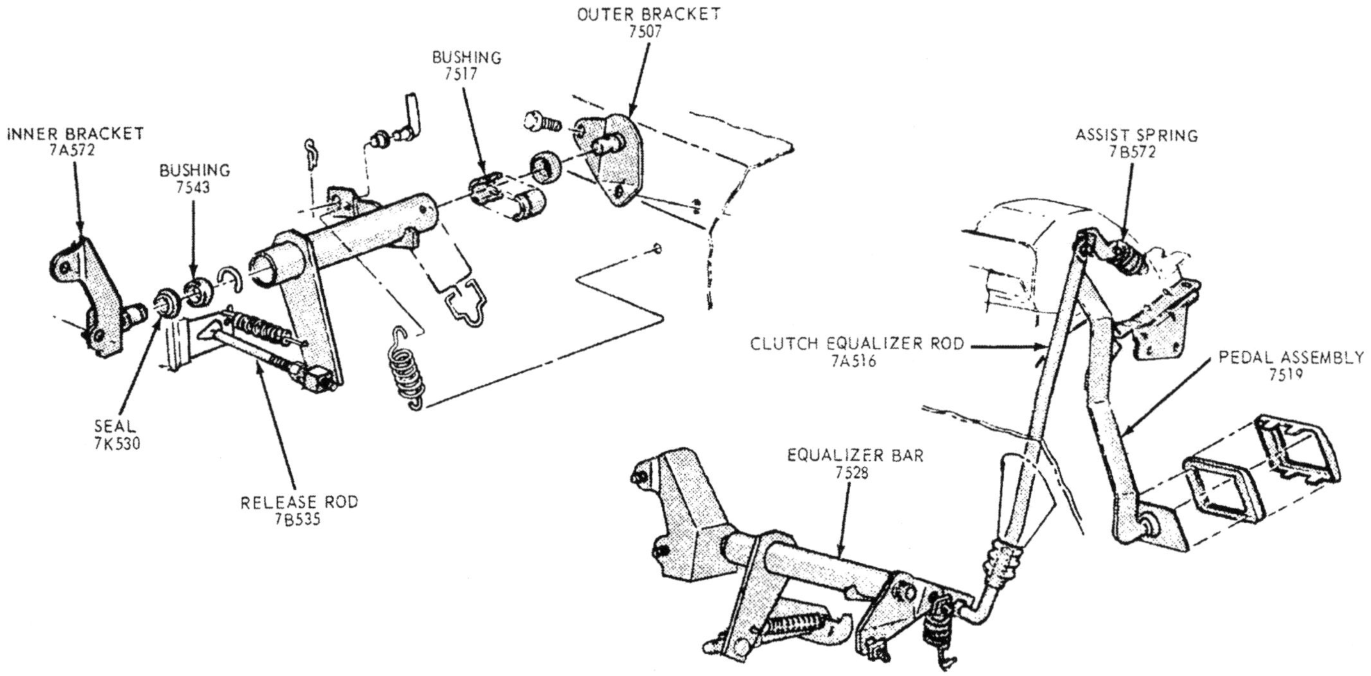

FIG 5:2 Clutch linkage on later models

the total travel within specification. Start the engine and leave it running at curb-idle speed. Press down the clutch pedal by hand until the release bearing is just felt to contact the levers in the clutch cover and note the free travel of the pedal. A typical adjustment point is shown in **FIG 5 : 6**. Slacken the nuts and move them in the direction required to set the correct pedal free travel. Tighten both nuts to the trunnion at the correct setting. Start the engine again and run it up to 3000 rev/min. With the engine at speed check that the minimum pedal free travel is $\frac{1}{2}$ inch. If the free travel at speed is not sufficient, the free travel at idling must be increased.

On later models the check is carried out in a similar manner but the adjustment is easier. Disconnect the return spring from the clutch release lever. Slacken off the nuts and pull the release lever rearwards until the release bearing can just be felt to touch the release levers in the clutch and hold the main release lever in this position. Press the release rod into the release lever so that they are in contact, and hold them in this position. Fit a .136 inch feeler gauge between the adjusting nut and swivel. Tighten down the adjusting nut until it just nips the feeler gauge. Hold the adjusting nut and tighten the locknut to keep the adjusting nut in position. Check the pedal free travel with the engine at idle and at 3000 rev/min, as for the earlier models.

5 : 3 The flywheel housing

When removing the flywheel housing from the engine, collect and carefully store any shims between them. On reassembly install the shims into their original positions. The shims are fitted to align the housing with the engine.

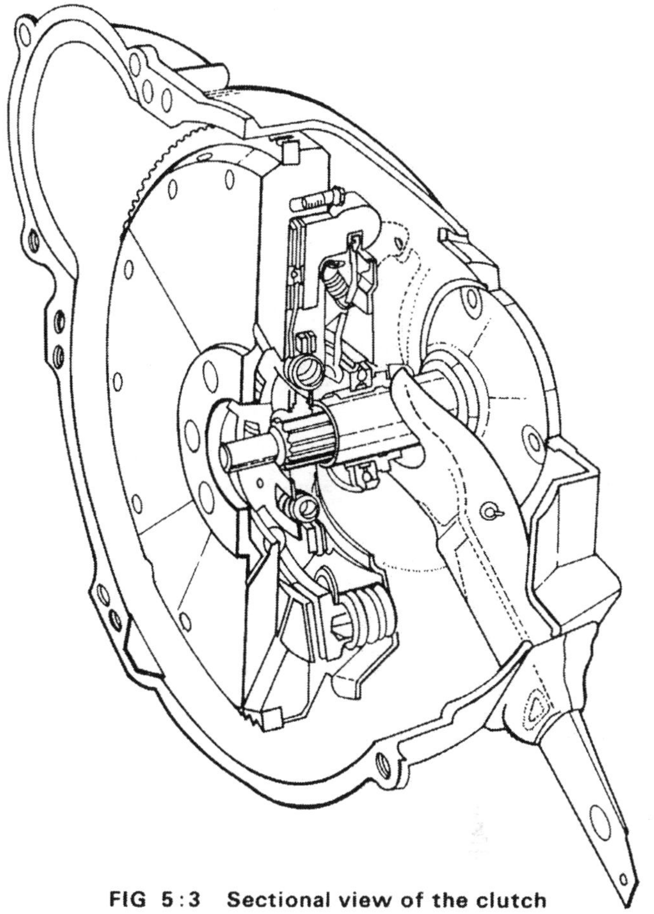

FIG 5 : 3 Sectional view of the clutch

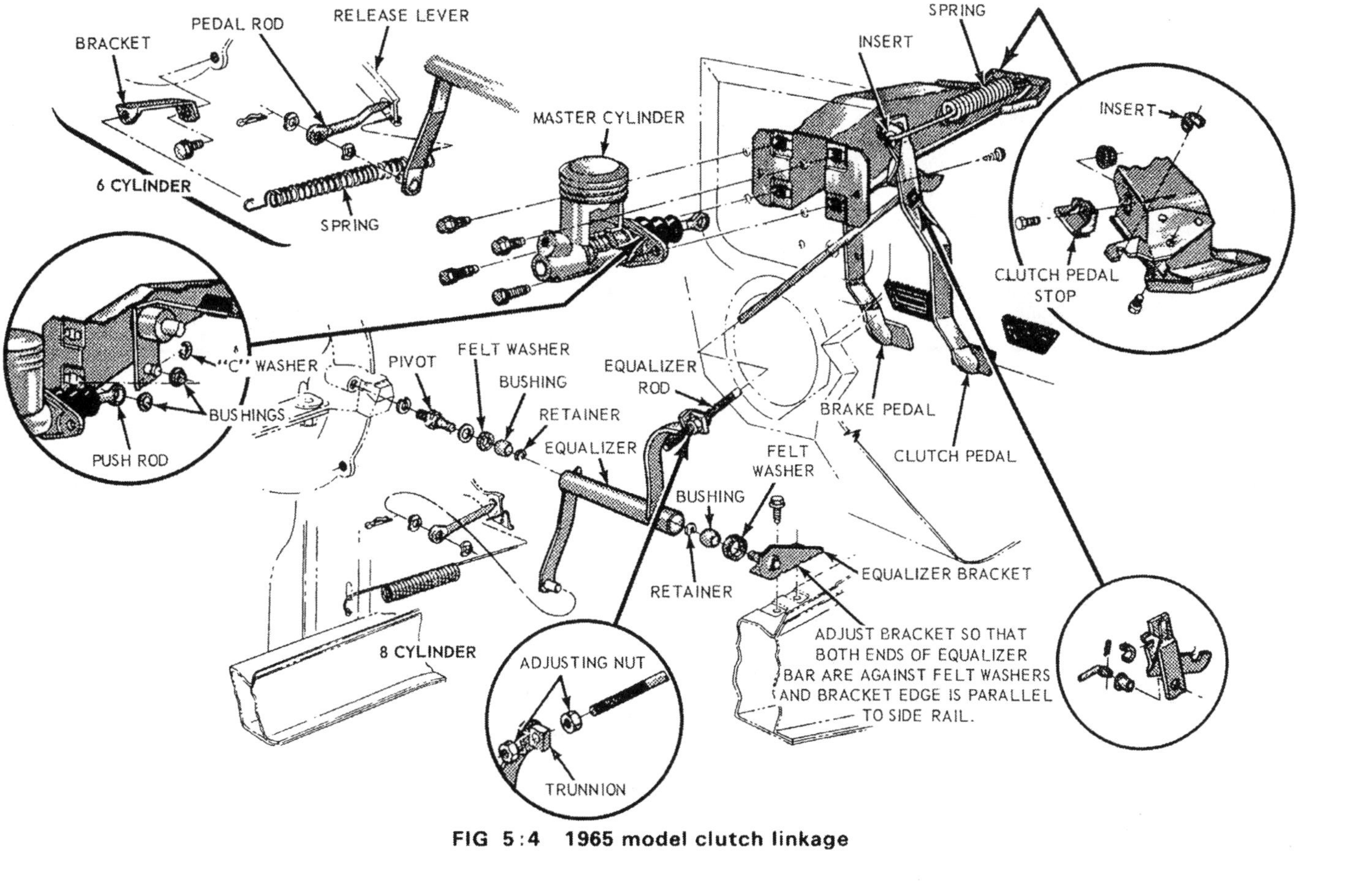

FIG 5 : 4 1965 model clutch linkage

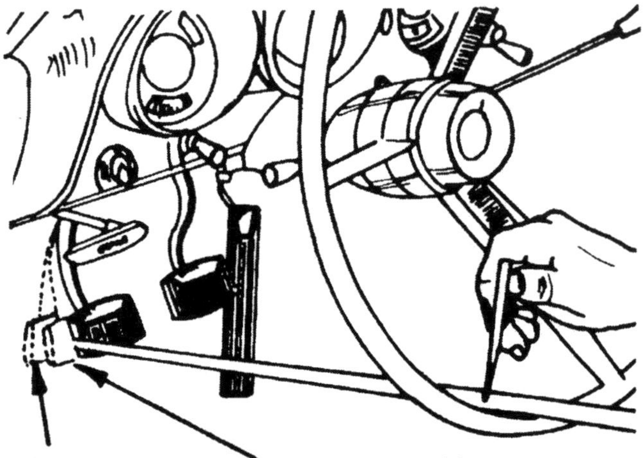

FIG 5:5 Checking the clutch pedal travel

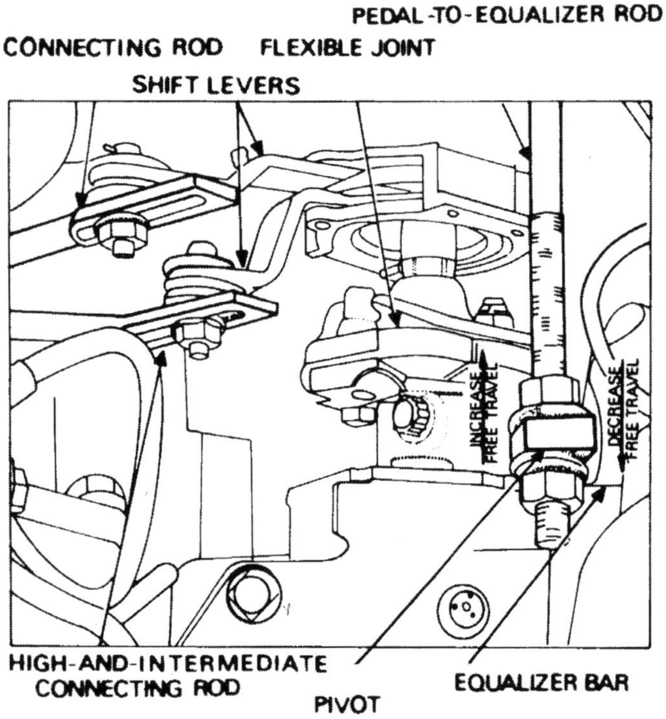

FIG 5:6 Typical clutch adjustment

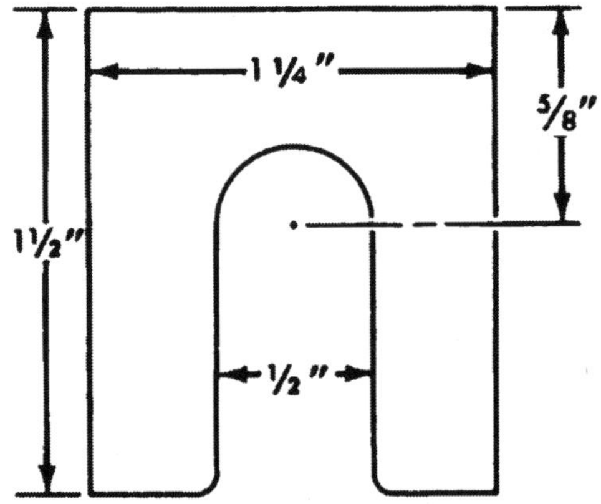

FIG 5:7 Dimension of the shims between flywheel housing and engine

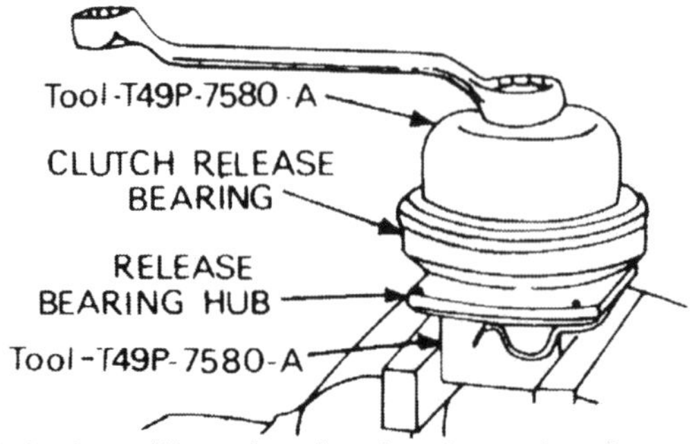

FIG 5:8 Installing the clutch release bearing onto its sleeve

If it is suspected that the alignment of the flywheel housing to the engine is incorrect, a dial indicator, pilot tool, and Dial-L-Igner gauge will be required. The clutch release mechanism must be removed from the housing and the housing freed from the transmission. The spark plugs should be taken out of the engine so that it can be rotated slowly by hand. Full instructions are given on the Dial-L-Igner gauge and shims are made up to the dimensions shown in **FIG 5:7**. **Shims exceeding .010 inch must not be fitted at any point between the housing and engine.** If the Dial-L-Igner gauge gives a reading of greater than .014 inch, the housing is not acceptable.

A certain amount of excessive misalignment can be corrected by machining, using new locating dowels and drilling faulty mounting holes oversize.

5:4 Release bearing and lever

The release bearing and lever can be removed from the flywheel housing after the transmission and housing have been removed from the vehicle (see **Chapter 6**).

The bearing is sealed and no further lubricant can be added to it. **Do not wash the bearing in solvent otherwise the lubricant in it may be removed.** If the

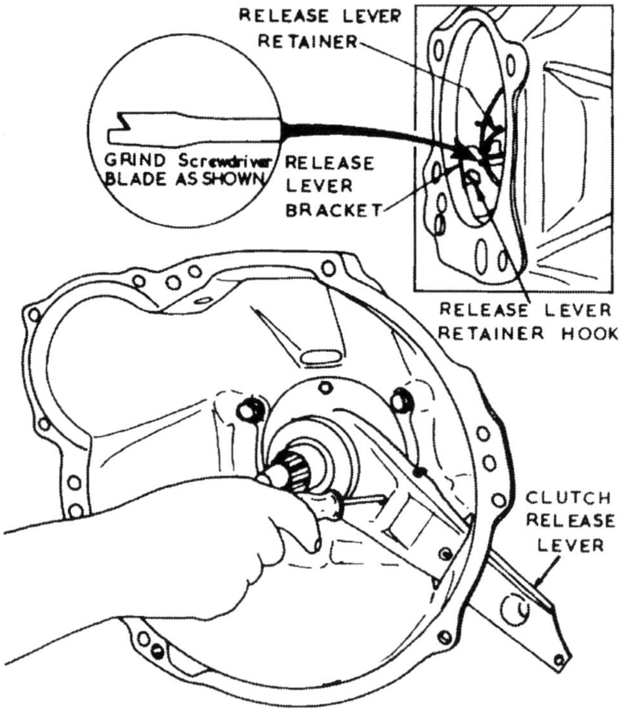

FIG 5:9 Removing or installing the clutch release lever

flywheel housing needs washing, first remove the release bearing. The method of freeing the release lever, using a reworked screwdriver to unhook the retainer, is shown in **FIG 5:9**.

Inspect the bearing for nicks or roughness in operation. Light external damage can be cleaned off using Crocus paper. Apply pressure to the bearing and rotate it while listening for noise. Reject the bearing if it is rough, noisy, or worn, and install a new bearing in its place. The old bearing can be carefully driven from the sleeve but when fitting a new bearing use the tools shown in **FIG 5:8** to draw it back into position.

Install the parts in the reverse order of disassembly. The pivot and bearing points on the release lever should be lightly lubricated with Lubriplate and the face of the release bearing, where it contacts the fingers of the clutch cover release levers, should be lightly lubricated with lithium-based grease. Avoid excessive lubrication otherwise grease may find its way onto the clutch friction plates.

5:5 Servicing the clutch

The removal and installation of the clutch is dealt with in **Chapter 1, Section 1:12**. It is essential to use the arbor when installing the clutch otherwise the disc will not be central and great difficulty will be experienced in trying to line up the transmission input shaft.

Damage or wear to the cover assembly or the clutch disc will require renewal of that component as neither can be dismantled for repairs or servicing.

Cover assembly:

Do not immerse the cover assembly in solvent or wash it with large amounts of solvent as all lubricant will then be washed out. Cleaning should be confined to brushing and blowing out loose dirt and dust and wiping oily surfaces with a cloth moistened with solvent, such as alcohol. Do not use petroleum based solvents.

Examine the cover for cracks or distortion, paying particular attention to areas around rivets and attachment holes.

Inspect the pressure plate surface for cracks, burn marks or scores. Though not advisable, light damage can be skimmed off. If the damage is deep, fit a new cover assembly.

Lay the assembly, pressure plate downwards, onto a clean floor or plank of wood. Check that the release fingers are all at the same height approximately and that no finger shows more wear than the others. Press each finger down in turn and release it smartly, checking that it returns freely and rapidly. If the fingers stick or are at uneven heights, and cleaning does not free the binding, install a new cover assembly.

Clutch disc:

Take great care not to drop the disc when the cover is removed or to allow grease, oil, or paint onto the friction linings.

Check the disc for wear, mechanical damage, or contamination and renew it if it is defective.

The surface of the linings must be well above the rivet heads that secure them. If the linings are worn down nearly to the rivets, install a new disc. Do not attempt to rivet new linings to the old disc.

Check that all rivets are tight and secure. Check that the hub is firmly attached and that none of the damper springs are loose or broken. Slide the disc back onto the input shaft of the transmission and check that it slides freely without excessive rotational play or movement in the damper springs.

Any oil leaks onto the clutch must be traced and rectified before the parts are installed. If all gaskets and seals on both engine and transmission are satisfactory the cause can be overfilling of the engine or transmission. A blocked vent in the transmission will also force oil out past the seals.

For maximum efficiency the linings should have a polished smooth finish through which the grain of the friction material is clearly visible. Small amounts of oil or grease will leave dark colored smears on the linings. Larger amounts will leave the whole surface covered with a dark glaze hiding the grain of the friction material. Excessive amounts of oil will be obvious from the free oil and the oil-soaked appearance of the linings. The disc can be considered fit for further service, provided that it is not mechanically defective or worn, as long as the grain of the friction material can still be seen. Do not attempt to wash off contaminant with solvents or attempt to bake it out.

5:6 Trouble shooting guide

(a) Drag or spin

1 Oil or grease on the linings
2 Incorrectly adjusted clutch pedal free travel
3 Misalignment between engine and flywheel housing
4 Binding pilot bushing in crankshaft
5 Distorted clutch disc
6 Warped or damaged pressure plate
7 Broken disc linings

(b) Fierceness or snatch

1 Check 1, 3, 4, 5, 6 and 7 in (a)
2 Excessively worn clutch disc linings
3 Worn or loose disc hub

(c) Slip

1 Check 1, 2 and 3 in (a) and 2 in (b)
2 Weak cover assembly springs

(d) Judder

1 Check 1, 2 and 7 in (a)
2 Contact area of friction linings not evenly distributed
3 Buckled disc
4 Faulty or loose engine mountings
5 Defective rear suspension or drive shaft

(e) Rattles

1 Worn release linkage
2 Worn release lever and bearing
3 Broken spring in disc or cover assembly

Inches	Decimals	Milli-metres	Inches to Millimetres Inches	mm	Millimetres to Inches mm	Inches
1/64	.015625	.3969	.001	.0254	.01	.00039
1/32	.03125	.7937	.002	.0508	.02	.00079
3/64	.046875	1.1906	.003	.0762	.03	.00118
1/16	.0625	1.5875	.004	.1016	.04	.00157
5/64	.078125	1.9844	.005	.1270	.05	.00197
3/32	.09375	2.3812	.006	.1524	.06	.00236
7/64	.109375	2.7781	.007	.1778	.07	.00276
1/8	.125	3.1750	.008	.2032	.08	.00315
9/64	.140625	3.5719	.009	.2286	.09	.00354
5/32	.15625	3.9687	.01	.254	.1	.00394
11/64	.171875	4.3656	.02	.508	.2	.00787
3/16	.1875	4.7625	.03	.762	.3	.01181
13/64	.203125	5.1594	.04	1.016	.4	.01575
7/32	.21875	5.5562	.05	1.270	.5	.01969
15/64	.234375	5.9531	.06	1.524	.6	.02362
1/4	.25	6.3500	.07	1.778	.7	.02756
17/64	.265625	6.7469	.08	2.032	.8	.03150
9/32	.28125	7.1437	.09	2.286	.9	.03543
19/64	.296875	7.5406	.1	2.54	1	.03937
5/16	.3125	7.9375	.2	5.08	2	.07874
21/64	.328125	8.3344	.3	7.62	3	.11811
11/32	.34375	8.7312	.4	10.16	4	.15748
23/64	.359375	9.1281	.5	12.70	5	.19685
3/8	.375	9.5250	.6	15.24	6	.23622
25/64	.390625	9.9219	.7	17.78	7	.27559
13/32	.40625	10.3187	.8	20.32	8	.31496
27/64	.421875	10.7156	.9	22.86	9	.35433
7/16	.4375	11.1125	1	25.4	10	.39370
29/64	.453125	11.5094	2	50.8	11	.43307
15/32	.46875	11.9062	3	76.2	12	.47244
31/64	.484375	12.3031	4	101.6	13	.51181
1/2	.5	12.7000	5	127.0	14	.55118
33/64	.515625	13.0969	6	152.4	15	.59055
17/32	.53125	13.4937	7	177.8	16	.62992
35/64	.546875	13.8906	8	203.2	17	.66929
9/16	.5625	14.2875	9	228.6	18	.70866
37/64	.578125	14.6844	10	254.0	19	.74803
19/32	.59375	15.0812	11	279.4	20	.78740
39/64	.609375	15.4781	12	304.8	21	.82677
5/8	.625	15.8750	13	330.2	22	.86614
41/64	.640625	16.2719	14	355.6	23	.90551
21/32	.65625	16.6687	15	381.0	24	.94488
43/64	.671875	17.0656	16	406.4	25	.98425
11/16	.6875	17.4625	17	431.8	26	1.02362
45/64	.703125	17.8594	18	457.2	27	1.06299
23/32	.71875	18.2562	19	482.6	28	1.10236
47/64	.734375	18.6531	20	508.0	29	1.14173
3/4	.75	19.0500	21	533.4	30	1.18110
49/64	.765625	19.4469	22	558.8	31	1.22047
25/32	.78125	19.8437	23	584.2	32	1.25984
51/64	.796875	20.2406	24	609.6	33	1.29921
13/16	.8125	20.6375	25	635.0	34	1.33858
53/64	.828125	21.0344	26	660.4	35	1.37795
27/32	.84375	21.4312	27	685.8	36	1.41732
55/64	.859375	21.8281	28	711.2	37	1.4567
7/8	.875	22.2250	29	736.6	38	1.4961
57/64	.890625	22.6219	30	762.0	39	1.5354
29/32	.90625	23.0187	31	787.4	40	1.5748
59/64	.921875	23.4156	32	812.8	41	1.6142
15/16	.9375	23.8125	33	838.2	42	1.6535
61/64	.953125	24.2094	34	863.6	43	1.6929
31/32	.96875	24.6062	35	889.0	44	1.7323
63/64	.984375	25.0031	36	914.4	45	1.7717

UNITS	Pints to Litres	Gallons to Litres	Litres to Pints	Litres to Gallons	Miles to Kilometres	Kilometres to Miles	Lbs. per sq. In. to Kg. per sq. Cm.	Kg. per sq. Cm. to Lbs. per sq. In.
1	.57	4.55	1.76	.22	1.61	.62	.07	14.22
2	1.14	9.09	3.52	.44	3.22	1.24	.14	28.50
3	1.70	13.64	5.28	.66	4.83	1.86	.21	42.67
4	2.27	18.18	7.04	.88	6.44	2.49	.28	56.89
5	2.84	22.73	8.80	1.10	8.05	3.11	.35	71.12
6	3.41	27.28	10.56	1.32	9.66	3.73	.42	85.34
7	3.98	31.82	12.32	1.54	11.27	4.35	.49	99.56
8	4.55	36.37	14.08	1.76	12.88	4.97	.56	113.79
9		40.91	15.84	1.98	14.48	5.59	.63	128.00
10		45.46	17.60	2.20	16.09	6.21	.70	142.23
20				4.40	32.19	12.43	1.41	284.47
30				6.60	48.28	18.64	2.11	426.70
40				8.80	64.37	24.85		
50					80.47	31.07		
60					96.56	37.28		
70					112.65	43.50		
80					128.75	49.71		
90					144.84	55.92		
100					160.93	62.14		

UNITS	Lb ft to kgm	Kgm to lb ft	UNITS	Lb ft to kgm	Kgm to lb ft
1	.138	7.233	7	.967	50.631
2	.276	14.466	8	1.106	57.864
3	.414	21.699	9	1.244	65.097
4	.553	28.932	10	1.382	72.330
5	.691	36.165	20	2.765	144.660
6	.829	43.398	30	4.147	216.990

CHAPTER 6

MANUAL TRANSMISSION

6:1 Description

The standard manual transmission fitted to six cylinder Mustangs is a threespeed floor shift unit.

The power flow through a typical threespeed transmission is shown in **FIG 6:1**. The input shaft is splined to the clutch disc and is therefore driven by it. Synchromesh engagement is used on all three forward speeds. The input shaft is supported by a bearing in the casing and by a needle roller bearing around the front of the mainshaft, and it is shown in **FIG 6:2**. The gear integral on the input shaft is in constant mesh with the front gear of the countershaft gear cluster. The countershaft gear cluster, which has its remaining two gears in constant mesh with those on the mainshaft, revolves about the countershaft on needle roller bearings, one at either end as shown in **FIG 6:3**. The main gears on the mainshaft are free to revolve about the shaft without driving it and all the gears mentioned up to now are helical cut. Synchronizer units are fitted to the mainshaft, shown in **FIG 6:4**, to select the drive. The hub of each synchronizer is splined to the mainshaft while the outer sleeves are splined to the hubs. When the shift mechanism is moved, forks slide the appropriate synchronizer sleeve in the direction shown in **FIG 6:1**. The blocking rings press against the cone of the gear so that friction speeds or slows the shaft to the same speed as the gear. Once the speed is correct, the sleeve moves further and makes the drive positive by meshing with the dog teeth on the gear so that drive is transmitted from the gear through the synchronizer to the mainshaft. Top speed is selected by joining the mainshaft to the input shaft so that drive is direct. An idler, shown in **FIG 6:4**, is in permanent mesh with a gear on the mainshaft and reverse is engaged by sliding the sleeve of the rear synchronizer back until the two spur cut gears mesh and a reverse drive is transmitted. There is no synchromesh on reverse as this speed is only selected when the vehicle is stationary.

A typical transmission case and related parts are shown in **FIG 6:5**.

Some earlier models have an optional fourspeed Dagenham transmission (**FIG 6:6**). It is basically similar in operation to the threespeed unit, but with an extra pair of meshing gears on the mainshaft and countershaft (**FIG 6:7**). The two synchronizer sleeves on the mainshaft select the four forward speeds and a separate selector moves the idler gear into mesh to select reverse.

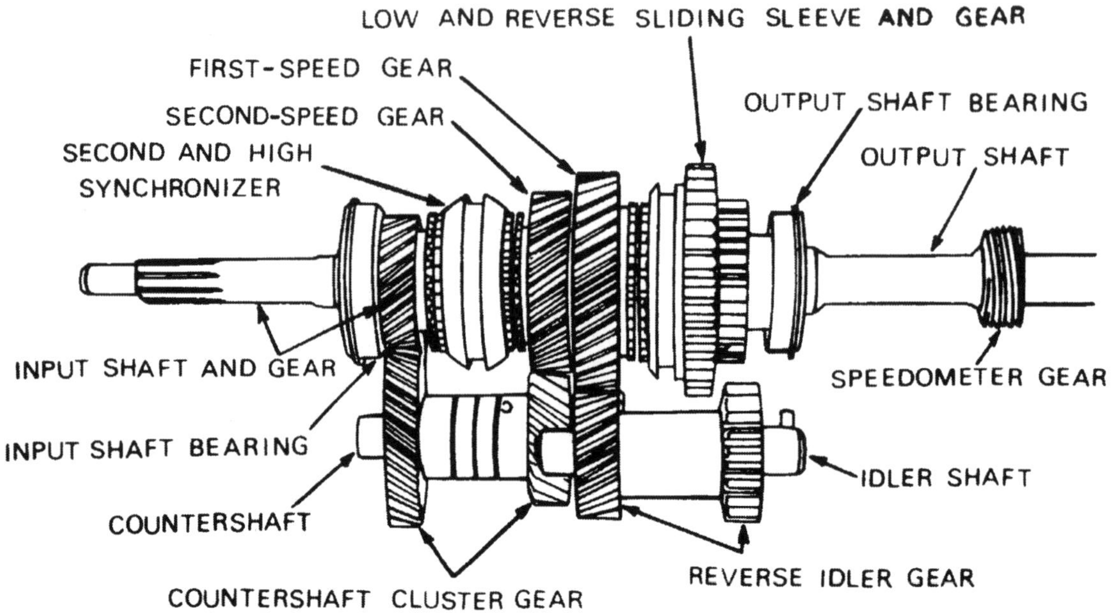

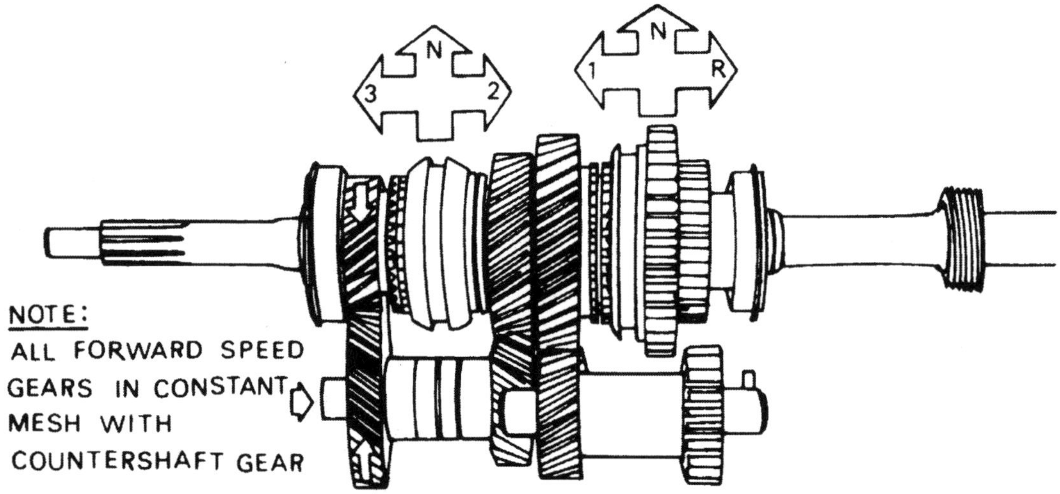

FIG 6:1 **Power flow through a threespeed transmission**

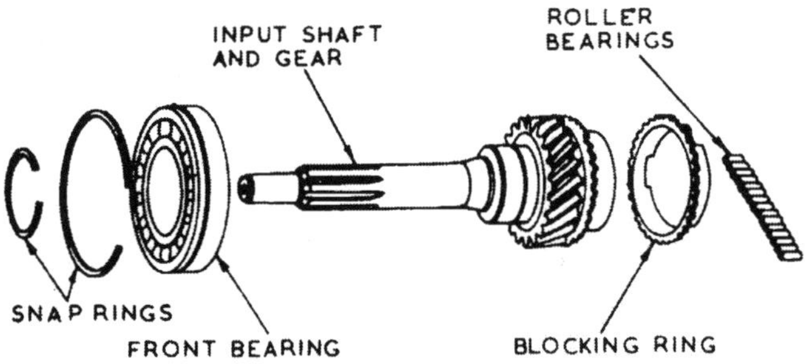

FIG 6:2 **Input shaft assembly**

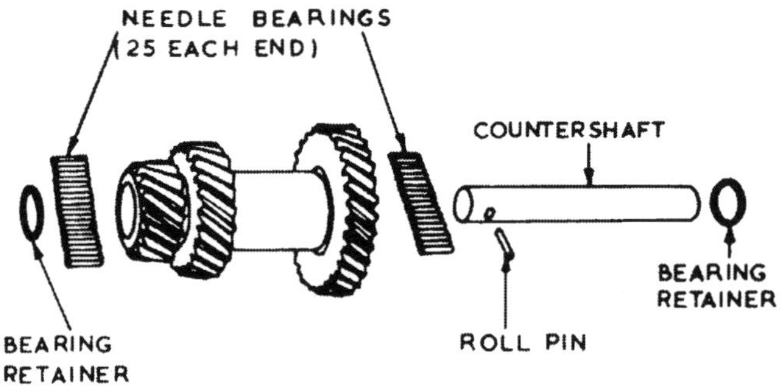

FIG 6:3 **Countershaft assembly**

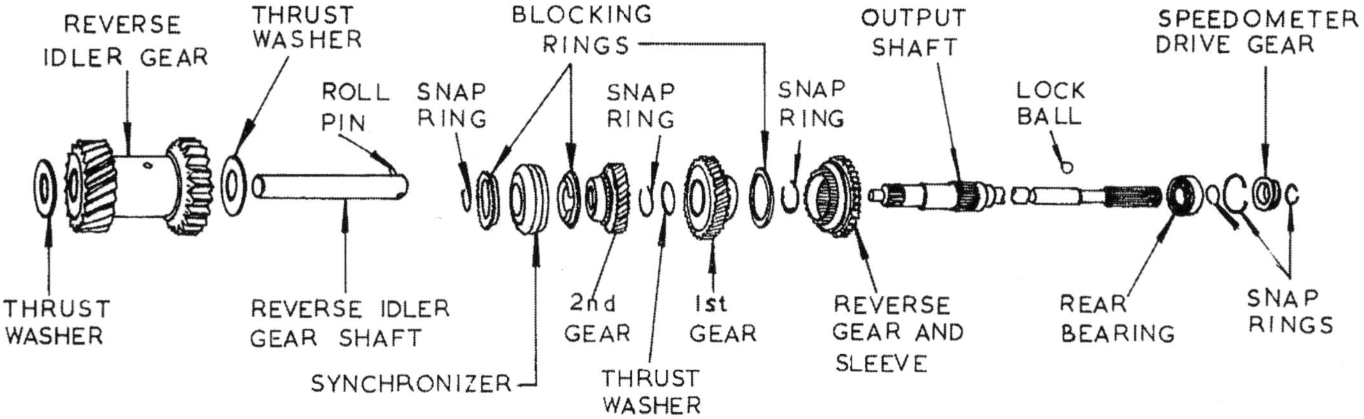

FIG 6:4 Mainshaft and reverse idler parts

6:2 Maintenance

Seasonal or regular oil changes are not required. However when removing the transmission, drain it and refill with fresh oil after installation.

The oil level should be checked at intervals of 6000 miles. The casing is fitted with a combined filler and level plug on the side. Park the vehicle on level ground and if it has recently been driven, leave it for a while to allow the oil to settle. Wipe around the area of the filler plug to remove road dirt and unscrew the filler plug. If the level is correct it will be just at the bottom of the filler aperture. If the level is low, top up using Ford Standard Transmission Fluid C3RZ-19C547-B to the correct level. **Do not overfill the transmission: allow any surplus oil to drain out of the filler hole before installing the filler plug.**

If the oil requires to be drained out, most models are fitted with a drain plug under the casing. If a drain plug is not fitted, drain by removing the lowest bolt that secures the rear extension to the casing. The drain plug is magnetic to trap ferrous metal chips.

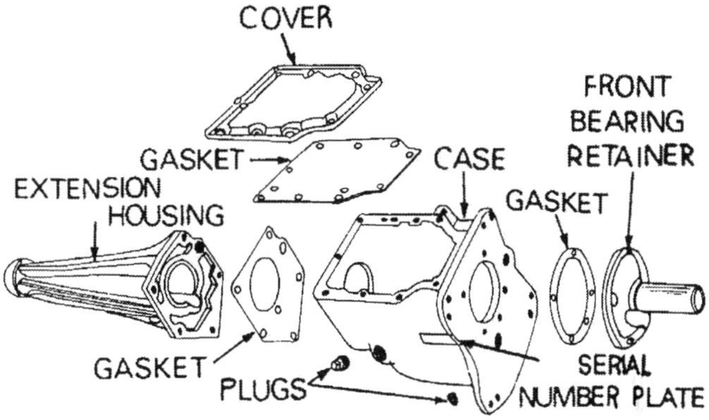

FIG 6:5 Transmission case and related parts

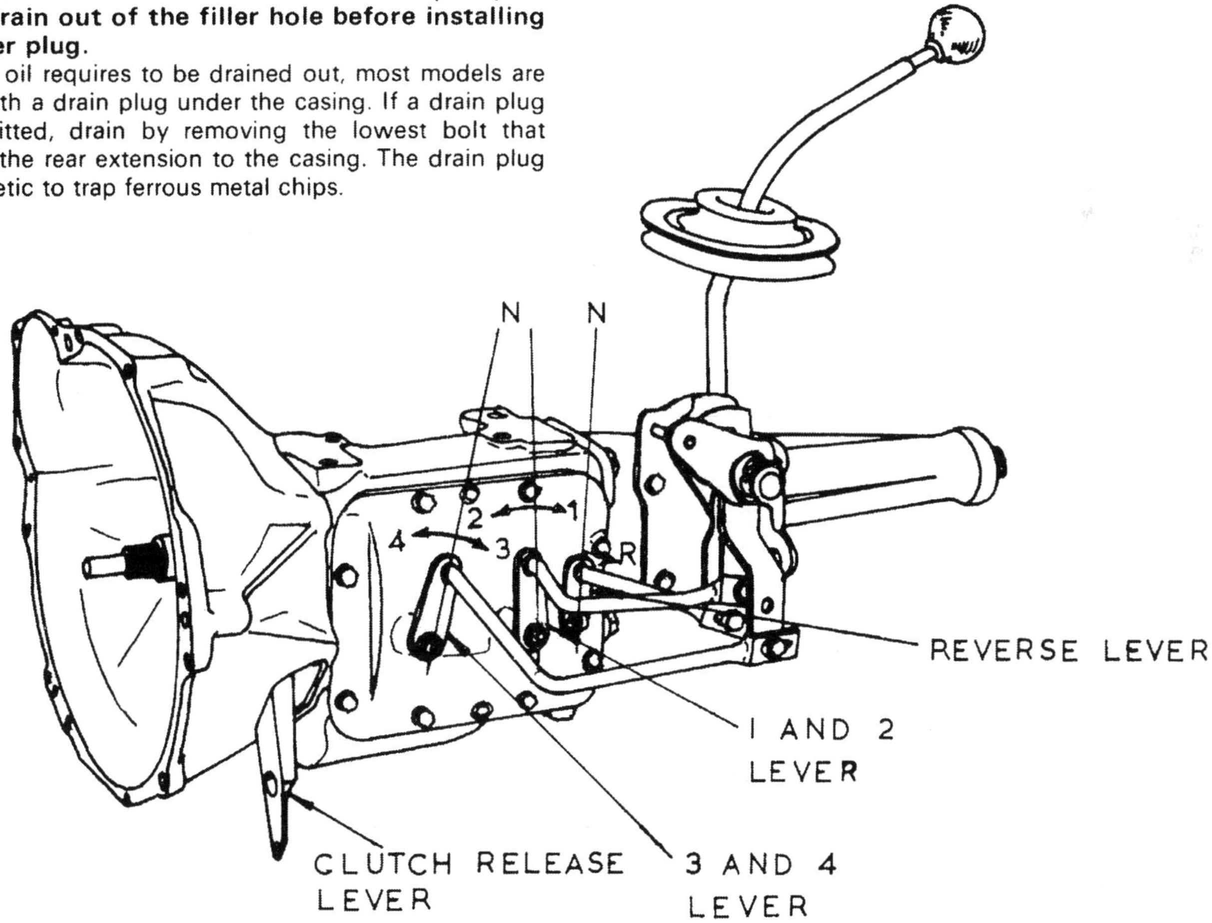

FIG 6:6 Dagenham fourspeed transmission

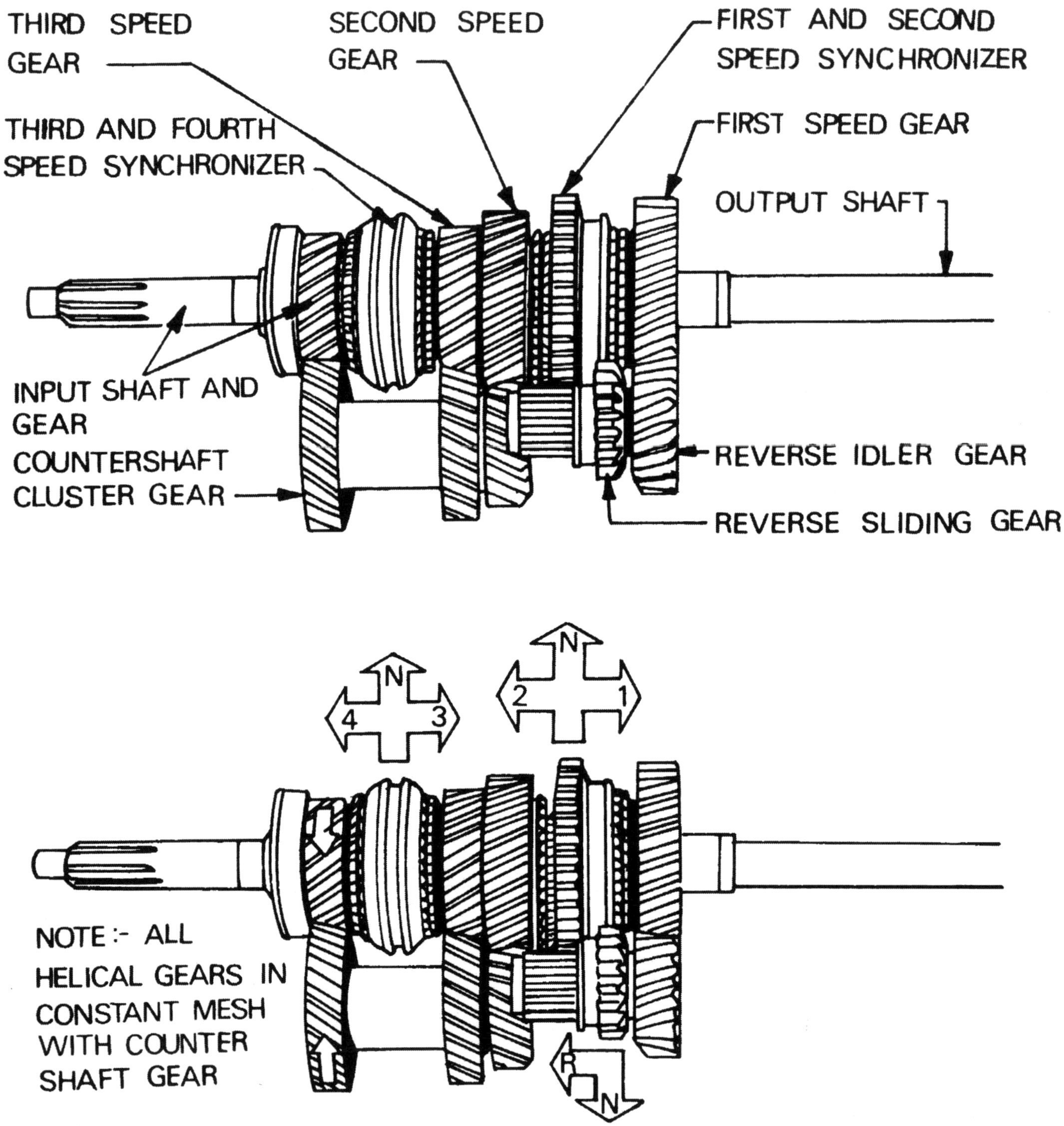

FIG 6:7 Power flow through a fourspeed transmission

6:3 Shift linkage adjustment

A typical adjustment is shown in **FIG 6:8** and the method is the same for all models. On earlier models the alignment pin shown is made by bending a piece of $\frac{1}{4}$ inch diameter rod to an L-shape but from 1969 $\frac{3}{16}$ inch rod should be used instead. Set the gearshift lever to the neutral position and slacken the nuts that secure the shift rods. Move the shift levers on the transmission to their neutral positions and insert the aligning tool as shown in the figure. If the neutral positions of the shift levers are uncertain, disconnect the shift rods from them and move each lever through its three detent positions, leaving it in the central position (which is the neutral one). With the levers set and the pin in place, tighten the adjusting nuts on the shift rods. Remove the pin and check that all speeds select smoothly. Check the adjustment on a road test.

6:4 Transmission removal

Removal procedure on all models is fairly similar with only minor variations between types. The work will be easiest to carry out if the vehicle is raised on a hoist or

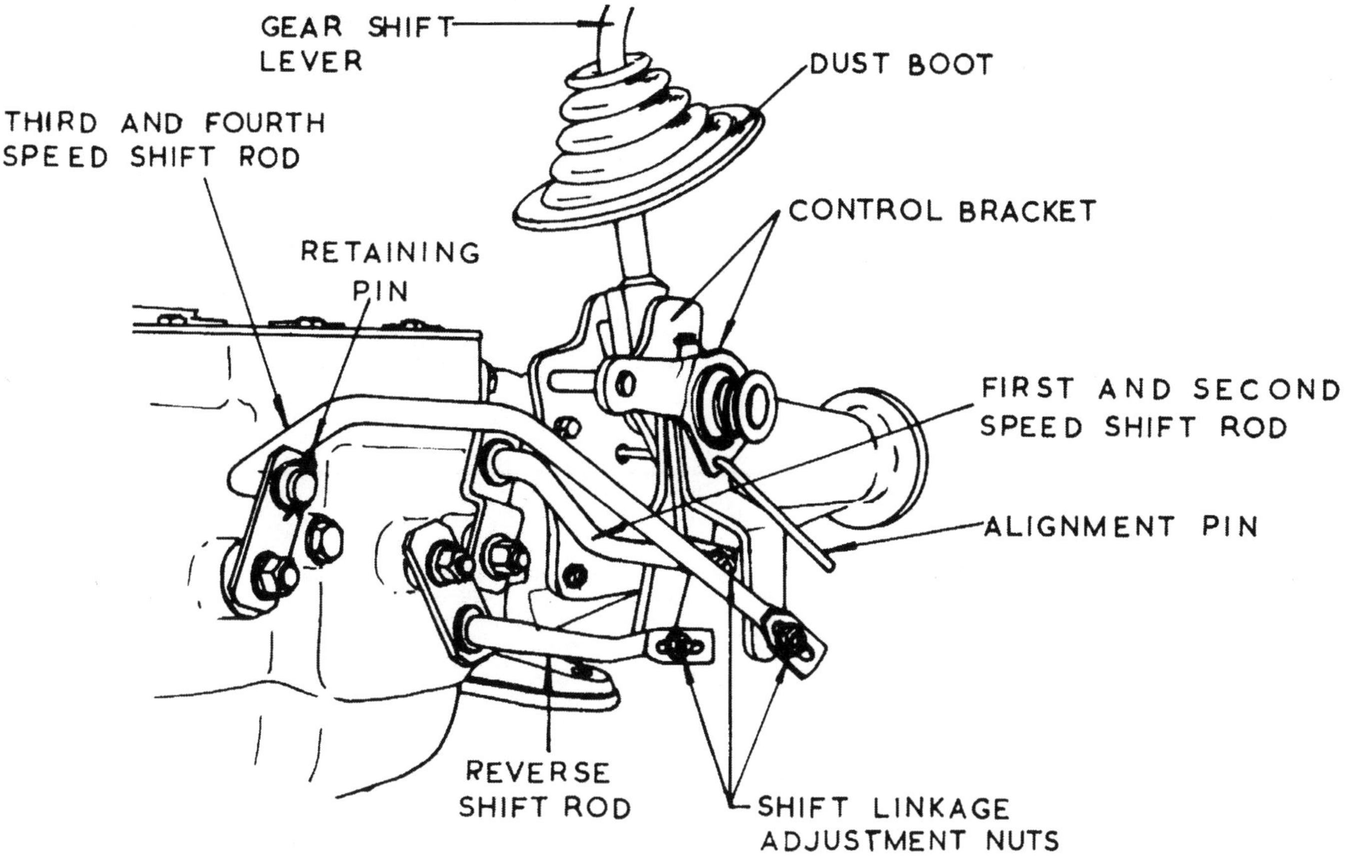

FIG 6:8 Adjusting the shift linkage

over a pit. If neither of these is available raise the car on stands, making sure that the stands are firmly based and capable of supporting the weight.

1 Drain the oil from the transmission. Mark the drive flange of the rear axle and drive shaft, then remove the U-bolts that secure the drive shaft to the rear axle. Lower the end of the drive shaft and withdraw it from the transmission. Blank off the extension housing on the transmission to prevent dirt from entering, lubricant leaking out, or damage being caused. Fuller instructions on drive shaft removal are given in **Chapter 8**.

2 On earlier models only, remove the dust boot from inside the vehicle and take out the two bolts securing the gearshift lever so that the lever can be removed. Take care not to lose the two trunnions or springs. The remainder of the shift linkage can then be left attached to the transmission.

3 On later models, disconnect the shift rods and remove the three bolts that secure the shift selector assembly to the rear extension, leaving the parts hanging from the gearshift lever.

4 Remove the cap screw and retainer that secure the speedometer drive cable to the transmission and withdraw the cable. If a back-up light switch is fitted, free the switch and lay it safely out of the way.

5 Support the engine with a jack and pad of wood at the rear of the oil pan, and lift it to take the weight off the rear mounting. Free the mounting from the transmission and raise the rear of the engine slightly, about the front mountings, so that the mounting crossmember can be removed from the frame.

6 On some models it will be necessary to disconnect the parking brake linkage so that the transmission can be lowered down and out of the vehicle.

7 Support the transmission with another jack and pad of wood. Pads of wood should always be used on a jack to protect the casting or oil pan from damage. Remove the bolts that secure the transmission. On some models the transmission is removed, leaving the flywheel housing attached to the engine. If the bolts securing transmission to housing are not accessible, disconnect the clutch linkage and remove the bolts that secure the flywheel housing to the engine. **If the flywheel housing is left installed, do not operate the clutch pedal while the transmission is out,** otherwise the clutch disc will fall out of alignment and have to be recentralized with an arbor.

8 Support the weight of the transmission and draw it rearwards until the input shaft is well clear from the clutch. The transmission can then be lowered down and out of the vehicle. **Do not allow the weight of the transmission to hang on the input shaft in the clutch otherwise the clutch may be seriously damaged.**

Flywheel housing:

This can be removed from the engine, by taking out its attachment bolts, after the transmission has been removed and the clutch mechanism freed from it. Carefully collect and store any shims fitted between the housing and engine, installing them into their original positions on reassembly.

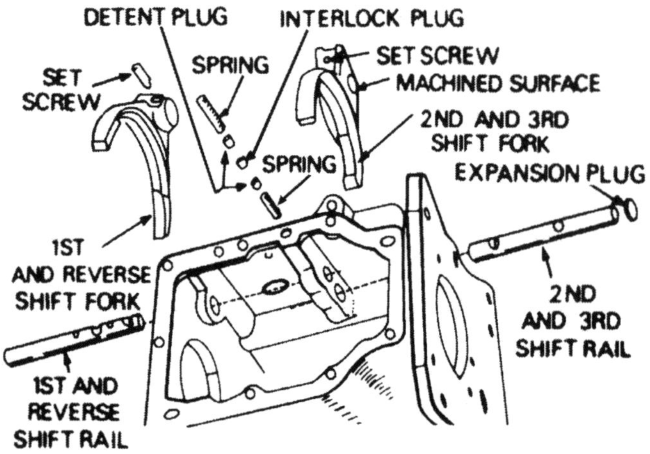

FIG 6:9 Shift rails and forks

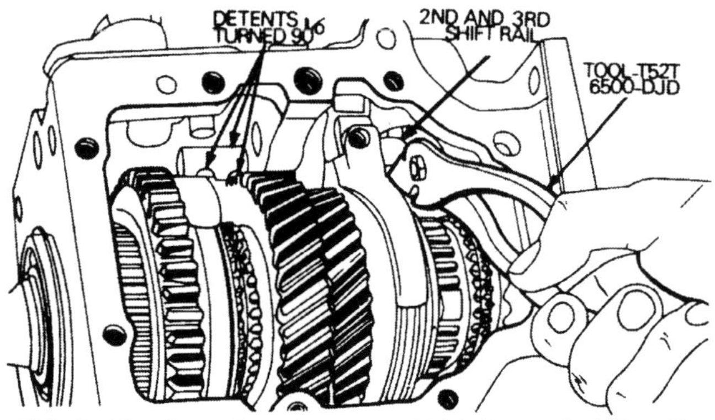

FIG 6:10 Rotating the second/third speed shift rail

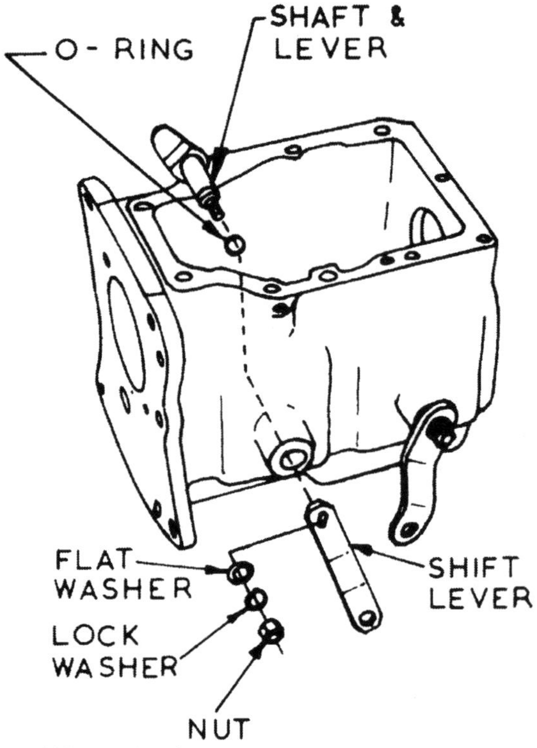

FIG 6:11 Shift levers and shaft parts

The housing is installed in the reverse order of disassembly, after making sure that all the mounting faces are clean and free from damage or burrs. The method of aligning the housing with the engine is outlined in **Chapter 5, Section 5:3**, if this is required.

Installation:

The transmission is installed in the reverse order of removal. The task will be made easier if two long studs are fitted to the lower mounting holes and the transmission is slid along these. Alignment of transmission to engine is important, otherwise difficulty will be found in pressing home the input shaft. Do not use force and if the input shaft will not pass through the clutch hub, select a gear and rotate the output shaft to align the splines of the input shaft with those of the clutch hub. A little grease should be applied to the splines. Support the weight of the transmission until it is fully in place and held by bolts.

Install and connect the drive shaft, speedometer cable and shift linkage and rear mounting. If the parking brake linkage was disconnected, reconnect it and set the park brake correctly (see **Chapter 11**). Similarly reconnect and adjust the clutch linkage if necessary (see **Chapter 5**).

Wipe the magnetic drainplug clean and install it into the bottom of the case. Fill the transmission slowly through the filler hole to the bottom of the filler hole and install the filler plug.

Check the shift selector linkage adjustments and road test the vehicle.

6:5 Servicing the transmission

Special tools and an arbor press will be required for full dismantling and reassembly of the mainshaft components, so many owners will find it beyond their capacity to carry out a full strip down of the unit.

Drain the oil and remove the transmission from the vehicle as described in **Section 6:4**.

Disassembly:

1 Remove the cap screws securing the cover so that the cover and its gasket can be taken off. Carefully remove the upper spring from the case and withdraw the detent with a small magnet. The springs and detents are shown in **FIG 6:9**. Remove the cap screws securing the rear extension to the case and pull off the rear extension housing with its gasket. Similarly remove the front bearing retainer and its gasket.

2 Remove the filler plug from the right side of the case. Use a $\frac{1}{4}$ inch punch through the plug opening to drive out the roll pin that secures the countershaft in place. A dummy countershaft of the same diameter as the correct shaft but shorter and only as long as the gear cluster itself is now required. Hold up the countershaft gear cluster with a hook of wire and insert the dummy shaft at the rear of the case and tap it gently forwards. The dummy shaft should always be kept in contact with the real one, to prevent the washers or bearings coming free, and it should be driven in until it is fully in the countershaft gear cluster so that the real shaft can be withdrawn. Lower the countershaft cluster, with the dummy shaft in it, to the bottom of the casing.

3 Remove the speedometer drive gear from the mainshaft. On models where the gear is secured by a snap ring, remove the snap ring and slide the gear off the shaft, collecting the drive ball fitted between shaft and gear. On the later version, depress the tang on the spring clip and slide off the gear, removing the spring clip from the shaft after the gear is off.

4 Remove the snap ring that locks the mainshaft rear bearings as well as the locating ring around the bearing. Use a suitable extractor to remove the bearing from the case and shaft.

5 Place both selector levers into their central neutral positions. Remove the setscrew that secures the first/reverse selector fork and slide the first/reverse shift rail out of the rear of the casing. The parts are shown in **FIG 6:9**. Slide the synchronizer as far forwards as possible and remove the selector fork by rotating it upwards. Move the second/third selector fork to the second speed position and remove its setscrew. Rotate the shift rail through 90 deg. as shown in **FIG 6:10**. Remove the interlock plug from the case using a small magnet. Tap on the end of the second/third shift rail to knock out the expansion plug and remove the rail. Extract the lower detent plug and spring from the case.

6 Pull the input shaft assembly forwards and remove it from the case, taking care to collect and label the blocking ring. Rotate the second/third selector fork upwards and remove it. Lift the output shaft assembly out through the top of the case.

7 Drive out the roll pin securing the idler gear shaft and withdraw the shaft from the front of the case. Lift out the reverse idler gear with its thrust washers from the case followed by the countershaft gear cluster assembly. The attachments of the shift levers are shown in **FIG 6:11**. If the shift levers are removed, the O-rings must be discarded and new ones, smeared with oil, fitted on reassembly.

Mainshaft assembly:

The parts are shown in **FIG 6:4**. Remove the snap ring from the front of the shaft. Slide off the second/third synchronizer unit as a complete assembly, followed by the second speed gear. Remove the next snap ring and tabbed thrust washer so that the first speed gear and blocking ring can be removed from the shaft. The blocking rings should be labelled as they are removed so that they will be installed into their original positions. The first speed synchronizer has a snap ring in front of it and the hub is a press fit onto the shaft. **An arbor press is essential for removing or installing the synchronizer hub** and if an attempt is made to drive it off or on with drifts the parts will be damaged.

Wash the parts in a suitable solvent, such as gasoline, and dry them with air or non-fluffy cloth. Check the gears for damaged or worn teeth, or worn cones. Examine the shaft for fretting or wear marks and make sure that the gears rotate freely about it. Renew all worn or defective parts. The bearing should be washed separately in clean solvent and renewed if it rotates roughly after thorough cleaning.

The parts are assembled in the reverse order of disassembly. The blocking rings should be installed back

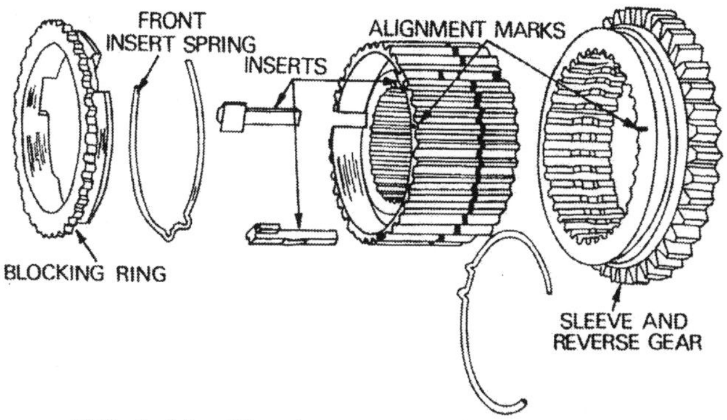

FIG 6:12 First/reverse synchronizer unit

into their original positions. Use new thrust washers if the old ones are worn. Lubricate the parts with transmission fluid.

The input shaft:

The parts are shown in **FIG 6:2**. A press must be used to remove or install the bearing onto the shaft. Clean the shaft and bearing and renew the shaft if it is worn or the gear teeth are damaged. Renew the bearing if it is worn or runs noisily.

Coat the bore in the input shaft with a **thin** film of grease and install the fifteen needle rollers into the grease. A thick layer of grease should not be used to hold the rollers as it will block the lubrication holes and prevent the oil from reaching the bearing.

Synchronizer units:

The parts of the first/reverse unit are shown in **FIG 6:12**, and the hub of this unit is a press fit on the mainshaft. The components of the second/third unit are shown in **FIG 6:13**.

When dismantling a unit, wrap it in cloth to catch all the parts and push the sleeve off the hub.

Clean the components and check them for wear or damage. **Do not intermix the parts between the two synchronizer units.**

Install the rear spring in the groove of the first/reverse synchronizer hub. Start the hub into the sleeve, making sure that the marks are aligned. Position the three inserts in the hub, making sure that the small end is over the spring and that the shoulder is on the inside of the hub. Slide the hub onto the sleeve until the detent engages. Install the front spring to hold the inserts against the hub. It is essential that the humps on the springs cover different inserts, as shown in **FIG 6:14**.

The second/third speed synchronizer is assembled in a similar manner.

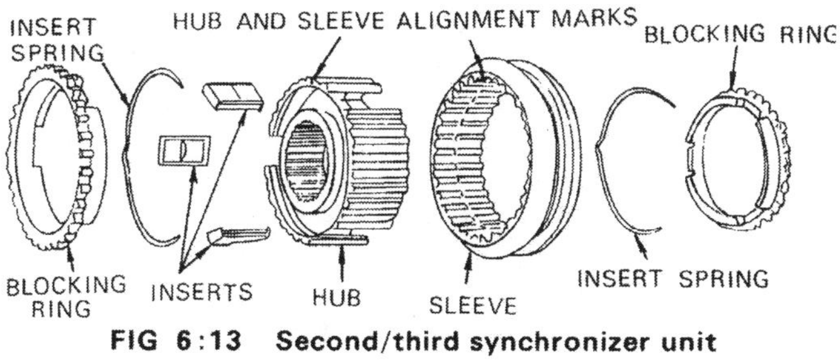

FIG 6:13 Second/third synchronizer unit

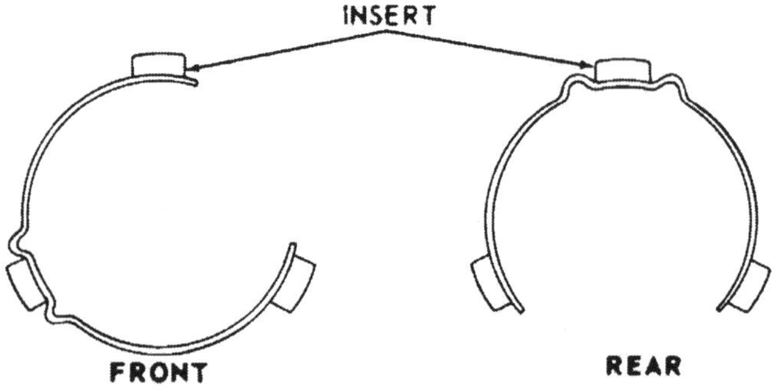

FIG 6:14 Correct installation of the springs in the synchronizer unit (first/reverse shown)

Countershaft assembly:

If necessary, remove the 25 rollers from each end of the cluster, after removing the dummy countershaft. Wash the parts in solvent and check them for wear or damage.

Slide the gear cluster and its thrust washers back into place in the case and use feeler gauges to measure the end play of the assembly in the case. If the play is not within the correct specification of .004 to .018 inch, discard the old thrust washers and check that new thrust washers bring the play within specification. Remove the parts from the case.

If the rollers have been removed, lightly smear the bore of the cluster with grease, insert the dummy countershaft and feed the 25 rollers back into each end.

Reverse idler shaft:

Check the shaft and gear for wear or damage. Install the parts, shown in **FIG 6:4**, back into the case and use feeler gauges to measure the end play on the gear in the case. If the end play is not within the correct specifications of .004 to .018 inch, fit new thrust washers. Remove the parts from the case.

Assembling the transmission:

Clean the casing parts and examine them for cracks, stripped threads, or damaged mating faces. Discard old gaskets and seals, using new ones on reassembly.

Reassembly is the reverse of the disassembly procedure. All splines and bearing surfaces should be lightly lubricated with transmission oil.

Install the reverse idler components. Lower the countershaft gear cluster assembly, with dummy shaft in place, down to the bottom of the case. The thrust washers can be held in place with a thin smear of grease. Install the mainshaft assembly through the top aperture, followed by the input shaft assembly through the front of the case. Make sure that the spigot on the mainshaft fits into the needle roller bearing in the input shaft, without disturbing or losing the rollers, and do not forget to install the blocking ring between the two shafts. The input shaft is driven in until the snap ring in the bearing locates against the case. The selector mechanism parts are then installed, not forgetting to fit the detent springs, detent plugs and interlock plug. The rear bearing is driven along the shaft until its snap ring is tight against the case. Raise the countershaft cluster assembly into position, making sure that the thrust washers are also in place. Press out the dummy countershaft with the correct one and lock the shaft in place with a new roll pin.

Make sure that a new expansion plug is fitted to the case for the shift rail. The cover, extension housing and bearing retainer are installed after lightly smearing the gaskets with sealer and dipping the threads of the bolts into sealer. The front bearing retainer must be fitted with the oil return slot downwards. Before fitting the cover, pour transmission oil over the complete gear train. Do not forget to install the upper spring and detent plug before fitting the cover.

Clean the drain and filler plugs and install them in the case. The drain plug is magnetic, so remove all metal chips from it.

Check the transmission for correct operation in all the gear selections. The input shaft can be rotated by hand while checking the transmission.

Fill the transmission to the level of the filler plug with fresh lubricant after it has been installed in the vehicle.

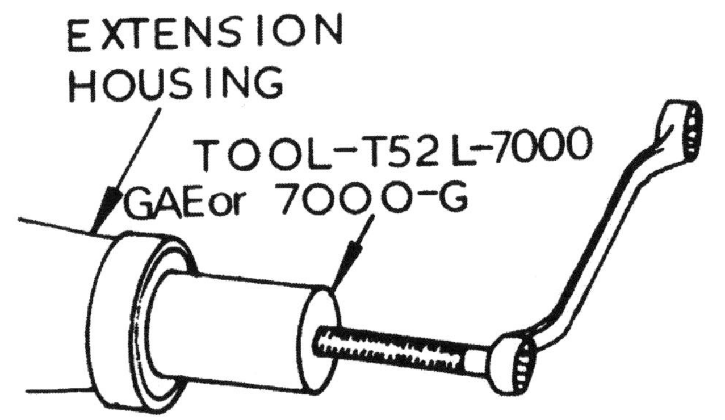

FIG 6:15 Removing the bush and oil seal from the extension housing

6:6 Rear seal and bush

The rear oil seal and bush in the extension housing of the transmission can be renewed without removing the transmission from the vehicle. Remove the drive shaft (see **Chapter 8**).

The seal on its own can be removed using a suitable internal extractor (preferably Tool 1175-A-B) and a suitable impact extractor. The new oil seal should be driven back into place using the special drift T61L-7657-A.

If the bush is worn or damaged, use the tool as shown in **FIG 6:15** to remove the bush and seal together. Insert the tool until it grips on the front side of the bush and then turn the screw clockwise until the bush and seal are drawn out of the housing. Drive a new bush back into place with the tool T57P-7697-B (or 7657-G) and then drive a new oil seal back into position. Install the drive shaft.

Check the transmission lubricant level by removing the filler drain plug on the righthand side of the case and, if necessary, top up until the level is at the bottom of the hole.

6:7 Trouble shooting guide

(a) Jumping out of gear

1 Shift mechanism incorrectly adjusted
2 Broken or weak detent spring
3 Worn detent grooves in shift rails
4 Worn blocking rings or synchronizer units
5 Shift fork loose on rail

(b) Noisy transmission

1 Insufficient or incorrect grade of lubricant
2 Excessive end play on shafts or parts (worn or broken thrust washers)
3 Worn or damaged bearings
4 Worn or damaged gear teeth

(c) Difficulty in engaging gear

1 Check 1, 4 and 5 in (a)
2 Stiff shift mechanism
3 Incorrect clutch adjustments
4 Defective spigot bushing in crankshaft

(d) Lubricant leaks

1 Damaged joint faces
2 Defective gaskets
3 Defective oil seals
4 Stripped threads
5 Cracked case

Inches	Decimals	Milli-metres	Inches to Millimetres Inches	mm	Millimetres to Inches mm	Inches
1/64	.015625	.3969	.001	.0254	.01	.00039
1/32	.03125	.7937	.002	.0508	.02	.00079
3/64	.046875	1.1906	.003	.0762	.03	.00118
1/16	.0625	1.5875	.004	.1016	.04	.00157
5/64	.078125	1.9844	.005	.1270	.05	.00197
3/32	.09375	2.3812	.006	.1524	.06	.00236
7/64	.109375	2.7781	.007	.1778	.07	.00276
1/8	.125	3.1750	.008	.2032	.08	.00315
9/64	.140625	3.5719	.009	.2286	.09	.00354
5/32	.15625	3.9687	.01	.254	.1	.00394
11/64	.171875	4.3656	.02	.508	.2	.00787
3/16	.1875	4.7625	.03	.762	.3	.01181
13/64	.203125	5·1594	.04	1.016	.4	.01575
7/32	.21875	5.5562	.05	1.270	.5	.01969
15/64	.234375	5.9531	.06	1.524	.6	.02362
1/4	.25	6.3500	.07	1.778	.7	.02756
17/64	.265625	6.7469	.08	2.032	.8	.03150
9/32	.28125	7.1437	.09	2.286	.9	.03543
19/64	.296875	7.5406	.1	2.54	1	.03937
5/16	.3125	7.9375	.2	5.08	2	.07874
21/64	.328125	8.3344	.3	7.62	3	.11811
11/32	.34375	8.7312	.4	10.16	4	.15748
23/64	.359375	9.1281	.5	12.70	5	.19685
3/8	.375	9.5250	.6	15.24	6	.23622
25/64	.390625	9.9219	.7	17.78	7	.27559
13/32	.40625	10.3187	.8	20.32	8	.31496
27/64	.421875	10.7156	.9	22.86	9	.35433
7/16	.4375	11.1125	1	25.4	10	.39370
29/64	.453125	11.5094	2	50.8	11	.43307
15/32	.46875	11.9062	3	76.2	12	.47244
31/64	.484375	12.3031	4	101.6	13	.51181
1/2	.5	12.7000	5	127.0	14	.55118
33/64	.515625	13.0969	6	152.4	15	.59055
17/32	.53125	13.4937	7	177.8	16	.62992
35/64	.546875	13.8906	8	203.2	17	.66929
9/16	.5625	14.2875	9	228.6	18	.70866
37/64	.578125	14.6844	10	254.0	19	.74803
19/32	.59375	15.0812	11	279.4	20	.78740
39/64	.609375	15.4781	12	304.8	21	.82677
5/8	.625	15.8750	13	330.2	22	.86614
41/64	.640625	16.2719	14	355.6	23	.90551
21/32	.65625	16.6687	15	381.0	24	.94488
43/64	.671875	17.0656	16	406.4	25	.98425
11/16	.6875	17.4625	17	431.8	26	1.02362
45/64	.703125	17.8594	18	457.2	27	1.06299
23/32	.71875	18.2562	19	482.6	28	1.10236
47/64	.734375	18.6531	20	508.0	29	1.14173
3/4	.75	19.0500	21	533.4	30	1.18110
49/64	.765625	19.4469	22	558.8	31	1.22047
25/32	.78125	19.8437	23	584.2	32	1.25984
51/64	.796875	20.2406	24	609.6	33	1.29921
13/16	.8125	20.6375	25	635.0	34	1.33858
53/64	.828125	21.0344	26	660.4	35	1.37795
27/32	.84375	21.4312	27	685.8	36	1.41732
55/64	.859375	21.8281	28	711.2	37	1.4567
7/8	.875	22.2250	29	736.6	38	1.4961
57/64	.890625	22.6219	30	762.0	39	1.5354
29/32	.90625	23.0187	31	787.4	40	1.5748
59/64	.921875	23.4156	32	812.8	41	1.6142
15/16	.9375	23.8125	33	838.2	42	1.6535
61/64	.953125	24.2094	34	863.6	43	1.6929
31/32	.96875	24.6062	35	889.0	44	1.7323
63/64	.984375	25.0031	36	914.4	45	1.7717

UNITS	Pints to Litres	Gallons to Litres	Litres to Pints	Litres to Gallons	Miles to Kilometres	Kilometres to Miles	Lbs. per sq. In. to Kg. per sq. Cm.	Kg. per sq. Cm. to Lbs. per sq. In.
1	.57	4.55	1.76	.22	1.61	.62	.07	14.22
2	1.14	9.09	3.52	.44	3.22	1.24	.14	28.50
3	1.70	13.64	5.28	.66	4.83	1.86	.21	42.67
4	2.27	18.18	7.04	.88	6.44	2.49	.28	56.89
5	2.84	22.73	8.80	1.10	8.05	3.11	.35	71.12
6	3.41	27.28	10.56	1.32	9.66	3.73	.42	85.34
7	3.98	31.82	12.32	1.54	11.27	4.35	.49	99.56
8	4.55	36.37	14.08	1.76	12.88	4.97	.56	113.79
9		40.91	15.84	1.98	14.48	5.59	.63	128.00
10		45.46	17.60	2.20	16.09	6.21	.70	142.23
20				4.40	32.19	12.43	1.41	284.47
30				6.60	48.28	18.64	2.11	426.70
40				8.80	64.37	24.85		
50					80.47	31.07		
60					96.56	37.28		
70					112.65	43.50		
80					128.75	49.71		
90					144.84	55.92		
100					160.93	62.14		

UNITS	Lb ft to kgm	Kgm to lb ft	UNITS	Lb ft to kgm	Kgm to lb ft
1	.138	7.233	7	.967	50.631
2	.276	14.466	8	1.106	57.864
3	.414	21.699	9	1.244	65.097
4	.553	28.932	10	1.382	72.330
5	.691	36.165	20	2.765	144.660
6	.829	43.398	30	4.147	216.990

CHAPTER 7

AUTOMATIC TRANSMISSION

7:1 Description

A C4 automatic transmission can be fitted in place of the manual unit described in the previous chapter.

A sectional view of the unit is shown in **FIG 7:1**, and the main components are shown separated in **FIG 7:2**. A torque convertor is bolted to the engine drive plate and takes the place of the mechanically operated clutch. From the convertor the drive is passed through an epicyclic gearbox in which the ratios are selected by clutches and brake bands operating on hydraulic pressure from the control system.

The converter:

A sectional view of the converter is shown in **FIG 7:3**. The case unit is sealed and there is a constant flow of transmission fluid through it so that all the parts operate in fluid. The impeller and its vanes are integral with the case, and as this is bolted to the engine drive plate they revolve with the engine. The vanes of the impeller drag the transmission fluid around with them setting up a concentric flow around the converter. Because of centrifugal force there will be a greater pressure at the tip of the impeller vanes than at the inner edge, and a rotational flow is set up in the plane of the cross-section. The two flows of the fluid combine so that the fluid follows the path of an imaginary coil spring laid concentrically between impeller and turbine. This flow applies torque to the turbine, causing it to rotate. When the turbine is stationary or rotating at slow speed compared with the impeller, the flow from the inner edge of the turbine is deflected by the front of the blades of the stator in such a way that the fluid impinges on the impeller blades to assist its rotation. This effect gives a torque multiplication on the turbine of up to 2:1. As the turbine speeds up the torque multiplication factor drops towards 1:1. From rest therefore, the converter acts as an infinitely variable gearbox between the ratios 2:1 and 1:1. At a turbine speed which is 90 per cent of the impeller speed the flow has changed so that it impinges on the back of the stator blades. The stator is fitted on a one-way clutch so that it is now free to revolve with the fluid and prevent power losses. From this point the converter acts as a fluid flywheel giving direct drive.

There will always be a slight speed differential, because of frictional losses, but under optimum conditions this loss is normally in the order of only 2 per cent. Frictional losses are dissipated as heat, the larger the slip the more heat produced. The fluid from the converter passes through

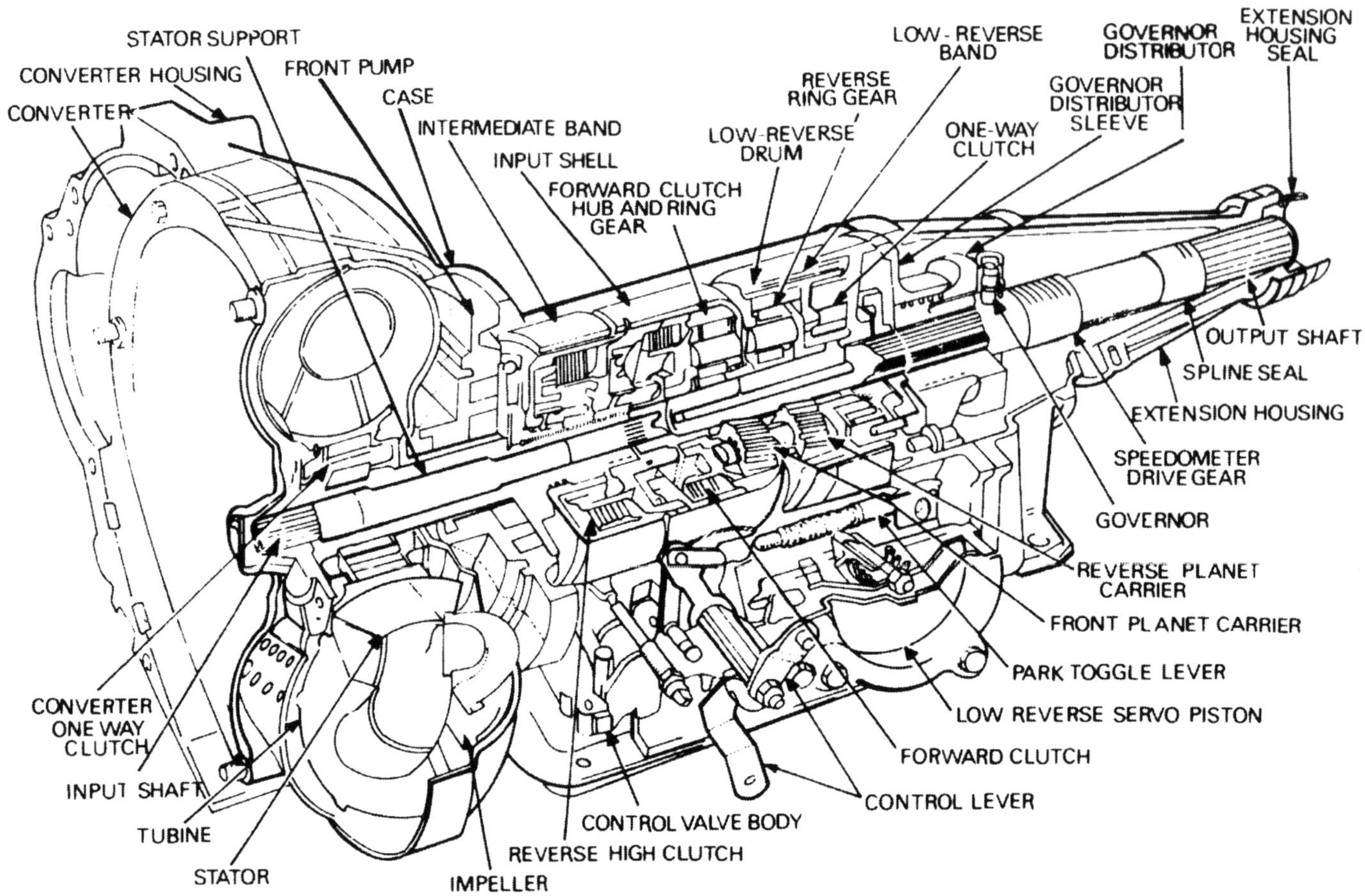

FIG 7:1 Sectional view of the automatic transmission

the oil cooler before being returned to the oil pan of the transmission unit. If the turbine is held stalled, by selecting drive and holding the car on the brakes, large amounts of heat will be rapidly produced which will damage the transmission fairly quickly.

Epicyclic gearbox:

An epicyclic gear train consists of a sun gear and outer gear with planet gears on a carrier revolving between the two. By holding any one component stationary and driving through another, drive can be taken from the third component. The ratio can be varied and the drive reversed, using a single epicyclic gear train.

Two epicyclic gear trains in series are used in the gearbox to give the three forward speeds and one reverse. The power flow through the transmission in the various selections is shown in **FIG 7:4**. The clutches are operated by pistons and the drums are held as required by brake bands around them, again operated by pistons and hydraulic pressure.

Special tools are essential for disassembly of the gearbox and such work should only be carried out by specialists.

Hydraulic control system:

A schematic layout of the system for the automatic transmission is shown in **FIG 7:5**. It must be emphasized that the figures are purely schematic and that many of the lines shown are integral passages in the components.

The fluid is contained in the oil pan under the unit. A gear-type pump is fitted to the input shaft and draws the fluid from the sump through an oil strainer. Normally the pressure is held to 60 lb/sq in by the regulator valve. Excess oil is allowed to pass through the converter, oil cooler and back to the oil pan while if this is not sufficient to reduce the pressure, more oil is released directly into the oil pan. When operation of the transmission requires a greater pressure than 60 lb/sq in, the regulator valve is held back by the pressure booster valve.

Internal control is from the governor, which is attached to the output shaft and therefore senses road speed. Throttle position is fed into the unit by the vacuum connection and the other two controls required are the kick-down link and selector link.

Tests on the operation of the system or any repair work, beyond that given in this chapter, must be left to experts using special equipment.

7:2 Maintenance

This is confined to checking that the fluid level is corrected and that the linkages operate correctly and freely, and cleaning dirt from the unit externally. At intervals adjust the brake bands. Seasonal or routine fluid changes are not required.

Cleanliness:

All operations involving the automatic transmission must be carried out with great care and

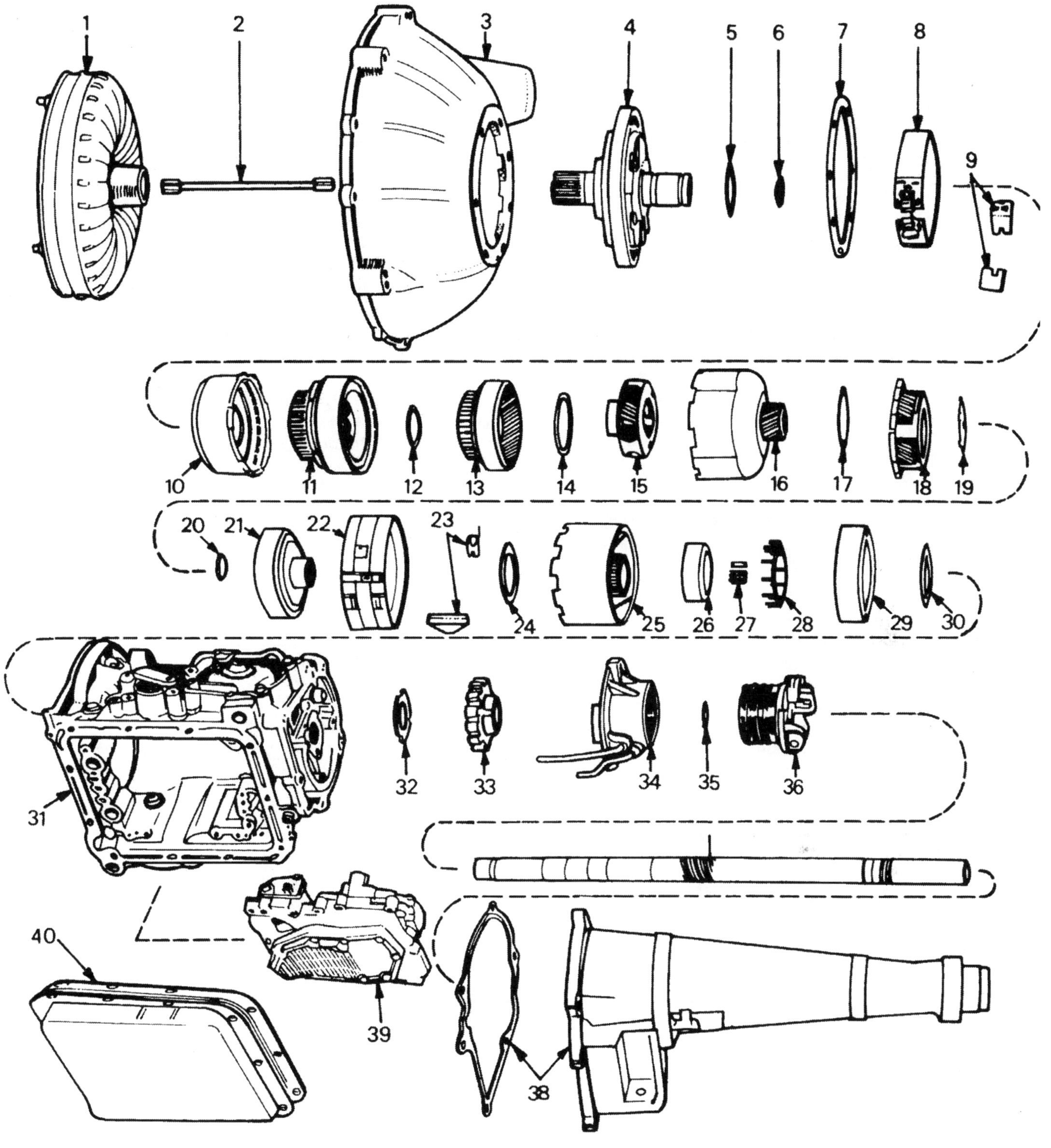

FIG 7:2 Main sub-assemblies of the transmission

Key to Fig 7:2 1 Converter 2 Input shaft 3 Converter housing 4 Front pump 5/6 Thrust washers 7 Front pump gasket 8 Intermediate band 9 Band struts 10 Reverse and high clutch drum 11 Forward clutch and cylinder 12 Thrust washers 13 Forward clutch hub and ring gear 14 Thrust washer 15 Front planet carrier 16 Input shell, sun gear and thrust washer 17 Thrust washer 18 Reverse planet carrier 19 Thrust washer 20 Snap ring 21 Reverse ring gear and hub 22 Low and reverse band 23 Band struts 24 Thrust washer 25 Low and reverse drum 26 One-way clutch inner race 27 Twelve rollers and springs 28 Spring and roller cage 29 One-way clutch outer race 30 Thrust washer 31 Case 32 Thrust washer 33 Parking gear 34 Governor distributor sleeve 35 Snap ring 36 Governor valves and distributor 37 Output shaft 38 Extension housing and gasket 39 Control valve body 40 Oil pan and gasket

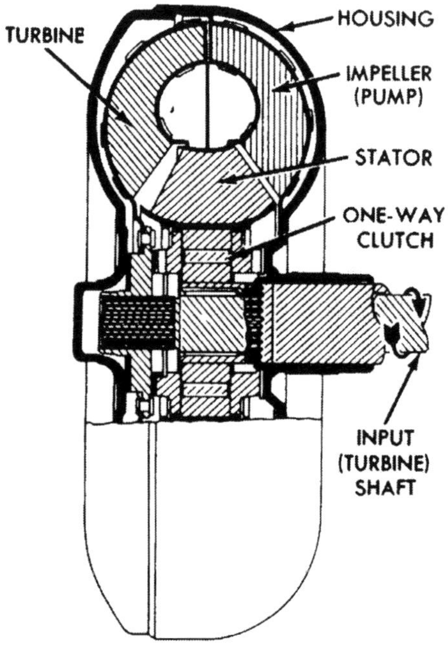

FIG 7:3 Sectional view of the convertor

extreme cleanliness. This includes wiping the filler tube and dipstick clean before removing the dipstick and using only clean fluid of the correct grade to top up with.

The outside of the unit should also be occasionally cleaned down, using a suitable proprietary cleanser and hose or brushing away road dirt. Check that air vents are free. Dirt on the exterior of the unit will act as an insulator and as the transmission operates at a high temperature insulation on the outside may cause it to overheat particularly in high ambient temperatures.

Fluid level:

This must always be checked with the transmission at its operating temperature, the engine running at idle and Park selected. If the level is checked with the transmission cold or the engine not running, a false reading will be given.

Drive the vehicle or run the engine until the transmission has reached its normal operating temperature. Drive the car onto level ground and briefly make each selection in turn. Leave the engine at idle speed. Wipe the dipstick and filler tube—in the engine compartment—and withdraw the dipstick. Wipe the dipstick clean with a piece of leather or lint-free cloth. Insert the dipstick fully into the filler tube and immediately withdraw it again. Check the level and top up with Ford Automatic Transmission fluid to the upper mark on the dipstick.

Do not overfill the transmission. Overfilling will lead to frothing of the fluid and incorrect operation of the transmission.

Linkage:

Check that the selector linkage operates fully and freely. Lightly lubricate the pivot points.

Draining:

This is only required when removing the transmission or carrying out repairs. If the fluid is drained out while the transmission is hot, take great care as the fluid will be hot enough to cause serious scalding.

Remove the cover from the bottom of the converter housing and crank the engine until the two converter drain plugs can be removed. A converter drain plug and attachment nut are shown in **FIG 7:6**.

If the remainder of the unit requires draining, slacken off the screws that secure the oil pan and tilt the pan so that the fluid can be poured out over one corner. It is advisable to remove the oil pan and check the strainer to the inlet of the pump.

If the fluid comes out blackened and has an obnoxious smell, the clutch discs are being burnt. The vehicle or transmission unit should be taken to a service garage for a strip examination of the transmission. A strip examination should also be carried out if metal or lining particles are found in the oil pan.

Oil cooler:

The oil cooler does not require routine maintenance. If damage has occured to the transmission unit, the oil cooler and its lines must be flushed through.

If transmission fluid is found in the radiator header tank, the oil cooler is defective. A defective oil cooler will also allow coolant into the transmission. Water in the fluid will cause a rise in the level and it may also cause fluid to froth out through the filler tube as the water turns to steam when the transmission becomes hot.

The oil cooler cannot be repaired by the owner as it is built into the radiator tank.

Brake band adjustments:

These normally only require adjustment at intervals of 36,000 miles. A special tool is made for adjusting the brake bands. When the band adjustment is tightened, the tool clicks when the correct torque load of 10 lb ft is reached. If the correct tool is not available, a torque wrench must be used instead and the adjusting screw tightened to the correct torque load.

A sectioned view through a reverse-low servo and brake band is shown in **FIG 7:7** and the intermediate band and servo are very similar. The figure gives an indication of the method of operation.

The method of adjusting the intermediate band is shown in **FIG 7:8**. Clean away all the dirt from the adjustment screw. Remove and discard the adjustment screw locknut and install a new locknut, leaving it slack. Tighten the adjustment screw with the tool shown until the correct torque is reached. Back off adjustment screw by exactly 1¾ turns. Hold the adjustment screw with a wrench to stop it turning and tighten the locknut to secure it.

The method of adjusting the low-reverse band is shown in **FIG 7:9**. The method is exactly the same as for the intermediate band, except that when the adjustment screw has been tightened to the correct torque it should be backed off by exactly 3 turns.

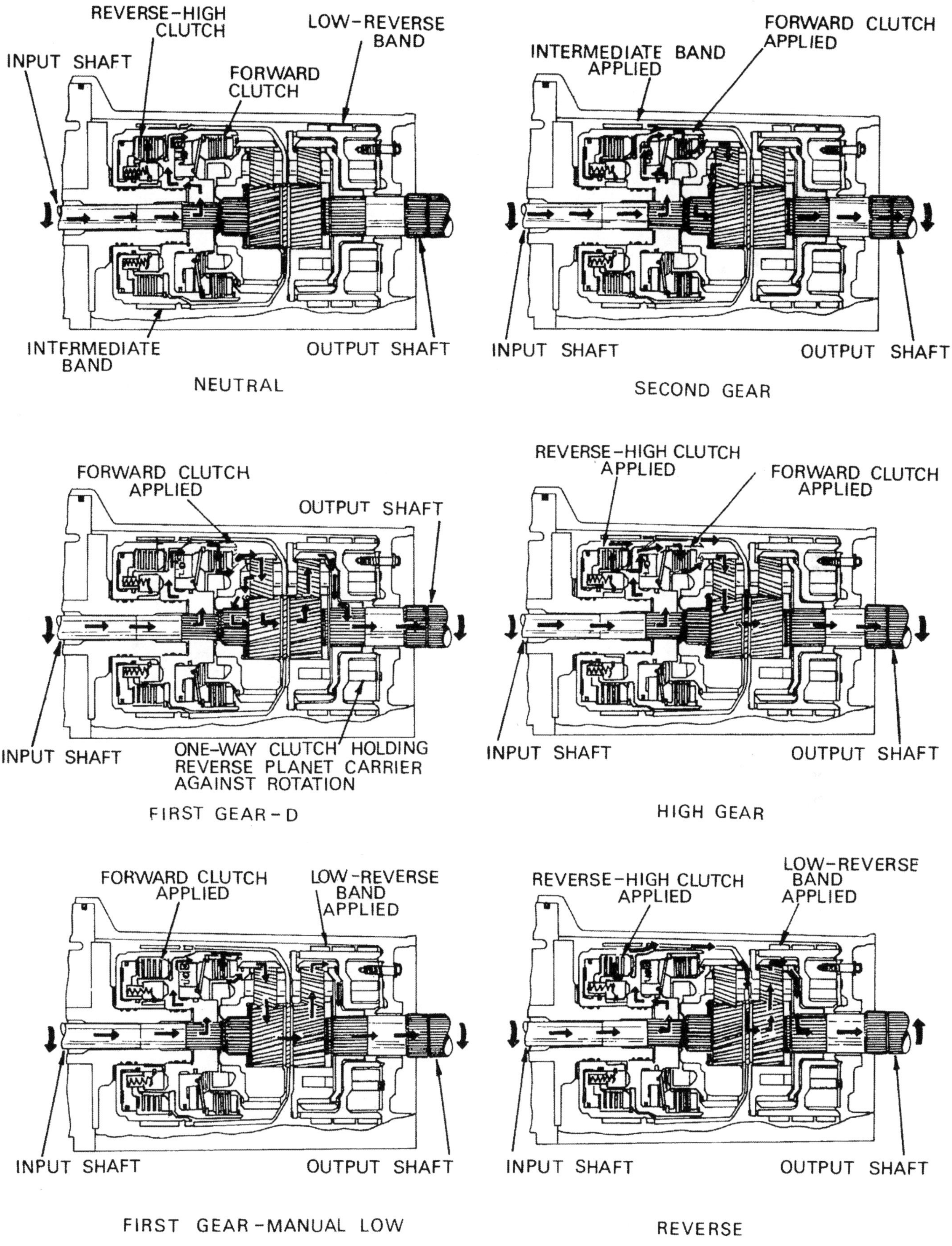

FIG 7:4 Power flow in the various selections

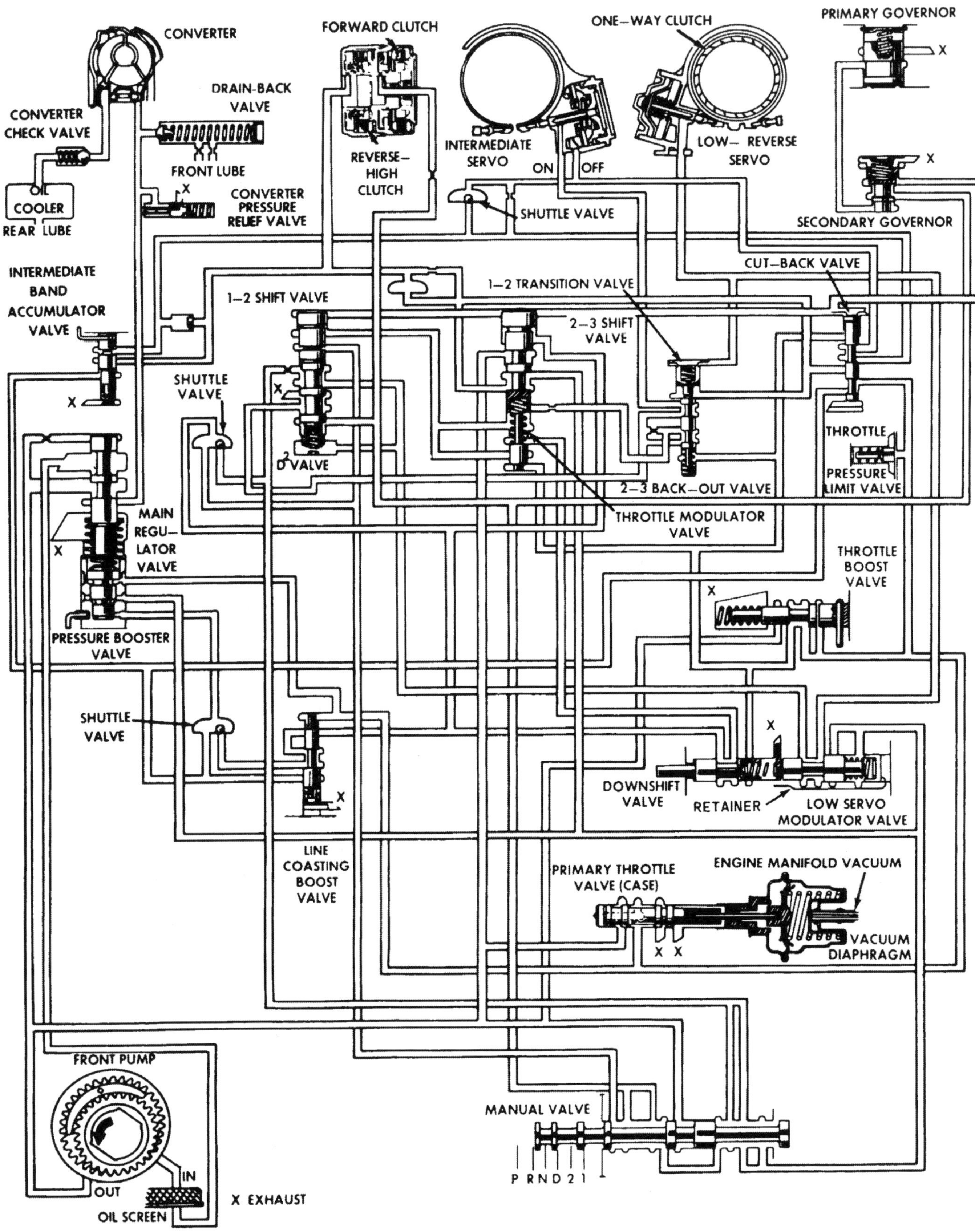

FIG 7:5 Hydraulic control system for the automatic transmission

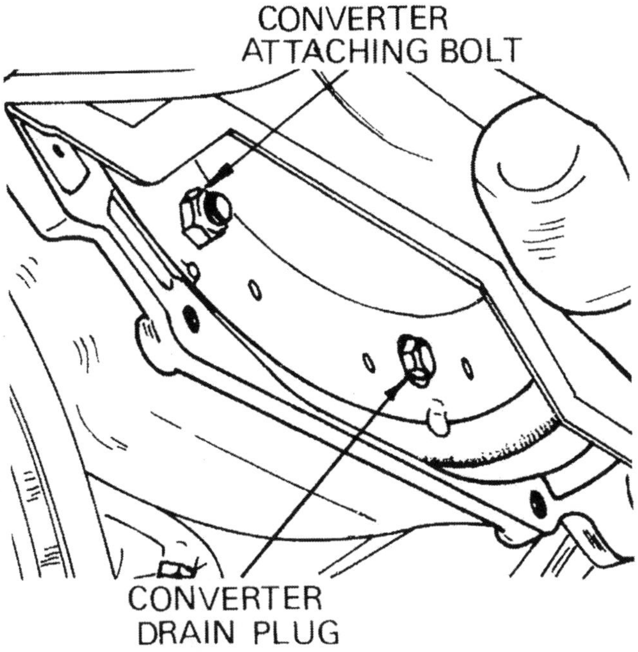

FIG 7:6 Convertor drain plug and attaching bolt

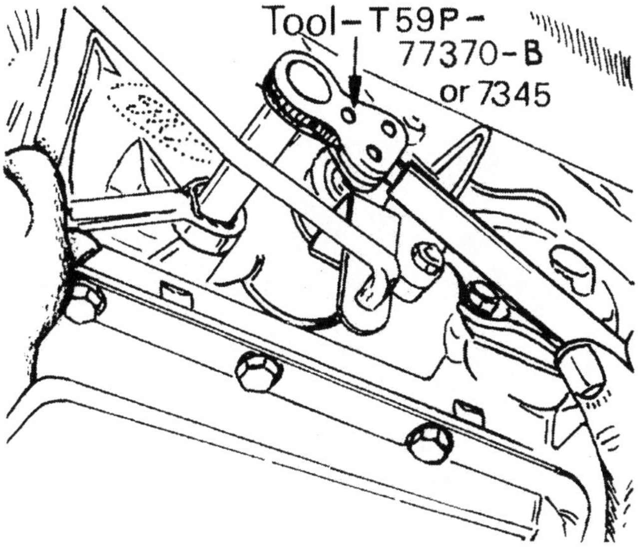

FIG 7:8 Adjusting the intermediate brake band

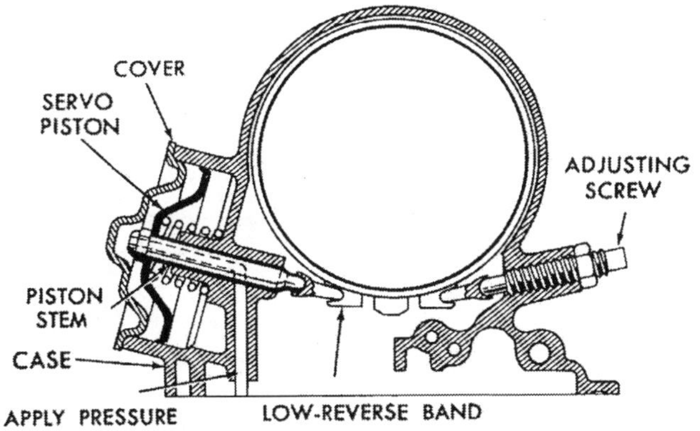

FIG 7:7 Section through the low-reverse servo and band

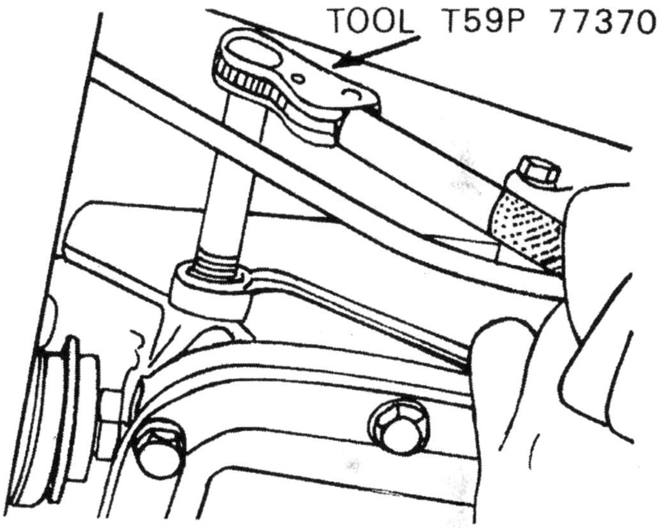

FIG 7:9 Adjusting the low-reverse brake band

7:3 Removing the transmission unit

The automatic transmission weighs considerably more than the manual-shift transmission and a jack and support are essential for supporting the transmission and for lowering it down and out of the car. Special heads for the jack are made to take the automatic transmission and one should be used if possible. The head is fitted with a chain which can be tightened round the transmission so that it cannot fall off the jack, and if the correct head is not used steps should be taken to lash down the transmission to the jack.

1 Drain the transmission by removing the two drain plugs, shown in **FIG 7:6,** from the converter and taking off the oil pan. Install both drain plugs and oil pan after the fluid has been drained. Free the oil filler tube from the transmission and free any attachments on the engine so that the tube can be removed. Disconnect the oil cooler lines from the transmission, after safely raising the vehicle onto stands. Blank off the lines and adaptors to prevent dirt from entering.

2 Disconnect the selector shift and 'kick-down' rods from the transmission. Free the speedometer drive cable from the transmission and disconnect the vacuum line from the vacuum unit, as well as freeing it from its clip on the transmission. Disconnect the drive shaft (see **Chapter 8**) and fit a plug to the rear extension housing of the transmission to prevent the entry of dirt or damage to the bushing.

3 If necessary free the park brake cable from the equalizer. The exhaust pipes will also have to be freed.

4 Disconnect the battery ground cable. Disconnect the cables from the starter motor and remove the motor. Disconnect the neutral start switch wires. If necessary, label all wires so that they will be correctly connected on installation.

5 Remove the bolts that secure the converter to the engine drive plate (see **FIG 7:6**). Remove the two bolts that secure the engine rear support to the extension housing of the transmission. Use a jack and support to raise the transmission slightly so that the

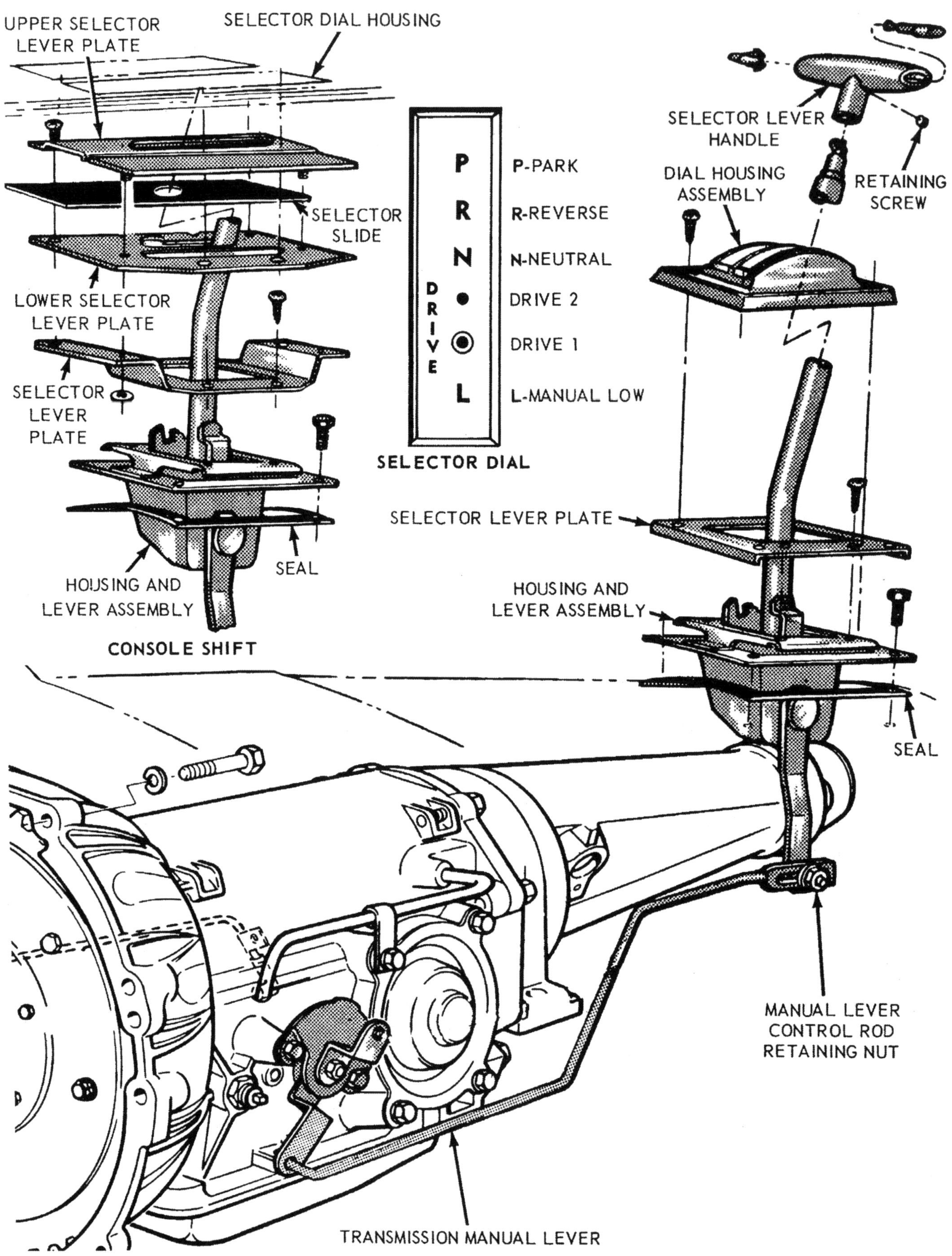

FIG 7:10 Earlier selector linkage

weight is just off the crossmember. Remove the attachments that secure the rear support crossmember to the frame and take out the rear support assembly.

6 Support the engine with a jack and pad of wood under the oil pan. Remove the bolts that secure the transmission to the engine. Draw the transmission slightly rearwards to clear it from the engine and then lower it down and out from the vehicle.

Installation:

The transmission is installed in the reverse order of removal. Tighten all bolts to the correct torque specifications. Check and adjust all linkages that have been disturbed. Fill the transmission to the upper mark on the dipstick and start the engine. Keep adding more fluid as it is drawn into the converter until the level stabilizes at the lower mark on the dipstick. Top up fully when the transmission has reached its normal operating temperature.

Oil cooler flushing:

If dirt has entered the oil cooler, or the transmission has been defective, flush through the oil cooler to remove any dirt or particles trapped in it.

With the transmission filled to the correct level, disconnect the return line from the cooler at the rear of the transmission. Start the engine and let at least two quarts of fluid drain into a suitable container.

If the fluid flow is sluggish, switch off the engine and disconnect the line to the oil cooler at the transmission. Blow through in a reverse direction to the normal flow with compressed air at a maximum pressure of 100 lb/sq in.

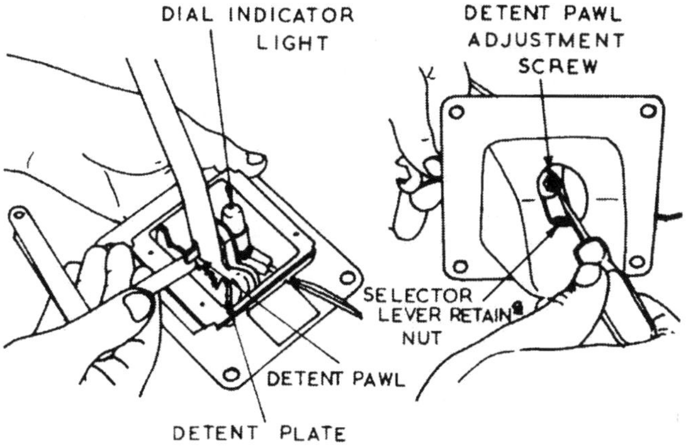

FIG 7:11 Checking and adjusting the selector lever detent pawl

If necessary disconnect the lines from the cooler so that they can be blown through separately. If the blockage cannot be removed with compressed air, new lines or oil cooler (engine radiator complete) must be fitted. **Do not attempt to cure leaks by bypassing the cooler or plugging the lines.**

Torque converter:

Once the transmission has been taken out of the vehicle, the converter can be removed from the transmission. The converter is sealed on manufacture so if it is defective a new unit must be fitted.

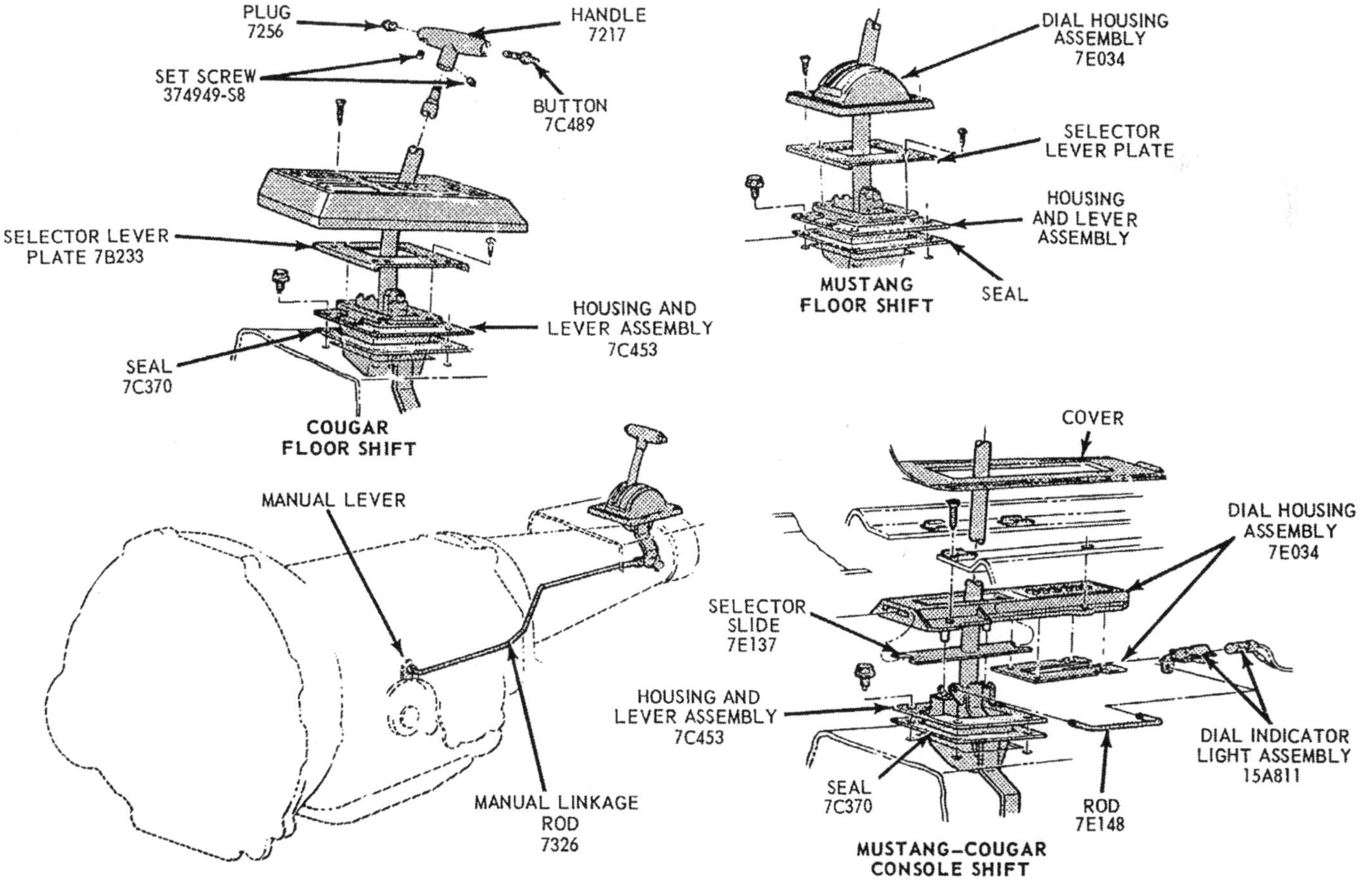

FIG 7:12 Later selector linkage

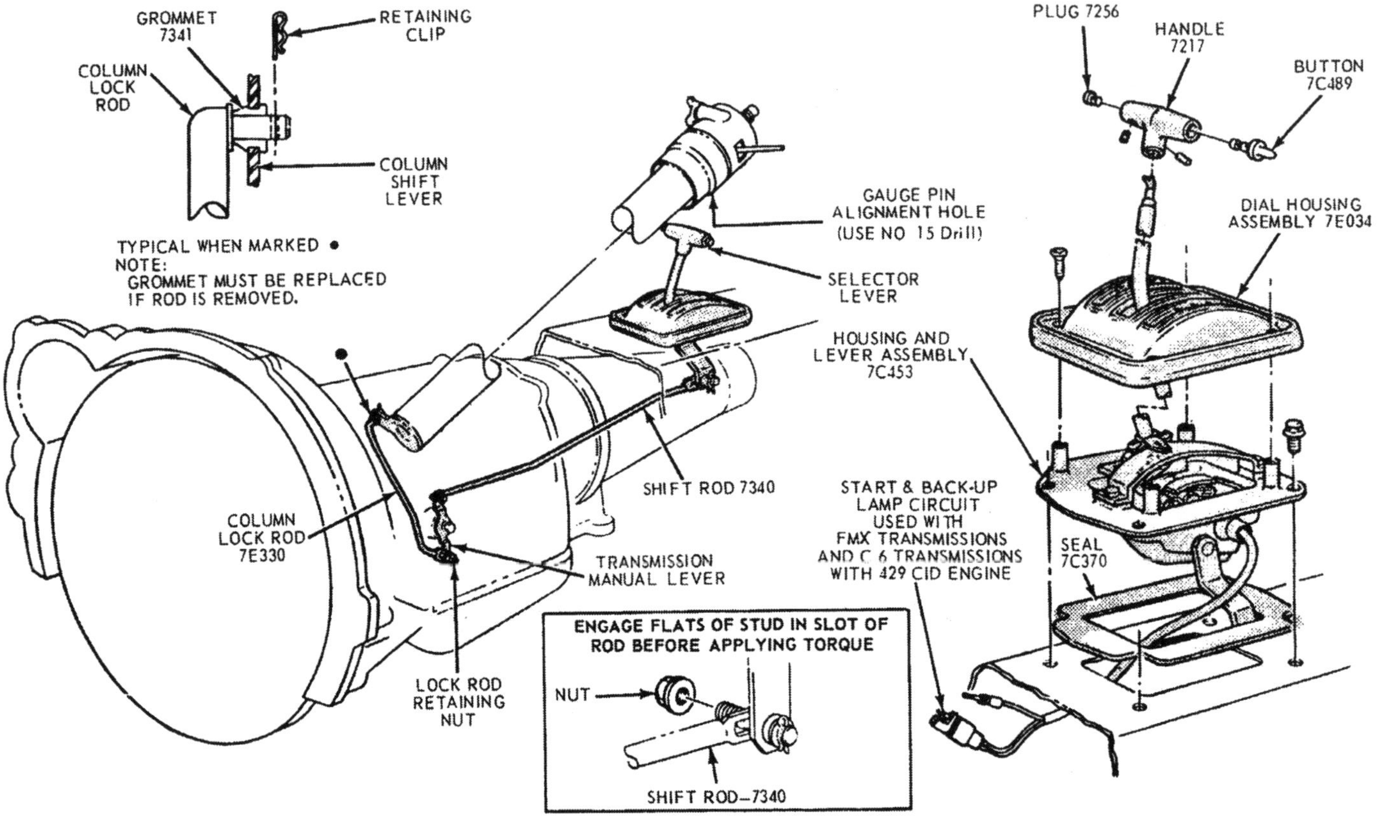

FIG 7:13 A more recent modification of the selector linkage–console or floor shift

Special tools are used for checking the converter. The tools are used for pressure testing the unit, as well as checking the interference fit of the stator to impeller and stator to turbine. Using the tools in conjunction with a torque wrench allows the stator one-way clutch to be checked. If it is suspected that faults in the converter are the cause of transmission faults, take the converter to a service garage holding the tools for checks.

If it is suspected that there is dirt or sludge in the system, drain out all the fluid from the converter by tilting it around with both drain plugs removed. Install one drain plug and fill the unit with kerosene or suitable cleansing solvent and then install the other drain plug. Preferably rotate the turbine but if this cannot be done, because of lack of special tools, shake the converter around to swill the inside. Remove both drain plugs and drain out all the solvent. Repeat the procedure as many times as are required to remove the dirt and sludge.

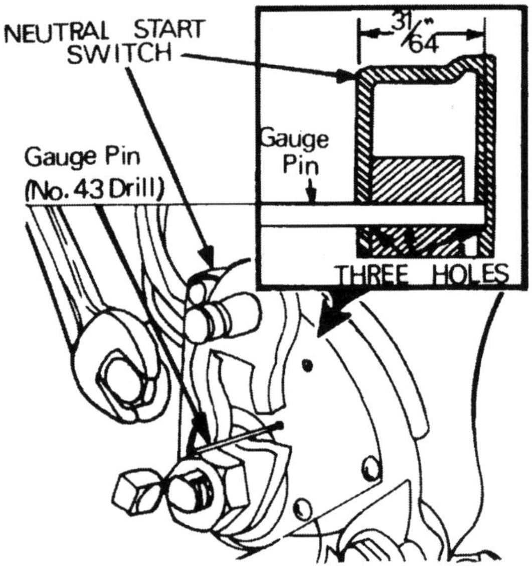

FIG 7:14 Adjusting the neutral start switch

7:4 Stall test

This test will give a rapid guide to the condition of the transmission. **Be sure that the transmission is correctly filled.**

Connect an accurate tachometer to the ignition circuit and fit the instrument so that it can readily be seen from the driver's seat. Start the engine and run it until the transmission has reached its normal operating temperature. Apply the park brake fully and also hold the vehicle on the service brake. Select a drive speed and floor the accelerator pedal. Note the maximum engine speed and release the accelerator pedal within 5 seconds. Run the engine at 1000 rev/min in Neutral for at least 15 seconds to allow the fluid to circulate and cool the transmission before repeating the test in the next selection. When all selections have been checked, run the engine at 1000 rev/min in Neutral until the transmission has cooled down. If the normal stall speed in any selection is exceeded, stop the test immediately as it is an indication of slip. **If the times and speeds given are not adhered to, the transmission may overheat to a serious degree as large amounts of heat are produced by the churning of the fluid.** The correct stall speed for all selections is 1400 to 1700 rev/min for a 200 engine, 1600 to 1800 rev/min for a 250 engine.

Low stall speed:

This can be caused by a lack of power from the engine, so make sure that the engine is satisfactory and that there are no faults which could cause a power loss.

A defective converter will also cause a low stall speed. If the stall speed is 300 to 400 rev/min low, take the vehicle for a road test. If the vehicle cruises normally but acceleration, particularly from standstill, is poor and starting on gradients is difficult then the stator one-way clutch is slipping.

If acceleration and starting on gradients are satisfactory but the vehicle drags at cruise and maximum speed (particularly in top) is limited while the transmission overheats easily, then the stator one-way clutch is seized.

In both cases, remove the transmission so that the torque converter can be checked at a service garage.

High stall speed:

A high stall speed indicates slip in the gearbox. Carefully check each selection in turn and note the selections that give high stall speeds. Excessive speed in only forward selections indicates forward clutch slippage. Excessive speed in R selection indicates either reverse-high clutch slippage or low-reverse band slippage. Excessive speed in only D1 indicates gear train one-way clutch slippage.

7:5 Adjustments

The adjustments of the brake bands have already been dealt with in **Section 7:2**.

For full and accurate checks of the transmission special equipment is required so that the various operating pressures can be noted and set. The vehicle should therefore be taken to a service agent for checking and diagnosis of faults that are beyond the powers of the owner to correct.

Selector linkage:

The earlier linkage is shown in **FIG 7:10**. Move the selector lever into the Drive 1 position against the large dot. Raise the vehicle and slacken the nut that secures the transmission manual to the selector. Move the manual ever on the transmission to the D1 position. The last position is manual low and the second from back position is D1. Check that the selector lever has not moved and tighten the control rod retaining nut to a torque of 20 to 25 lb ft.

The method of removal of the earlier selector mechanism is straightforward, but before installing the parts check that the clearance between the detent pawl and plate is .005 to .010 inch, as shown in **FIG 7:11**. The method of adjusting the height is also shown in **FIG 7:11**.

The later linkage is shown in **FIG 7:12**, and where a console shift is incorporated as an alternative see **FIG 7:13**. The adjustment of this linkage is exactly the same as for the earlier type, as is the setting of the pawl.

Neutral start switch:

The attachments of the neutral start switch are shown in **FIG 7:14**. Slacken the two nuts that secure the switch in place. Move the selector lever to the Neutral position. Insert a gauge pin (No. 43 drill shank) through the switch, as shown, and rotate the switch slightly until the gauge pin passes through the three holes **to its full depth**.

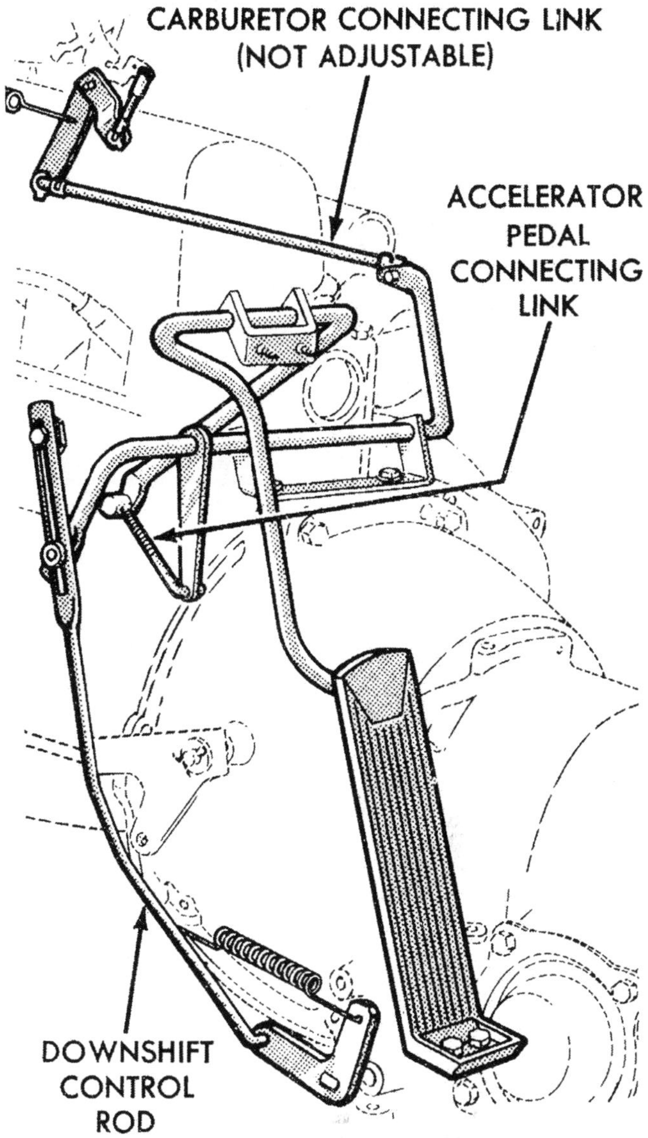

FIG 7:15 Earlier throttle linkage

Tighten the nuts that secure the switch and remove the gauge pin. Check that the starter motor will only operate in the Park and Neutral positions without operating in a drive selection.

Throttle linkage:

Typical throttle linkages are shown in **FIGS 7:15** and **7:16**.

Ensure that the idling speed is correctly set with the engine at normal working temperature (see **Section 2:6**) and that the fast idle system is not in operation.

Check that the floor covering is correctly in place and not obstructing throttle pedal movement. Check the height of the pedal from the floor pan (see **FIG 7:17**) and if it is outside limits adjust the pedal connecting link (earlier linkage) or the rod on the engine (later linkage).

Turn the engine off and have the throttle pedal held fully down. On models fitted with downshift rod, adjust the rod so that the lever on the transmission is in the kickdown detent position with approximately $\frac{1}{16}$ inch of free travel on the bellcrank assembly. On models fitted with a downshift cable, disconnect the cable. Rotate the downshift lever until it is against the internal stop.

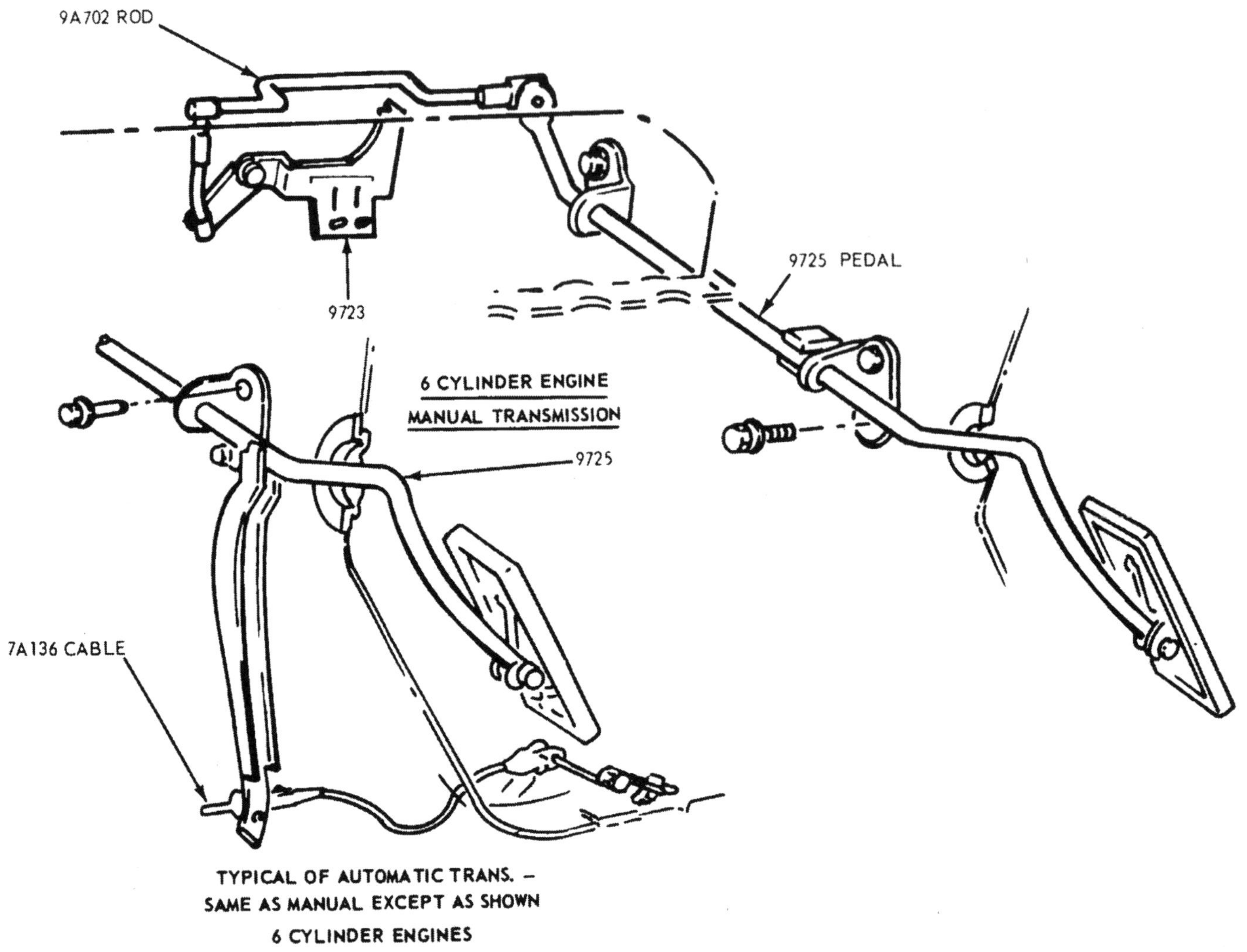

FIG 7:16 Later throttle linkage

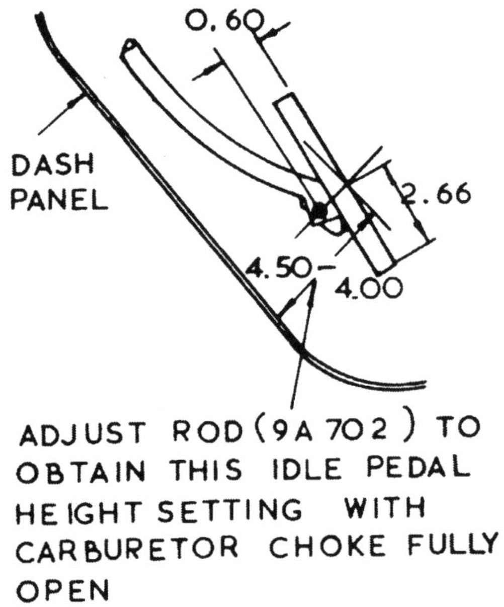

FIG 7:17 Throttle pedal setting

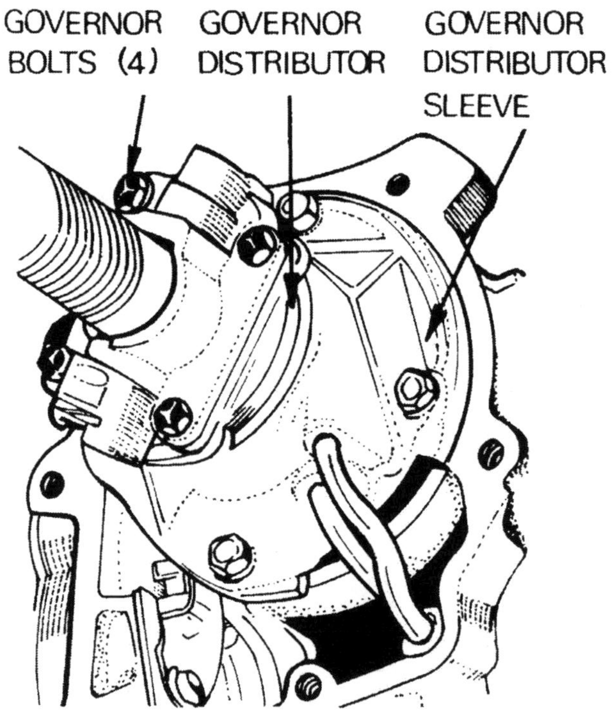

FIG 7:18 Governor in position

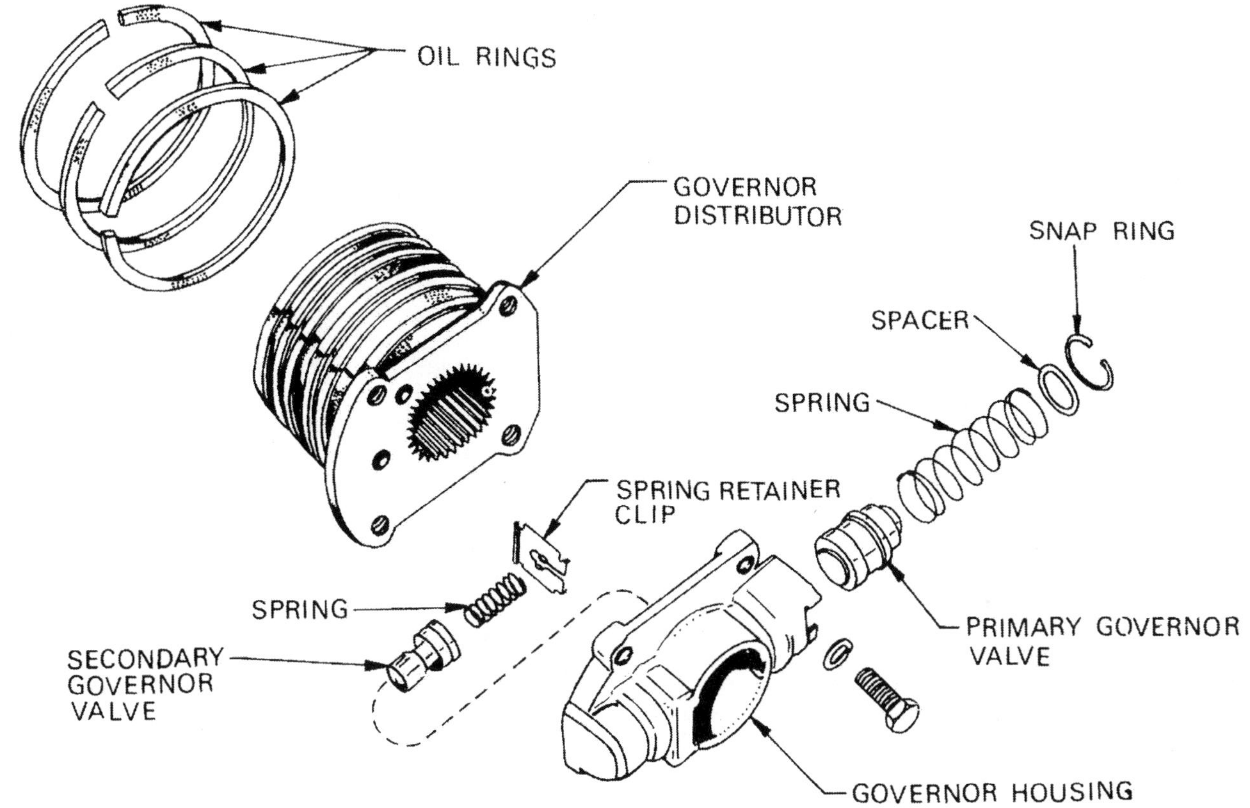

FIG 7:19 Governor and oil distributor components

Hold the lever in this position and adjust the length of the cable so that it is just taut when its trunnion slides easily into the lever on the throttle linkage. Turn the adjuster one additional turn clockwise, and reconnect the cable.

7:6 The governor and rear extension housing

The seal and bushing in the extension housing can be renewed in a similar manner to the equivalent parts on the manual-shift transmission (see **Chapter 6**).

Seal and bushing:

Special tools are made for removing and installing the parts. The seal is withdrawn using the three-legged internal extractor No. 1175.AB in conjunction with the impact tool T50T.100.A or B.160. An extractor No. T.57P.7697.A or 7000.G or AD is fitted behind the bushing and the bushing is then drawn out by tightening down the screw of the tool onto the output shaft of the transmission. Drift No. T57P.7697.B or 7000.HF is made for driving in the new bushing and drift T61L.7657.A, B, AA, or AB is made for driving in the new seal.

Before installing new parts, check that the counterbore in the extension housing is free from burrs or nicks, using Crocus paper to remove light damage. Similarly check the yoke of the drive shaft and either renew it or smooth it down with Crocus paper if damage is found.

The inner diameter of the seal and drive shaft yoke should be lightly lubricated with B8A.19589.A lubricant.

The governor:

This is exposed after removal of the extension housing. Drain the transmission (see **Section 7:2**) and remove the drive shaft (see **Chapter 8**). Remove the attachment bolts, after disconnecting the speedometer drive cable, and withdraw the rear extension. The attachments of the governor are shown in **FIG 7:18**. Remove the four bolts that secure the governor and withdraw the governor.

The components of the governor are shown in **FIG 7:19**. Remove the snap ring securing the primary valve and take out the primary valve parts. Remove the retaining clip so that the secondary valve parts can be removed.

Wash the components in a suitable solvent, such as trichlorethylene or denatured alcohol, to remove all dirt and any gum. Check that the valves slide freely in

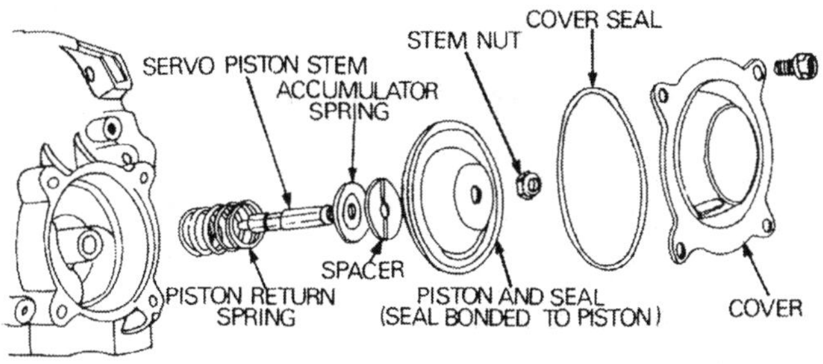

FIG 7:20 The low-reverse servo disassembled

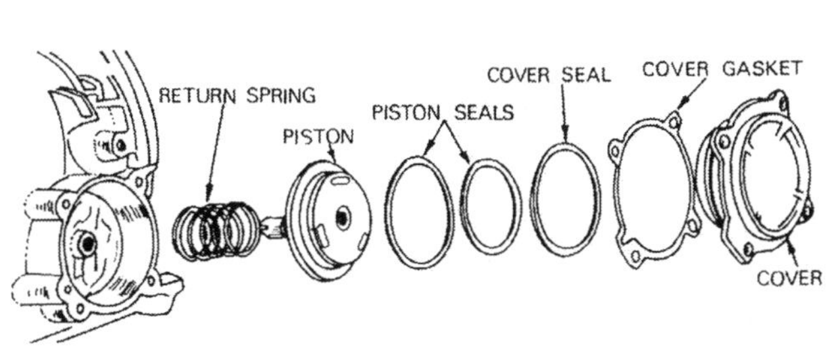

FIG 7:21 The intermediate servo disassembled

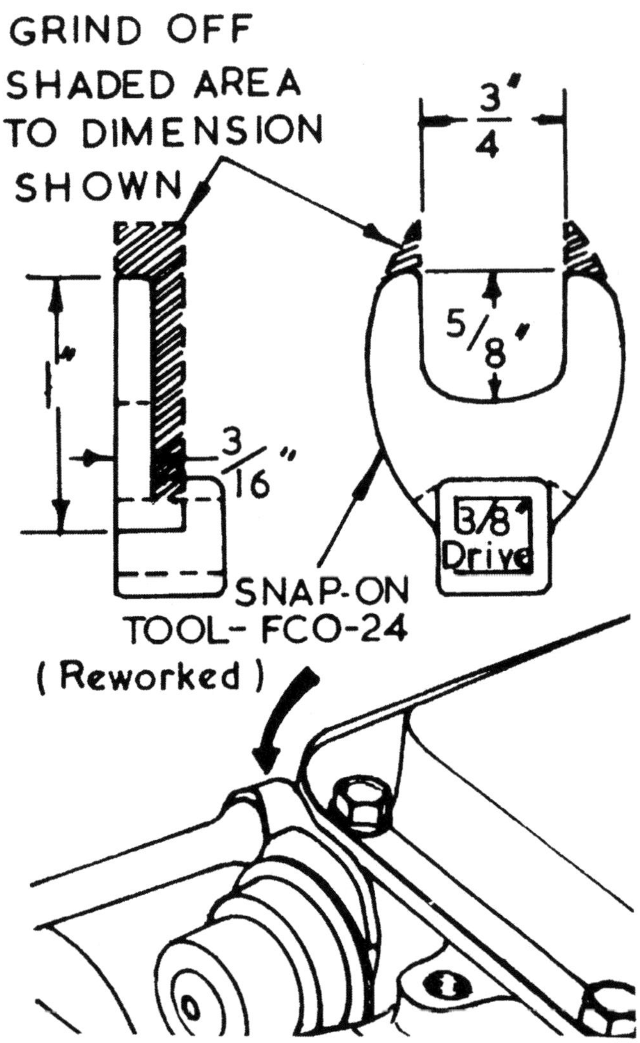

FIG 7:22 Removing or installing the vacuum valve unit

their bores. If the valves stick, they may be lightly polished with Crocus cloth but **take great care not to round off the sharp edges on the valves.** Install new parts if the old ones are badly scored or worn.

Install the parts in the reverse order of removal. Lightly lubricate the governor parts with transmission fluid. When installing the secondary valve parts, be sure that the retaining clip is fitted with its small concave surface downwards so that it locates the spring. Install the rear extension using a new gasket. When the drive shaft has been installed, fill the transmission to the correct level with fresh fluid, start the engine, and keep topping up until the converter is full. Run the engine until the transmission is at its normal operating temperature and top up correctly.

7:7 Servo assemblies and vacuum unit

The components of the low-reverse servo are shown in **FIG 7:20** and those of the intermediate servo in **FIG 7:21**. New seals and pistons can be fitted to both servos if the old parts are defective. When installing the cover for the intermediate servo, use two $\frac{5}{16}$-18 bolts $1\frac{1}{4}$ inch long to guide the cover into position and do not forget the transmission identification tag that is secured by one of the correct servo bolts.

The method of removing or installing the vacuum unit, after the vacuum line has been disconnected, is shown in **FIG 7:22**. If the vacuum unit is suspect, first check the vacuum line to be sure that it is clear and not split. Either have the vacuum unit tested at a service garage or install a new valve.

7:8 Trouble shooting guide

(a) Transmission overheats

1 Excessive dirt or mud on outside of unit
2 Oil cooler or lines plugged
3 Oil cooler defective
4 Convertor stator clutch seized
5 Incorrect brake band adjustment
6 Brakes binding or other drag on vehicle causing convertor to slip excessively

(b) Noisy operation

1 Incorrect fluid level or incorrect fluid used
2 Incorrectly adjusted shift linkage
3 Oil screen to pump intake choked

(c) Incorrect shift speeds

1 Check 1 in (b)
2 Defective vacuum unit
3 Incorrectly adjusted brake bands or defective servos

(d) Poor acceleration

1 Check 4 and 5 in (a)
2 Convertor stator clutch slipping

(e) Jumps in engagement or rough engagement

1 Check 5 in (a); 2 in (b) and 2 in (c)
2 Incorrect curb idle speed

(f) Vehicle does not hold in Park selection

1 Check 2 in (b)

(g) Fluid frothing or being forced out from filler tube

1 Too high fluid level
2 Water in system (suspect oil cooler)
3 Air leaks on suction side of pump

This guide is not fully comprehensive and only covers faults that can be cured by the owner. For other faults take the vehicle to a service garage for full checks and rectification.

CHAPTER 8

THE DRIVE SHAFT, REAR SUSPENSION AND REAR AXLE

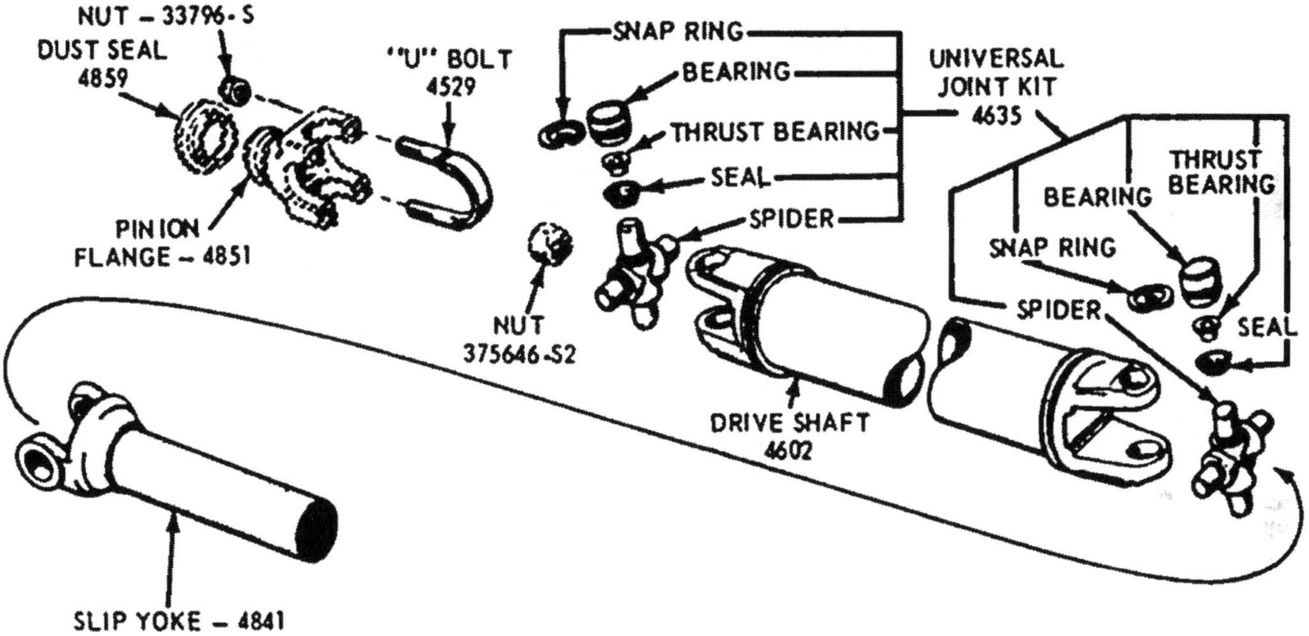

FIG 8:1 Drive shaft components

8:1 The drive shaft

A typical drive shaft assembly is shown in **FIG 8:1**. The pinion flange on the rear axle forms part of the rear universal joint and the two bearings on the spider are secured to the flange by U-bolts. The slip yoke on the front of the drive shaft slides into place in the rear extension housing of the transmission and fits onto the splines of the output shaft. The slip yoke is located by a bushing and sealed by an oil seal, which prevents transmission lubricant from leaking out, as shown in **FIG 8:2**.

Removal:

To preserve the balance of the shaft, the rear yoke of the drive shaft must be kept in the same relative position to the drive flange on the rear axle. When the parts are new alignment marks are made with yellow paint but if these have worn off or become covered in dirt, new marks should be made.

Remove the nuts and carefully drive out the U-bolts that secure the shaft to the pinion flange. **Wrap tape around the rear universal joint to prevent the bearings from falling off the spider.**

Lower the rear end of the shaft and withdraw the slip yoke from the transmission. Take care not to damage or nick the slip yoke surface and leading end as nicks or burrs will cause transmission fluid leakage. Light damage or burrs can be smoothed down with Crocus paper but if the wear or damage is excessive a new slip yoke must be fitted.

When the shaft has been removed, the aperture in the transmission rear housing should be covered over to prevent dirt from entering or lubricant from leaking out.

Install the drive shaft in the reverse order of removal, making sure that the marks are again in alignment. The splines in the slip yoke should be lubricated with B8A-A lubricant.

Maintenance:

Some universal joints are fitted with a removable plug for lubrication purposes. Remove the plug and thread a grease fitting into its place. Inject a few strokes of Ford Universal Lube through the grease fitting, remove the fitting and install the plug.

Vibration:

The drive shaft is balanced on manufacture by the addition of balance weights. If vibration is traced to the drive shaft, first check that the balance weights are

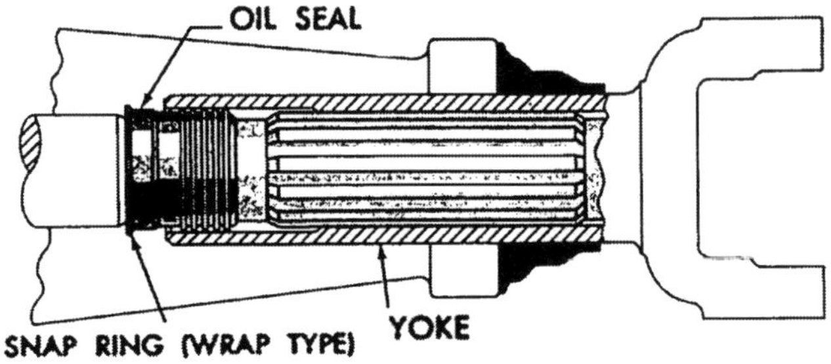

FIG 8:2 Drive shaft front attachment

present and have not fallen off, and then make sure that mud or underseal is not clinging to the shaft and upsetting the balance. Check the universal joints for wear. The joints are checked by smartly rotating them backwards and forwards by hand and trying to lift the shaft up by hand. Excessive wear will show up as excessive play or a 'clonk' as the direction of drive is reversed.

If there is no obvious cause of vibration, try rotating the shaft through 180 deg. relative to the pinion flange.

If rotating the shaft relative to the flange does not cure vibration, jack up the rear of the vehicle and place it **very securely** onto firm stands. **Chock the front road wheels both in front and behind so that there is no danger of the car coming off the stands.** Start the engine and select gear. Increase the engine speed until the speedometer is reading approximately 40 to

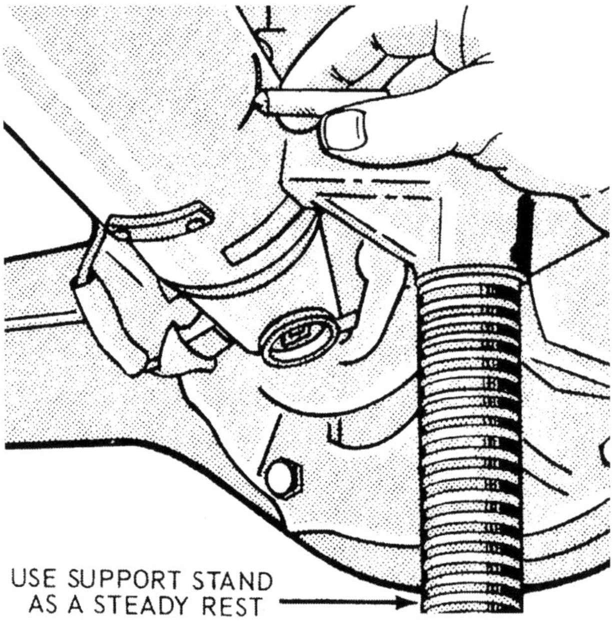

FIG 8:3 Marking the drive shaft to find the heavy side

50 mile/hr. From underneath the car bring a wax crayon into light contact with the rotating drive shaft, as shown in **FIG 8:3**, taking care to avoid the balance weights. Stop the engine and if the mark has been made lightly enough it will show the heavy side of the shaft. Install two Whittek clamps to the drive shaft so that their heads are both 180 deg. away from the chalk mark, as shown in **FIG 8:4**. Tighten the clamps to prevent them from moving. Start the engine and select gear. Speed up the engine until a speed of 65 to 70 mile/hr is indicated. If vibration is still present, stop the engine and rotate the two clamps so that their heads have an angle of 45 deg. between them, as shown in **FIG 8:5**. Repeat the test at speed and if vibration is still present, move the heads of the clamps slightly further apart and again repeat the test. Make small adjustments to the positions of the heads until vibration is eliminated.

Do not run the engine for long periods with the wheels off the ground as there is a danger of the transmission or engine overheating.

Servicing universal joints:

Kits of parts are available and these should be used. Individual parts of the universal joint should not be renewed on their own and all the parts supplied in a kit should be used. Before dismantling the shaft, make alignment marks so that the parts will be correctly reassembled and the balance not lost.

The method of removing the bearings from a universal joint is shown in **FIG 8:6**. On the rear universal, the two bearings that are secured to the drive flange can be pulled directly off the spider. Remove the snap rings that secure the bearings. If the snap rings stick, clean away all rust and enamel from around them. Mount the shaft in the padded jaws of a vice and use the tool to press the bearing on one side as far as possible. If the bearing cannot be removed fully with the special tool, press it out as far as possible and then use a pair of grips to withdraw it fully. In extreme cases, remove the shaft from the vice and use the vice to grip the bearing itself. Use a suitable drift and drive the shaft up and off the bearing, leaving the bearing retained in the vice. When all four bearings have been removed, the spider can be manoeuvered out of the yokes and shaft.

Before installing new parts, check the bores in the yoke arms. The bearings must pass through the bores with a light drive fit. If the bores are worn oval a new shaft or yoke must be installed.

Check that the needle rollers in the bearings are in place and held securely by grease. **If a needle roller is displaced and pressure is applied across it the needle will snap.**

Install the spider into the yoke and press a bearing into place so that it is $\frac{1}{4}$ inch below the surface. Install a snap ring onto the bearing. The method of pressing a bearing back in is shown in **FIG 8:7**. Start another bearing into the yoke and use the special tool to press it in until the first bearing contacts its snap ring. Remove the tool and secure the second bearing with a snap ring. Install the remaining bearings using the same method.

When all the bearings are in place, check that the joint articulates freely. Tap the ears of the yoke arms sharply to relieve binding. **It is essential that the joint moves freely before installing the shaft.**

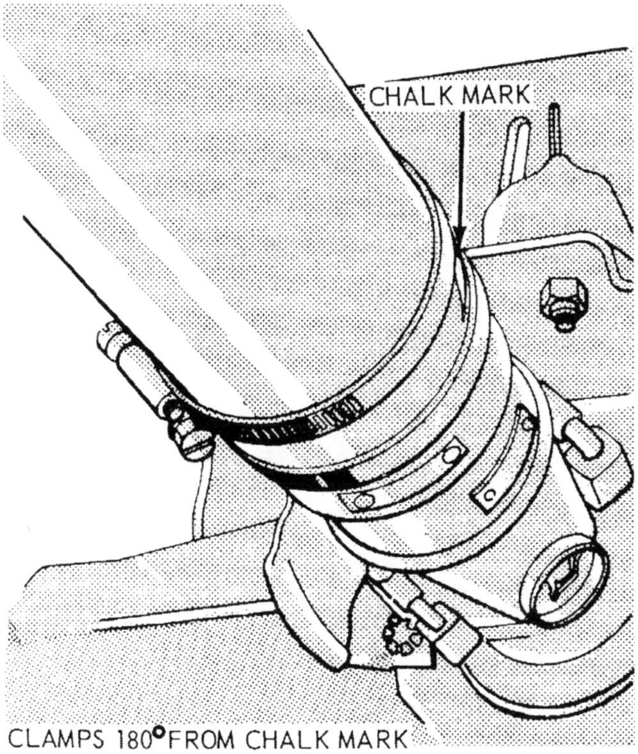

FIG 8:4 Whitteck clamps as installed initially

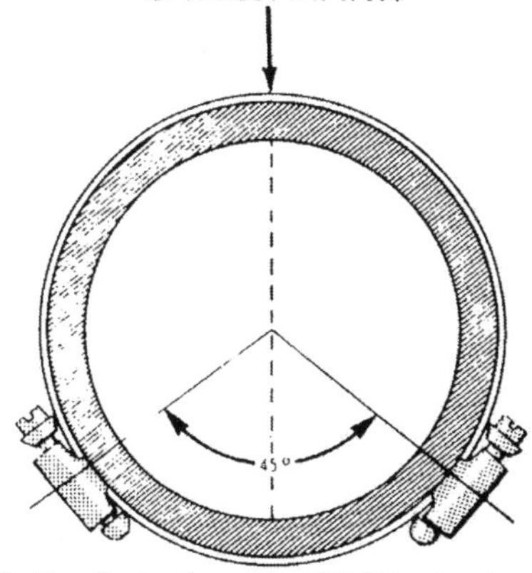

FIG 8:5 Rotating the Whitteck clamps

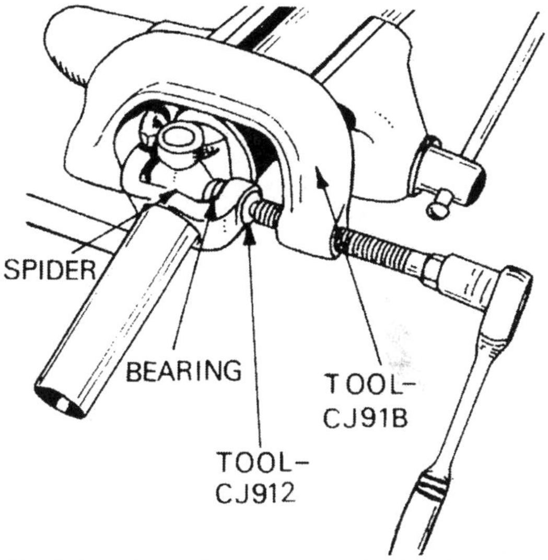

FIG 8:6 Removing a universal joint bearing

8:2 Rear suspension description

A typical rear suspension is shown in **FIG 8:8** and the components in **FIG 8:9**. Semi-elliptic springs and beam axles are used on all the models covered by this manual. Rubber insulated bushings are fitted into the eyes at the end of the springs, and insulators are fitted between the axle and spring, to ensure that road shocks and noise is not transmitted to the body. The front end of the spring is supported on a hanger mounting bolt but the rear is supported on shackles to allow the spring to flex. The axle is mounted above the spring, to lower the suspension height, and is secured by U-bolts and nuts.

Sealed telescopic shock absorbers are mounted between the frame and a plate under the spring, to control the movement of the suspension and to damp out oscillations. The shock absorbers are inclined inwards at the top to provide lateral location for the rear axle.

Each road spring is made up of several leaves. A clamp bolt secures the leaves at their centers and the ends of the leaves have insulators fitted between them. The springs should be kept clean and dry and do not require to be sprayed with any form of lubricant.

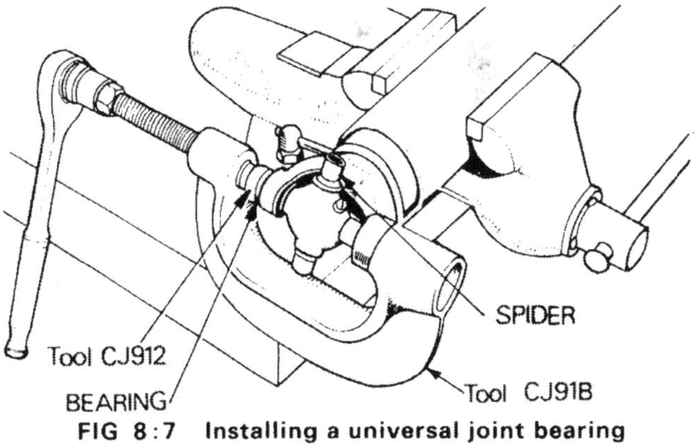

FIG 8:7 Installing a universal joint bearing

8:3 The shock absorbers

The owner has no means of accurately testing the shock absorbers but they should be suspected of being faulty if the vehicle continues pitching after a disturbance in the road or if the body rolls excessively on curves. The average life of a shock absorber is 36,000 miles so consideration should be given to renewing them after this mileage. As a rough test, bounce the corner of the vehicle and if it keeps pitching without rapid damping then the shock absorber is probably defective.

Once the shock absorber has been removed from the vehicle it can be checked for defects. Install a new shock

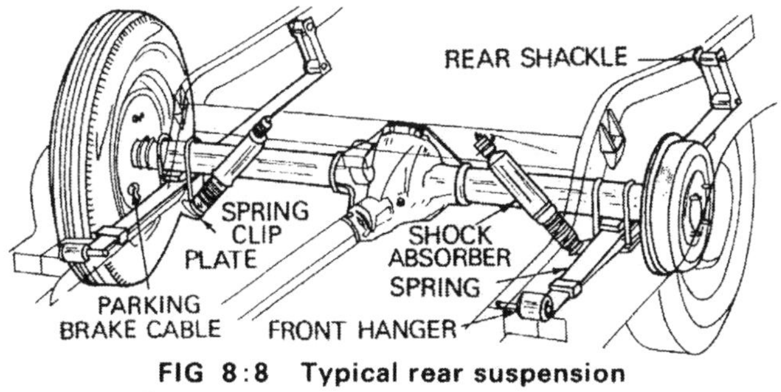

FIG 8:8 Typical rear suspension

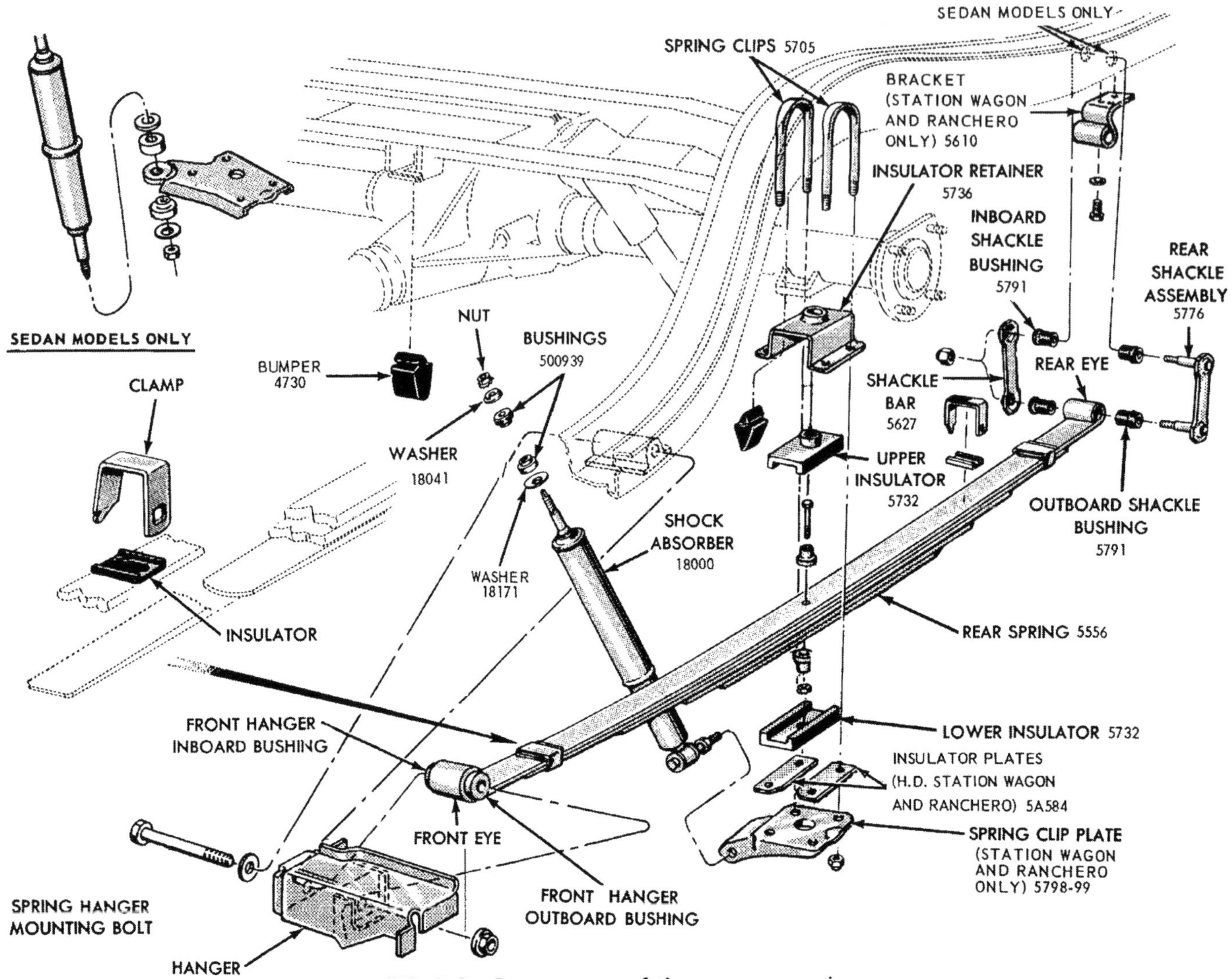

FIG 8:9 Components of the rear suspension

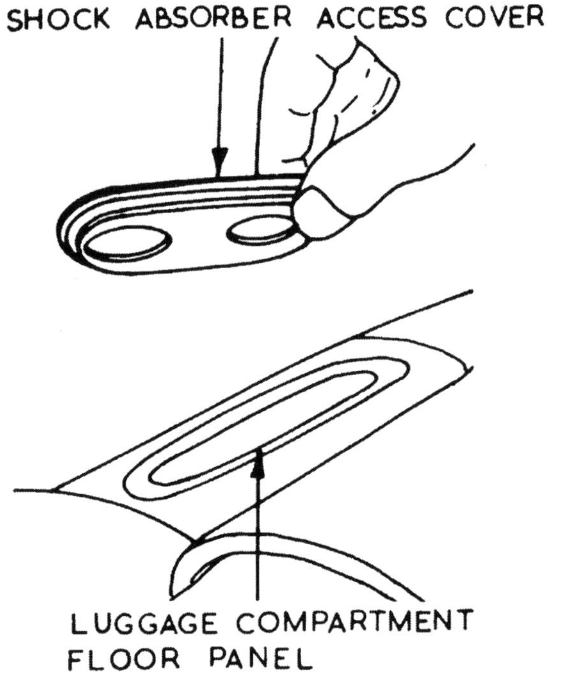

FIG 8:10 Access cover for the damper upper mounting

absorber if the old one has physical damage, such as a bent ram, dented body or fluid leaks. It is always advizable to renew shock absorbers in axle pairs.

Mount the shock absorber vertically with the lower eye in the padded jaws of a vice. Pump the shock absorber, starting with short strokes about the mid-point and gradually increasing the length until the strokes cover the full movement of the unit. There should be equal and uniform resistance to motion in either direction. If there are pockets of no or weak resistance, particularly when reversing stroke, the unit is defective and must be renewed. Similarly the shock absorber is defective if it operates noisily or leaks.

This test will ensure that any air or gas is bled to the top of the unit—if it has been in storage for some time—and it will also detect a damper that is defective. Even if a shock absorber passes the test this does not guarantee that it is operating at full efficiency.

Removal:

The task will be easier if the rear of the vehicle is jacked-up and the road wheel removed. Open the luggage compartment and if necessary remove the spare wheel. On convertible models, remove the rear seat and seat

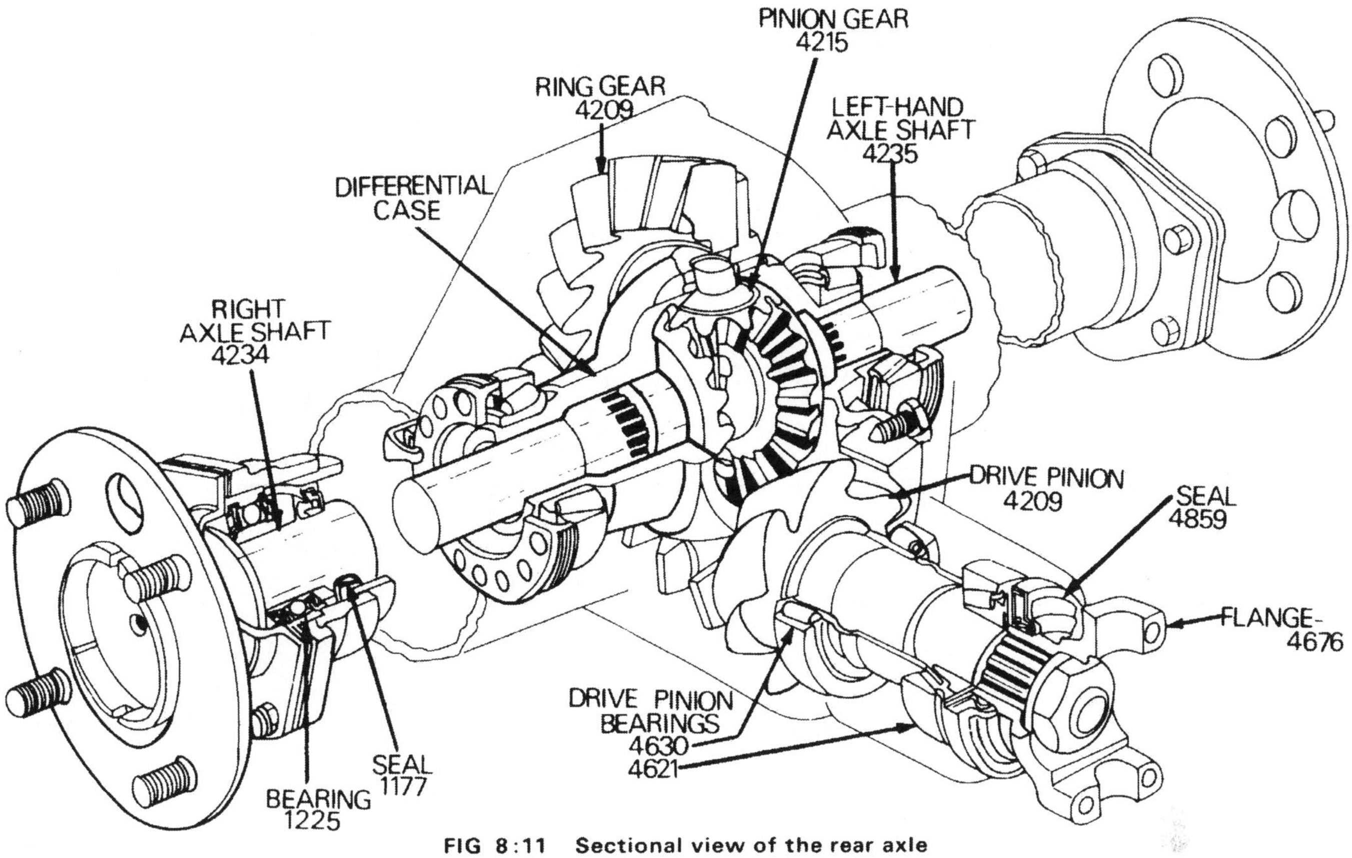

FIG 8:11 Sectional view of the rear axle

back. Fold back the mating and remove the access cover, as shown in **FIG 8:10**. Remove the nut, washer and bushing that secure the upper end of the damper to the floor pan. From underneath the car, remove the nuts and bushings that secure the shock absorber to the spring plate clip. Compress the shock absorber and remove it from the car.

Install the shock absorber in the reverse order of removal. Be sure that the spring plate clip is free from burrs or damage. Check the rubber bushings and renew them if they are defective.

8:4 The springs

Removal:

Raise the rear of the vehicle and place it securely onto stands. Two jacks will also be necessary for supporting the rear axle.

Support the rear axle and disconnect the shock absorbers from the spring plate clips by removing the nuts and bushings. Push the shock absorbers up so that they are out of the way. Lower the jacks under the axle until it is no longer raised.

Refer to **FIG 8:9**. Take off the nuts and remove the spring plate clip from underneath the spring, collecting the insulator and any plates fitted.

Raise the rear axle on the jack sufficiently to take the weight off the spring. Remove the U-bolts, insulator retainer and upper insulator.

Remove the nuts from the rear shackle bar and take off the shackle bar so that the inboard bushings can be removed. Support the spring either by hand or on a stand and withdraw the outer shackle and outboard bushings.

Remove the nuts and washer from the hanger bolt, then unscrew the hanger bolt to free the spring from the vehicle.

Examination:

Check the leaves for cracks and inspect the anti-squeak inserts between the leaves.

The spring can be disassembled by taking off the clamps and removing the center clamp bolt. When disassembling or reassembling the spring, clamp the leaves between the jaws of a strong vice to hold them together. **Carefully tighten or slacken the vice when the center bolt is out so as to keep control over the leaves.**

If the spring has been disassembled, scrub the leaves with a wire brush to remove dirt and corrosion. Examine the leaves carefully around the area for the center hole as this is one of the likeliest places for cracks to start.

When assembling the spring, use a long tapered rod through the center holes to align the leaves. Align the ends by hand and compress the spring together in the jaws of a strong vice, making sure that the anti-squeak inserts are fitted. Use a new center tie bolt and clamps to secure the spring. Be sure that the clamps are fully tight before installing the spring.

New shackle bushings should be fitted if the old ones are damaged or perished. The front bushings are a press fit in the spring eye and special tools are required to remove them and draw new ones back into place. If care is taken, an arbor press and suitable mandrels can be used for renewing the front bushes. If lubricant is required on any of the bushes, use only soapy water or soft soap as ordinary grease or oil will perish the material of the bush.

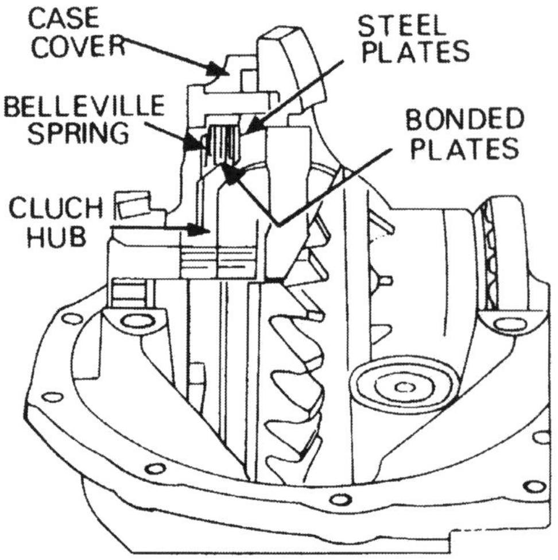

FIG 8:12 Typical limited slip differential

Installation:

The spring is installed in the reverse order of removal. Install it first at the hanger end and then at the shackle end. Lower the axle back into place—with the insulator and its retainer in place—and secure it to the spring using the U-bolts, spring clip plate and nuts. Leave the hanger and shackle bolts finger tight. When the spring is installed, raise the spring on a jack until it has reached its normal operating position, taking great care not to lift the vehicle off its stands. Tighten the hanger and shackle bolts in this position. This ensures that the rubber bushes are tightened into their normal working positions and will have the minimum of strain put onto them when the vehcile is back on its road wheels.

Reconnect the shock absorber and install the road wheel. Lower the vehicle back to the ground.

8:5 The rear axle

An integral carrier type axle is fitted to all the models covered by this manual and a sectioned view of the unit is shown in **FIG 8:11**. The owner should not attempt to dismantle the axle assembly completely as the differential parts must be set with accuracy to ensure long life and trouble-free operation. Special tools are required to set the mesh of the pinion and drive gear, and experience is also required as part of the operation involves examining the tooth marks.

Limited slip differential:

This can be fitted as an optional extra in place of the standard unit. A sectional view of the limited slip differential is shown in **FIG 8:12**. Under normal cornering forces the clutch will slip allowing the differential to operate normally. When one wheel loses traction (such as on mud or ice) the clutch transmits drive from the wheel that has no grip to the one that has grip, helping to prevent wheelspin and at the same time allowing the vehicle to be moved by the wheel that has grip. The unit also improves road holding and performance as if one wheel bounces clear of the road the other wheel will still transmit drive.

If the road wheels are to be driven with the vehicle stationary it is essential to raise both road wheels clear of the ground. Raising one wheel only, as is sufficient on standard rear axles, is dangerous as the other wheel will drive the vehicle off the stand.

The unit is checked with a torque wrench and suitable adaptor, in a very similar method to adjusting the park brakes. Raise **one wheel only** clear of the ground and fit an adaptor to the rear wheel studs so that the torque will be measured at the centre of the wheel. Fit a torque wrench, of at least 200 lb ft capacity, to the adaptor and use it to rotate the wheel—making sure that the brakes are not binding. The torque required to keep the wheel rotating must be at least 50 lb ft. The break-away torque to start the wheel rotating will be greater but once rotating the torque should not drop below the minimum.

Special lubricant must be used for the limited slip unit and the use of an incorrect lubricant can cause a high-pitched chattering noise on turns. To check the lubricant, drain out two tablespoonsful from the axle and mix it with an equal amount of white alcohol. Allow the mixture to stand for two minutes and then check its color. If the sample appears blue-tinted a Ford-approved lubricant is in use, but if the sample appears yellow-tinted then an incorrect lubricant is being used. Drain out the incorrect lubricant and refill with fresh Ford-approved. Flushing is not required but drive the vehicle in five tight circles in one direction followed by five more tight circles in the opposite direction to distribute the lubricant over the clutch plates. This treatment of five tight circles in either direction may also help to remove noise from the differential even if the correct type of lubricant is in use, as the clutch plates may partially dry off in normal use.

Maintenance:

No drain plug is fitted to the rear axle as seasonal changes of lubricant are not required. If, for major repairs, the unit has to be drained then slacken the bolts that secure the cover and allow the lubricant to drain out past the cover.

The level should be checked at intervals of 6000 miles. Clean the area around the filler plug on the rear cover and remove the plug. Top up as required to bring the level to the bottom of the filler hole using Ford-approved lubricant. Limited slip units must have a special lubricant. Allow the surplus lubricant to drain out from the filler hole before installing the filler plug. Normally Hypoid SAE.90 oil should be used but in climates where the temperature regularly drops below —25°F use SAE.80 oil.

At the same time as checking the lubricant level check that the axle vent is clear. A blocked vent will allow pressure to rise internally as the unit warms up with use and this can force lubricant past the oil seals.

Hubs and axle shafts:

The components of the axle are shown in **FIG 8:13**. Synthetic-material oil seals are installed on production but only leather seals are available as replacement parts. The whole length of the axle shaft is passed through the oil seal and any roughness on the shaft or carelessness in handling it will damage the oil seal. As a safety precaution against oil leaking into the brakes a new leather oil seal should be installed in place of the original seal every time that the axle shaft is removed.

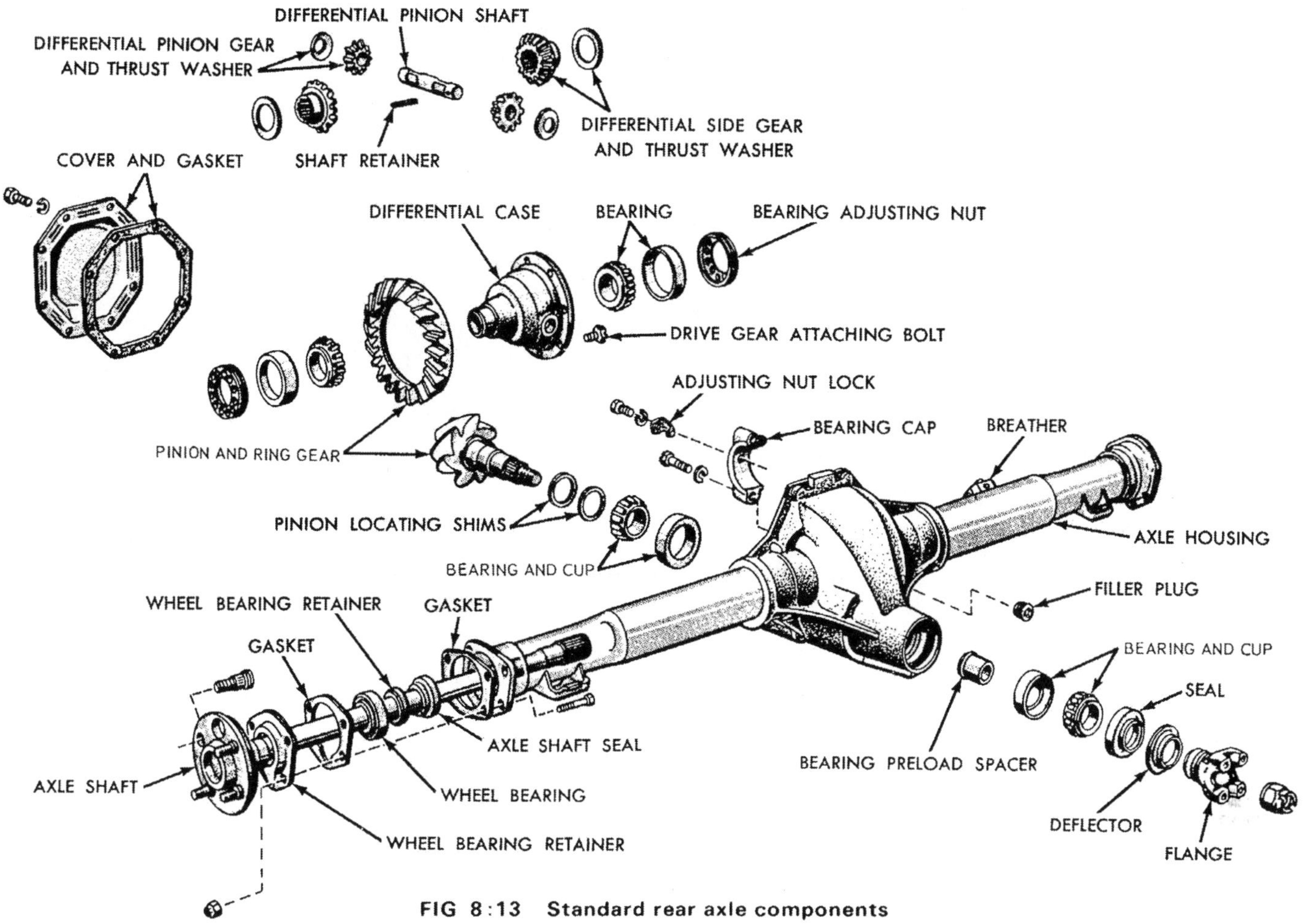

FIG 8:13 Standard rear axle components

Raise the rear of the car securely onto stands, after slackening the road wheel attachment nuts, and once the rear of the car is raised, remove the road wheel.

Take off the Timmerman nuts that secure the brake drum and pull the drum off. Rotate the axle shaft to a convenient position for working and remove the nuts that secure the wheel bearing retainer plate. Attach the special tools shown in **FIG 8:14** to the axle flange and use the impact hammer to draw the bearing and axle shaft out of the axle. When the axle shaft has been removed, prevent the brake carrier plate from moving by securing it with one nut.

While checking and cleaning the parts, put the new leather oil seal to soak in light SAE.10 oil as it requires at least $\frac{1}{2}$ hour soaking before installation.

Wash the shaft in a suitable solvent and check the bearing. Only if the bearing is defective should it be removed. Support the shaft on the padded jaws of a vice and use a cold chisel to nick the retaining ring deeply at several points around its circumference, as shown in **FIG 8:15**. **The aim is to nick the ring without actually cutting through it,** and once this has been done at several points the ring will slide easily off the shaft. The methods of removing and installing the bearing and new bearing retainer ring are shown in **FIG 8:16**. The bearing must be pressed right up against its shoulder on the shaft before attempting to press the new retainer into position.

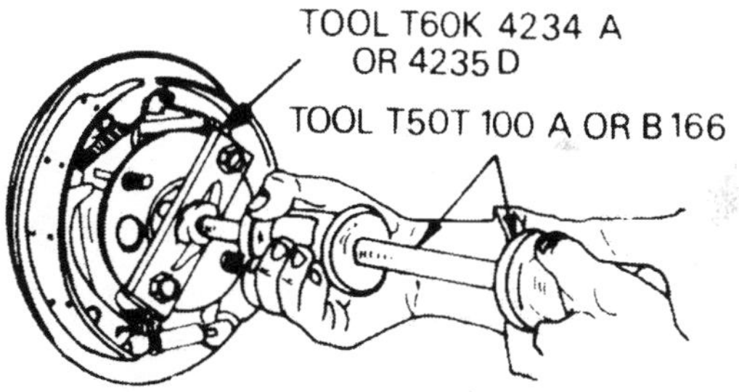

FIG 8:14 Withdrawing the axle shaft

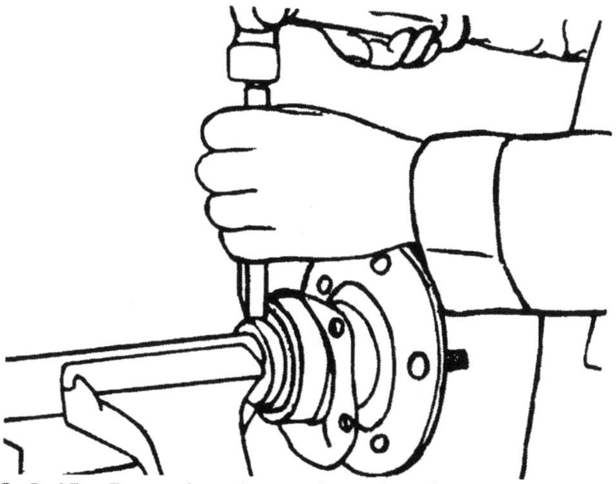

FIG 8:15 Removing the retainer ring that secures the axle shaft bearing

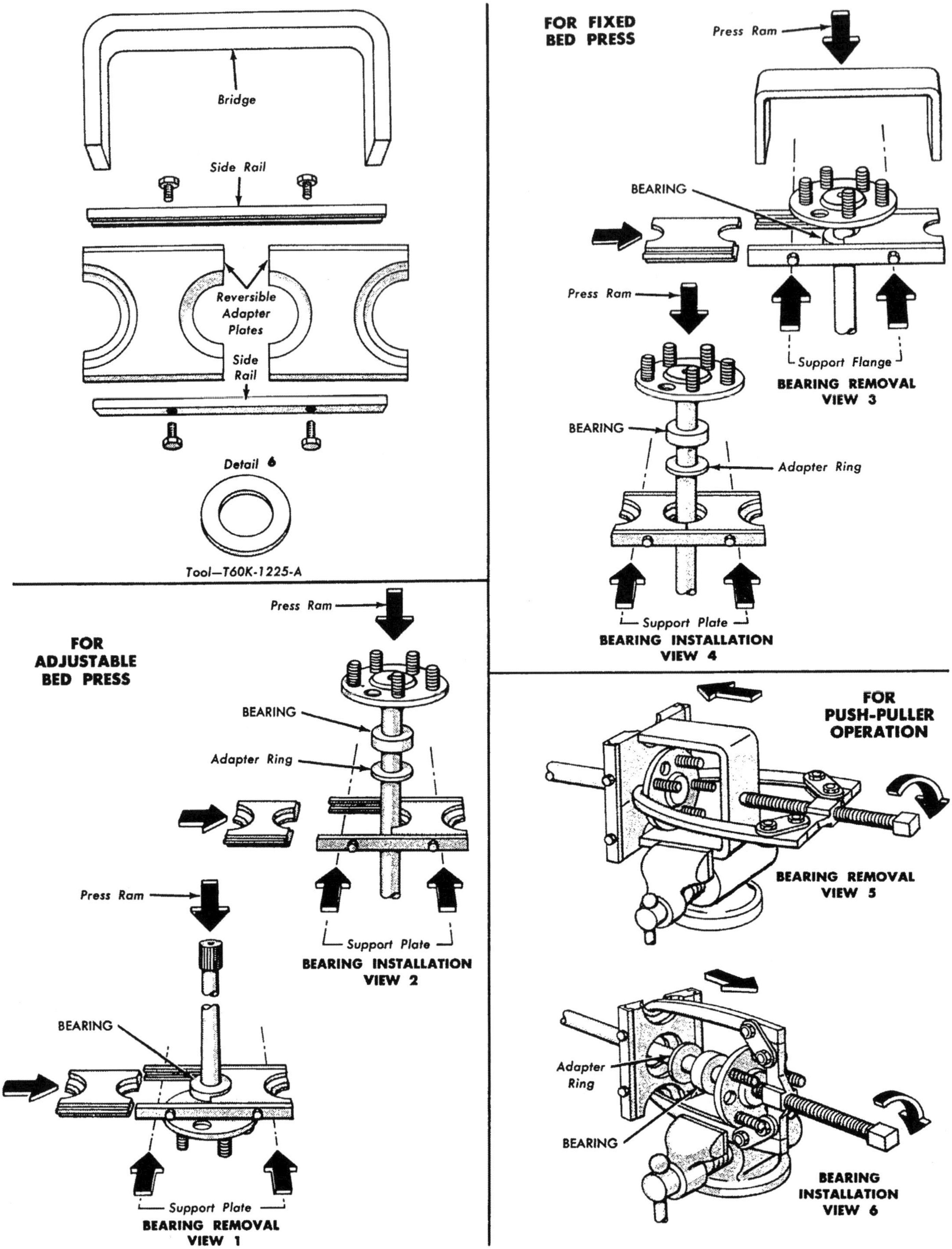

FIG 8:16 Removing and installing the axle shaft bearing

Remove the old oil seal from the axle tube, preferably using the tools shown in **FIG 8:17**. Wipe the outer circumference of the new oil seal free of oil and smear it lightly with sealing compound. Drive the new oil seal back into place using the special drift shown in **FIG 8:18**. It is essential to use the correct tool for installing the oil seal otherwise it may be damaged and leak in service.

Check the axle shaft for irregularities, burrs, or rough spots. Smooth down any damage if it is light and renew the parts if the damage is excessive.

Remove the nut holding the brake carrier plate, discard the old gaskets, and fit new ones on either side of the carrier plate. Very gently and carefully slide the axle shaft assembly back into the housing, **taking great care not to damage oil seal with the splines of the shaft.** Turn the shaft slightly if required to start the splines into the differential. Push the shaft in until the bearing bottoms in the housing and install the bearing retainer onto the bolts. Secure the retainer with the nuts. Install the brake drum and road wheel and check the brake adjustment. Lower the vehicle back to the ground and check the lubricant level in the rear axle.

Pinion oil seal:

This can be renewed without having to remove the axle or differential carrier from the vehicle. Make aligning marks, if the original yellow ones have worn off, and disconnect the drive shaft from the drive flange by removing the nuts and U-bolts. Wrap the rear universal joint of the drive shaft with adhesive tape to hold the loose pair of bearings in place on the spider.

1 Raise the rear of the vehicle clear of the ground and remove the road wheels and brake drums, to ensure that there is no drag on the axle from the brakes. Attach a torque spanner which reads in lb in centrally to the drive flange of the rear axle. Use the torque wrench to drive the axle and note the reading on the wrench required to keep the axle turning steadily. The break-away force required to start the axle rotating will be higher but should be ignored.

2 Hold the drive flange with a separate tool and remove the nut that secures it with a suitable socket wrench. Make aligning dot marks on the end of the pinion shaft and the drive flange. Pull off the drive flange, either by hand or with a suitable extractor if it is stiff.

3 Remove the old oil seal in a similar manner to that shown in **FIG 8:17**. Lubricant will flow out of the rear axle unless the front of the vehicle is raised higher than the rear, so place a suitable drip tray in position to catch spillage.

4 Check the splines on the pinion shaft for burrs or damage. Similarly check the surface of the drive flange. Light scores, nicks, or burrs can be smoothed down with Crocus paper but if the damage is deep then new parts must be installed.

5 Wipe away oil from the seating area. Earlier seals require a thin smear of sealing compound around the periphery (take great care not to allow any onto the lips themselves) but later seals already have sealing compound on them. Check with the dealer when obtaining the new seal. Install the new seal by driving it in with a suitable drift.

6 Lightly lubricate the splines of the drive flange and push it back into place so that the dots made earlier

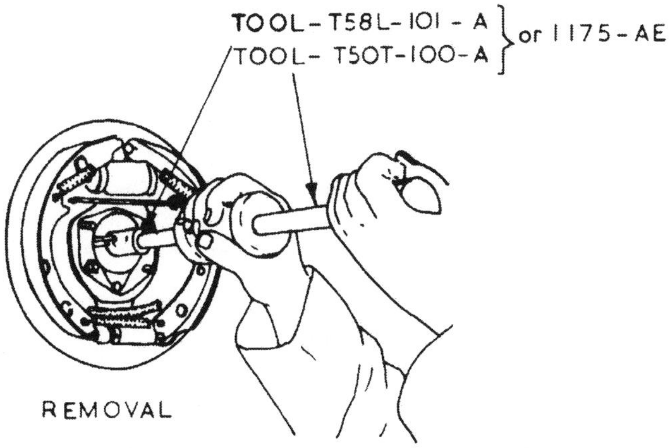

FIG 8:17 Removing the axle shaft oil seal

again align. Smear the washer part of a new retaining nut with oil and fit the nut back into position but do not tighten it.

7 Hold the drive flange and slowly tighten the flange securing nut. **At frequent intervals use a torque spanner to check the torque required to keep the axle rotating steadily.** Tighten the nut until the torque required is exactly the same as found before disassembly (see operation 1) then very carefully tighten the nut until the torque required to rotate the axle has increased by 6 to 12 lb in. **If the preload is exceeded then the pinion must be disassembled and a new spacer fitted between the bearings, as slackening the nut will leave the excessive clearance between the bearings.**

8 Install the drive shaft, brake drums and road wheels. Lower the vehicle back to the ground and check the lubricant level in the axle.

Axle removal:

If the axle complete with drive shafts and brake parts is being removed it will be necessary to disconnect the park brake cables from the brake carrier plates and axle, and to disconnect the main hydraulic line at the T-piece on the axle. The hydraulic system will than have to be bled after installation of the axle.

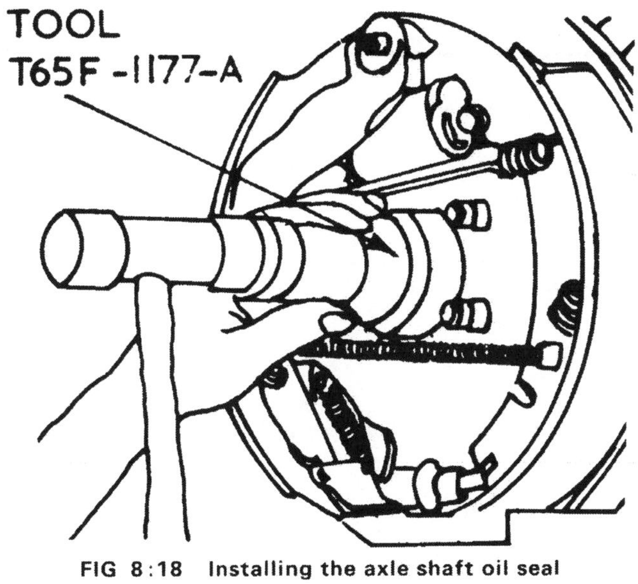

FIG 8:18 Installing the axle shaft oil seal

The method given here obviates the need to disconnect the braking system parts and as the axle will most likely have to be dismantled anyway it will save work in the long run.

1 Raise the rear of the vehicle and support it on stands under the frame. Remove both road wheels and brake drums and then withdraw both axle shafts. Suspend the brake carrier plates, still attached to the brake lines and park brake cables, safely out of the way.

2 Drain the unit by slackening the rear cover attachment bolts. Disconnect the vent pipe from the casing. Free the hydraulic lines and T-piece from the rear axle, without disconnecting the lines. Disconnect the shock absorber lower attachments from the spring plate clips.

3 Make alignment marks and free the drive shaft from the pinion drive shaft, carefully tapping the loose bearings in place on the drive shaft.

4 Remove the nuts that secure the spring plate clips and take off the spring plate clips, U-bolts, insulators, and insulator retainers. Manoeuvre the axle assembly out over the top of the road springs.

The axle is installed in the reverse order of removal. The points to watch for when reconnecting and installing the parts have been dealt with in the relevant sections.

8:6 Trouble shooting guide

(a) Noisy axle or suspension

1 Lack of lubricant or incorrect lubricant used
2 Defective shock absorbers
3 Defective shock absorber mounting bushings
4 Defective axle mounting insulators
5 Defective spring mounting bushings
6 Worn drive shaft universal joints
7 Defective differential unit
8 Worn bearings

(b) Vibration

1 Check 5 and 6 in (a)
2 Drive shaft out of balance
3 Drive shaft fitted 180 deg. out

(c) Lubricant leaks

1 Defective hub seals or gaskets
2 Defective pinion oil seal
3 Defective seals on universal joint spiders
4 Plugged breather on axle casing
5 Too high lubricant level in rear axle
6 Defective gaskets or damaged joint faces

CHAPTER 9

THE FRONT SUSPENSION AND HUBS

9 :1 Description

A typical front suspension assembly is shown in **FIG 9 :1**. The 1965 models differ in the attachments of the inboard ends of arms and this alters the method of setting the suspension geometry.

The upper and lower arms are attached to the frame by pivot bolts which allow them to swing vertically. The outboard end of the lower arm is located by a strut to prevent it from moving in the fore and aft direction. This strut is also used to set the suspension geometry. A spindle and stub axle assembly swivels between the outer ends of the arms in an upper and lower ball joint. The ball joints allow the spindle to turn for steering purposes and at the same time allow it to rise and fall vertically for the movement of the suspension.

A coil spring and concentric shock absorber are installed between the top arm and the frame to take the weight of the vehicle and movement of the suspension. To control these movements the shock absorber damps the oscillations of the suspension.

The road wheel is attached to a hub assembly which rotates about the stub axle on two opposed tapered bearings.

A stabilizer bar interconnects the lower suspension arms. When one suspension is at a different height from the other, the stabilizer bar is under torsion and exerts a restoring force between the two suspensions to bring them back to the same height. This transfers weight from the outer suspension to the inner one on curves and therefore improves road-holding as well as decreasing body roll.

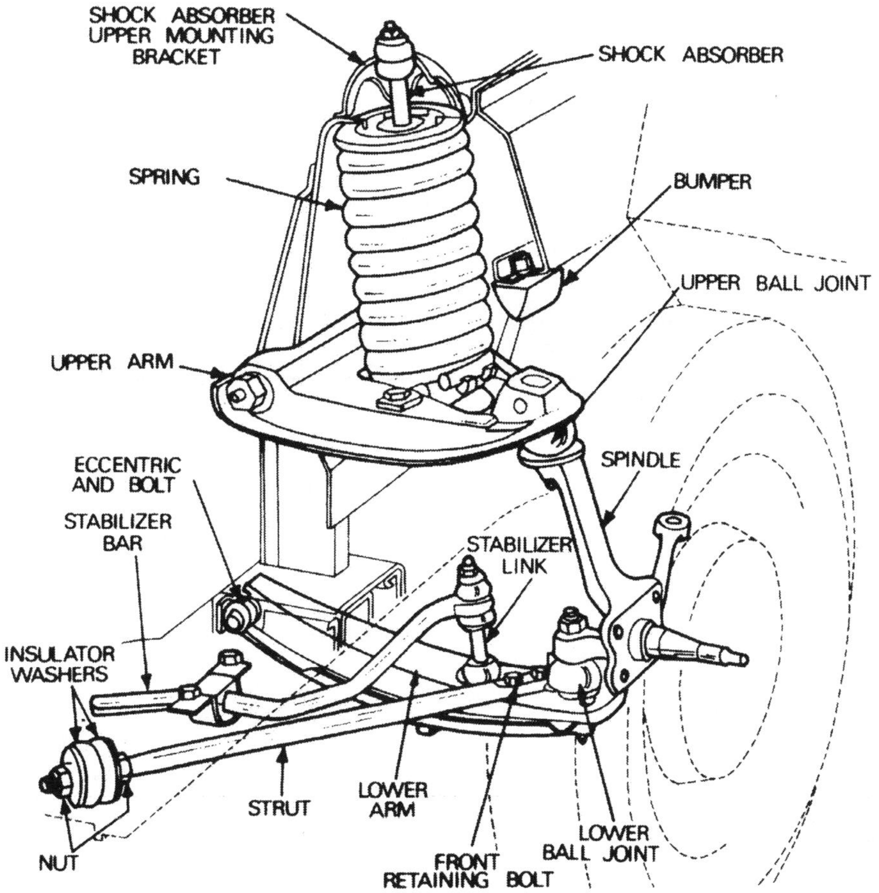

FIG 9 :1 Front suspension assembly

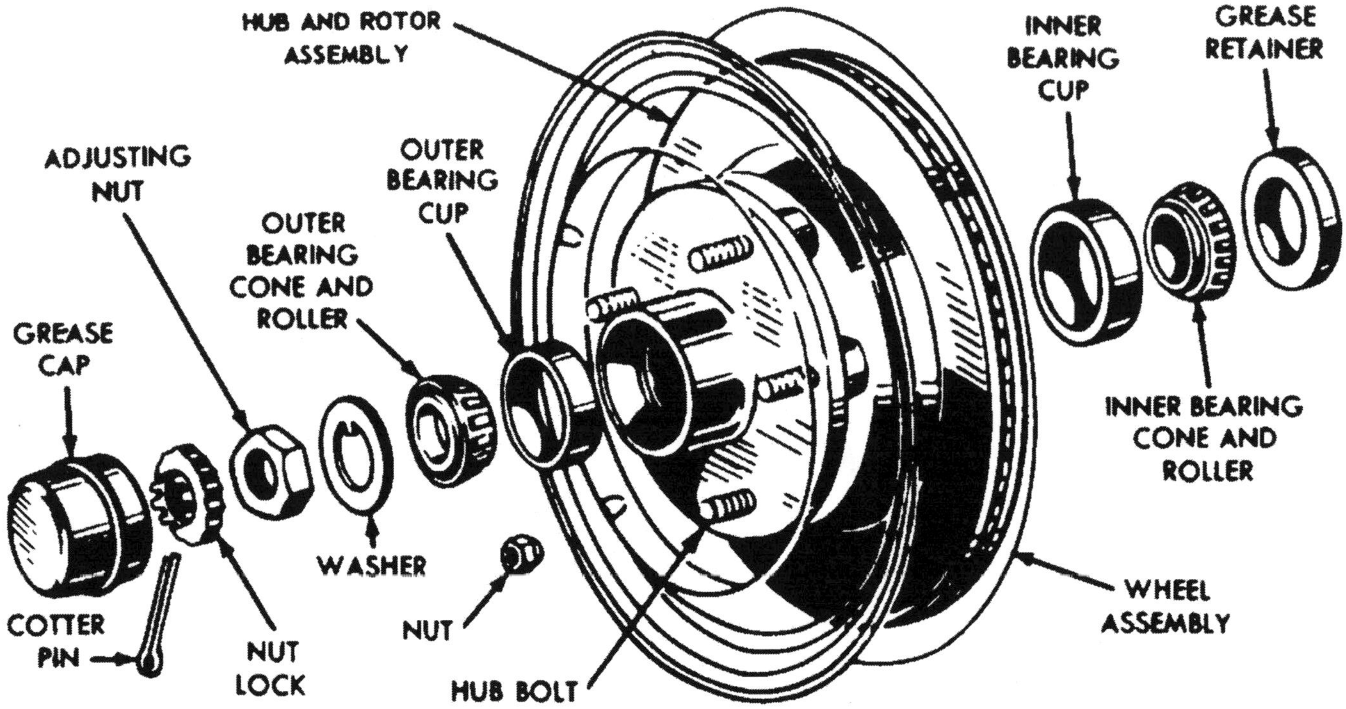

FIG 9:2 Wheel hub components

9:2 Maintenance

The suspension ball joints should be greased at intervals of 12,000 miles. Remove the plug from the ball joint and install a grease fitting. Inject grease through the fitting until the dust cover just starts to swell. Do not inject so much grease that it exudes past the dust cover, otherwise the weather seal properties will be damaged. Similarly grease all pivot points.

Check the dust covers for incipient splits or chafing. If a dust cover is found to be failing it must be renewed before it splits and allows dirt to enter the ball joint. If dirt has entered the ball joint, try removing it by pumping plenty of grease through the ball joint with the dust cover removed so as to flush out the dirt. If the dirt cannot be flushed out a new ball joint(complete lower arm in the case of the lower ball joints) must be installed.

After lubrication and inspection of the ball joints, remove the grease fittings and install the plugs.

At approximately 30,000 miles the wheel hub should be removed, old grease washed out and the hub installed after checking the bearings and packing with fresh grease. This operation will be dealt with in the next section.

9:3 The wheel hubs

Typical wheel hub components are shown in **FIG 9:2**. Whenever the front of the car is raised off the ground, briefly check the wheel bearings. Grip the tire at the nine and three o'clock positions and rock the wheel as though making it steer. If there is excessive play then it is likely that the wheel bearings require checking and adjusting. Do not confuse steering movement with play. Spin the wheel and check that it rotates freely without grinding noises. Do not confuse brake noise with that from a defective bearing. If the bearings are noisy, the wheel hub should be removed so that the bearings can be visually examined. **Packing in more grease on the outer bearing and tightening up the adjustment can be dangerous** as the inner bearing may be on the point of failure.

Removal:

Raise the front of the car and place it securely on stands. If the road wheel attachments are slackened before raising the car it will be found easier to remove the road wheel once the front end is up, as the wheel will spin freely once it is clear of the ground.

Remove the grease cap and extract the cotterpin that secures the locknut.

On models fitted with disc brakes, slacken back the hub nuts by three turns and then rock the hub firmly to press the friction pads and pistons back into the caliper. Remove the caliper as instructed in **Chapter 11**.

On models fitted with drum brakes, slacken back the brake adjusters to free the shoes from the drum.

Remove the locknut, nut and special washer. Pull the hub assembly firmly off from the stub axle, collecting the inner race of the outer bearing as it comes free.

Pry out the grease retainer so that the inner race assembly of the inner bearing can be removed from the hub. The bearing cups (outer races) can be left installed in the hub, unless the bearings are defective. To remove the outer races, drive them evenly out with a suitable drift made of soft-metal.

Cleaning and inspection:

Wipe away most of the old grease with rags. Wash the parts with gasoline or kerosene to remove the remainder of the grease and dirt. Rotate the inner races of the bearings under clean solvent to wash out dirt and grease. Dry the parts with an airline. **When drying the bearings, do not allow them to rotate under the air pressure or they will become chipped and damaged.** If an airline or tire pump is not available, use lint-free cloth on the main portions and allow the bearings to air-dry.

Brush away all dust and dirt from the brake carrier plate or disc brake dust shield and wipe the spindle with a piece of cloth wetted with solvent. Wipe down the stub axle and check it for cracks or damage.

Check the bearings for pitting, wear, or fret marks on their operating faces. If the bearings are visually satisfactory, lightly oil the inner races with thin oil and press them back into their cups. Press them firmly in and oscillate them backwards and forwards to check for any roughness in operation. If roughness is felt, wash the bearings again as specks of dirt may be the cause. If the bearings are found to be defective or rough in operation, both bearings must be completely renewed. Drive out the cups and drive the new cups back into place.

While cleaning and examining the parts, put a new grease retainer to soak in light oil as it requires at least $\frac{1}{2}$ hour soak before installation.

Installation and lubrication:

Wipe down the stub axle with an oily or greasy cloth to give it a film of lubricant which will prevent corrosion.

The points for applying grease in the hub are shown in **FIG 9:3**. Work ample grease into the inner races of both bearings and apply a light smear to the bearing cups. Pack the inside of the hub to the depth shown in the figure. Install the inner race of the inner bearing and secure it in place by installing the new grease retainer, after its $\frac{1}{2}$ hour soak in light oil.

Slide the hub assembly back into place, after cleaning the operating face of the brake with a harsh solvent (such as Trichlorethylene) to remove any grease or finger marks from the drum or disc. Take great care to slide the hub on so that the stub axle is central and does not damage the grease retainer.

Press the inner race of the outer bearing back into place and hold it there by lightly tightening the nut with the special washer in place.

Once the bearings have been adjusted, install the

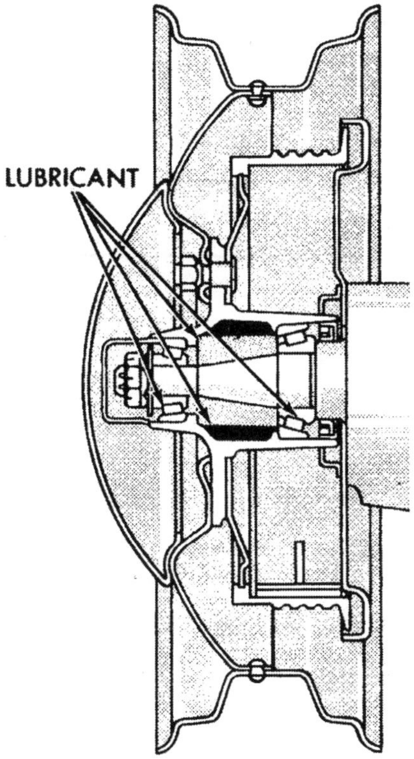

FIG 9:3 Wheel bearing lubrication

remainder of the parts in the reverse order of remova Check that the road wheel spins freely before lowerin the vehicle back to the ground. Adjust the brakes (se **Chapter 11**). On models fitted with disc brakes, adjus ment is automatically taken up by pumping the servic brake pedal hard several times.

Adjustment:

A method of adjusting the wheel bearings to the corre clearance is shown in **FIG 9:4**. If a torque wrench is n available, tighten the nut while spinning the hub un there is a definite resistance to rotation (showing th

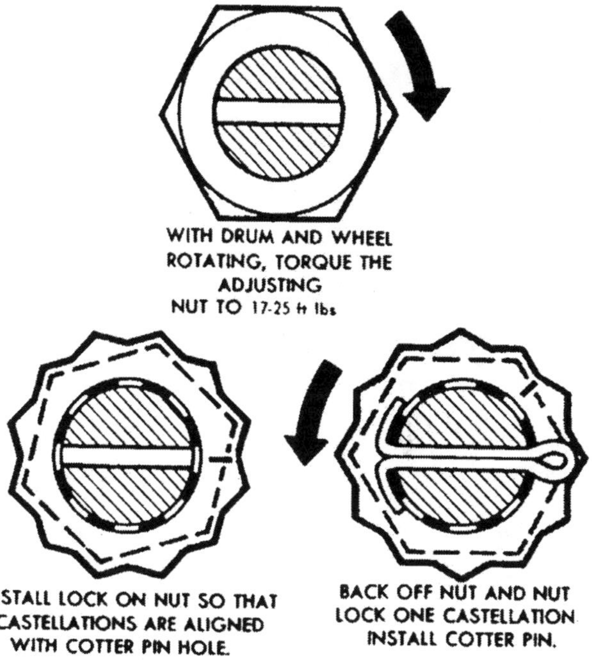

FIG 9:4 Adjusting the wheel bearings

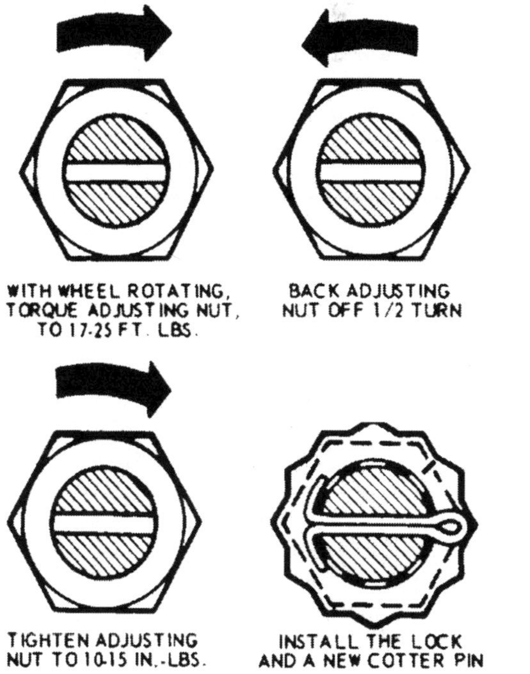

FIG 9:5 Adjusting the wheel bearings: alternate method

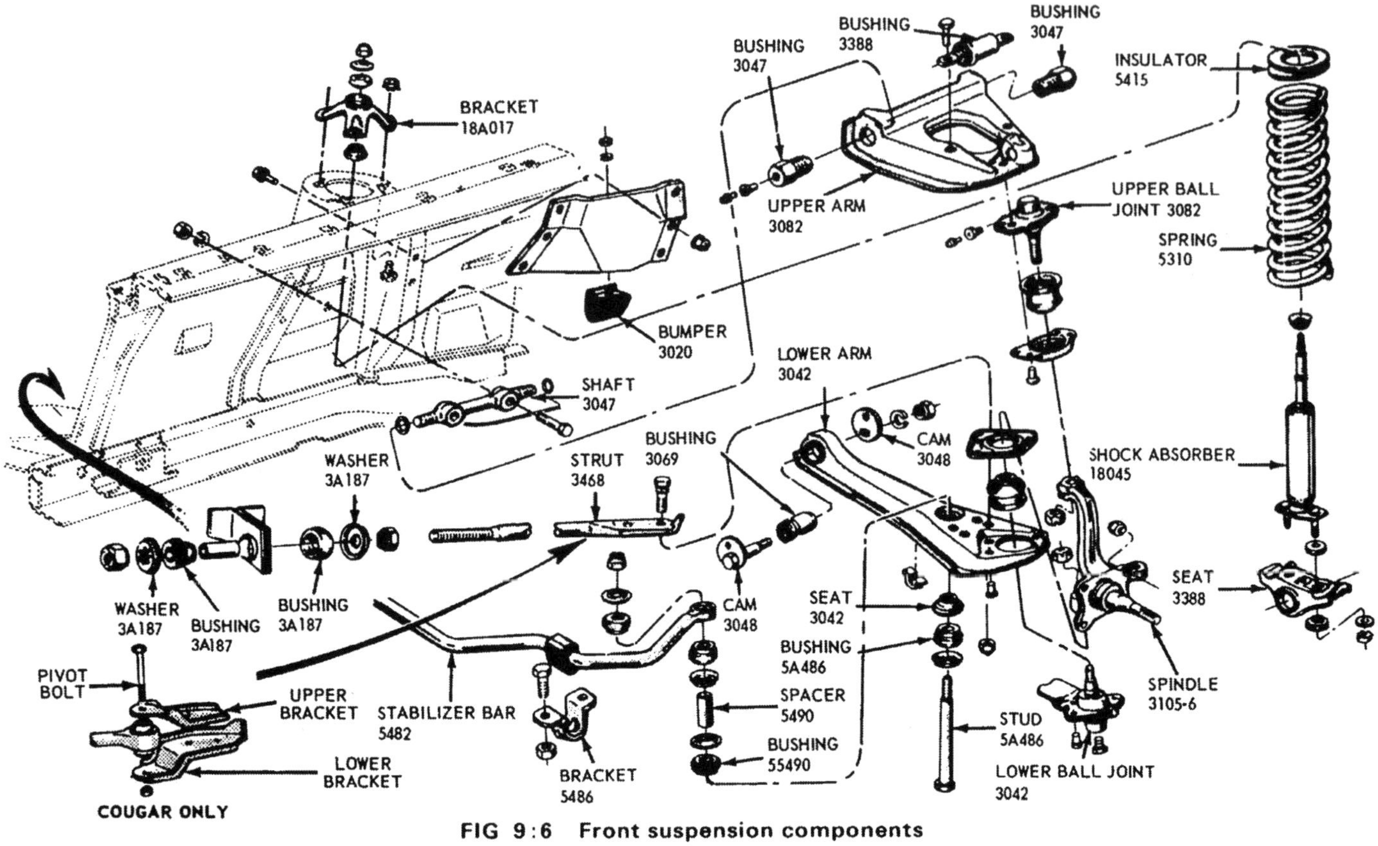

FIG 9:6 Front suspension components

the bearings are under slight preload) then slacken the nut again. Tighten the nut again using moderate pressure until all the end play in the bearing has been taken up and then install the locknut and turn both nuts back as shown in the figure.

An alternate method of setting the wheel bearings is shown in **FIG 9:5**. The alternative method is the better one to use, provided that suitable torque wrenches are available. When installing a new cotterpin to lock the nut it is essential that the legs are opened circumferentially as shown in the figures, otherwise the legs will damage the static pick-up point in the grease cap.

9:4 The shock absorbers

Testing of the shock absorbers for defects is the same as testing the shock absorbers fitted to the rear suspension (see **Chapter 8, Section 8:3**).

The components of a typical suspension are shown in **FIG 9:6** and a typical shock absorber assembly is shown in **FIG 9:7**.

Removal:

Raise the front of the vehicle onto stands and support the suspension with a jack under the lower arm. Keep the suspension low enough to gain access to the lower shock absorber attachments. Remove the road wheel to make the task easier.

Remove the nuts that secure the upper and lower attachments to the frame and suspension arm respectively. Lift the shock absorber and its attachment brackets up and out of the vehicle as shown and then separate the shock absorber from the bracket.

The shock absorber is installed in the reverse order of removal, after ensuring that all the rubber bushes are in good condition.

9:5 The stabilizer bar

The attachments of the stabilizer bar are shown in **FIG 9:6**, and also in **FIG 9:1**.

Raise the front of the vehicle and place it securely on stands. Remove the road wheels and support each suspension on a jack or small stand so that they are both at the same height.

Take off the nuts and withdraw the stabilizer link bolts from each lower arm, carefully collecting the spacer, bushings and washers as they come free.

Remove the bolts that secure the stabilizer brackets to the frame and withdraw the stabilizer bar from under the vehicle.

Before installing the stabilizer bar, check all bushings for wear, perishing, or deformation and renew any that are defective. Lubricate the bushings with Ruglyde or comparable lubricant.

Install the stabilizer bar in the reverse order of removal.

9:6 The suspension ball joints

The suspension components are shown in **FIG 9:6**.

Upper ball joint inspection:

Raise the front of the vehicle and support it on underframe stands so that both suspensions hang down freely at the end of their range of movement. This ensures that the upper ball joints are unloaded. Be sure that the wheel bearings are correctly adjusted (see **Section 9:3**) noting that this can be done with the road wheel in place, provided that the bearings spin correctly.

Mount a Dial Indicator Gauge on the upper arm and position it so that its plunger rests vertically on the underside of the spindle at the upper ball joint. Lift and lower the suspension assembly, noting the play on the dial indicator. The play should not exceed .200 inch. Mount

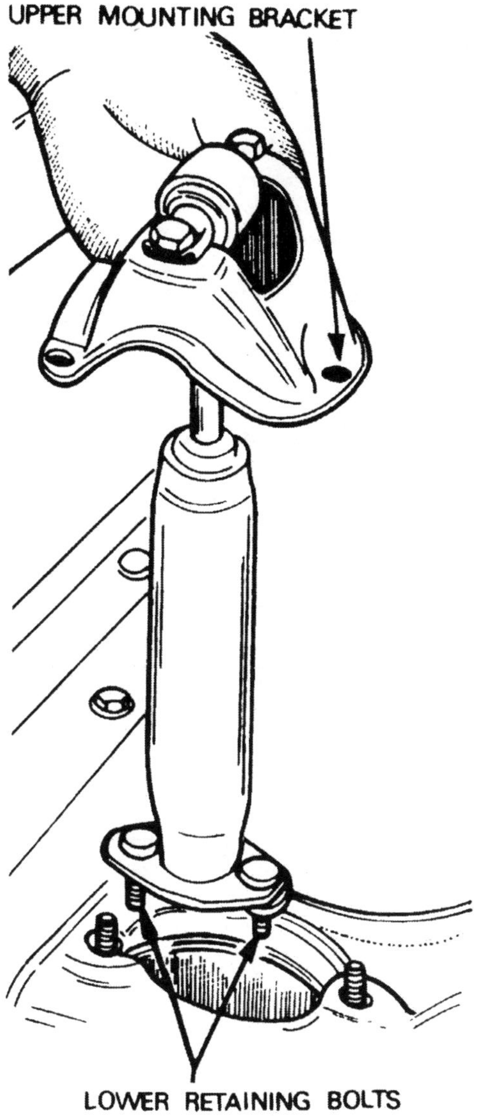

FIG 9:7 Typical shock absorber assembly

the dial indicator so that its plunger now rests against the inside wheel rim, vertical to the plane of the wheel. Grasp the top of the tire and rock it in and out so that the radial play of the ball joint can be measured on the dial indicator. The radial play must not exceed .250 inch. If the play is excessive in either direction, a new ball joint must be installed.

Lower ball joint inspection:

With the front of the vehicle still raised off the ground as for inspecting the upper ball joint, have an assistant rock the lower edge of the tire in and out, while watching the lower ball joint. Any movement between the lower end of the spindle and the lower suspension arm indicates that the ball joint is worn and a new complete lower suspension arm must then be installed.

Renewing upper ball joint:

A fabricated support, as shown in **FIG 9:8**, is required to take the pressure of the road spring. With the front of the vehicle off the ground, raise the suspension on a jack until the support can be inserted and the suspension lowered to hold it in place.

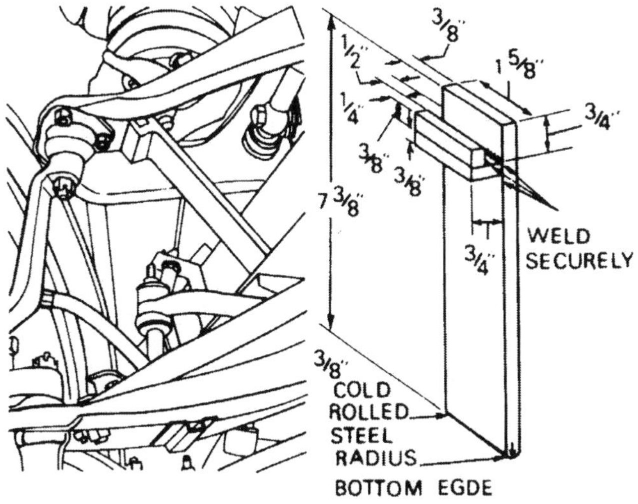

FIG 9:8 Method of supporting the upper suspension arm

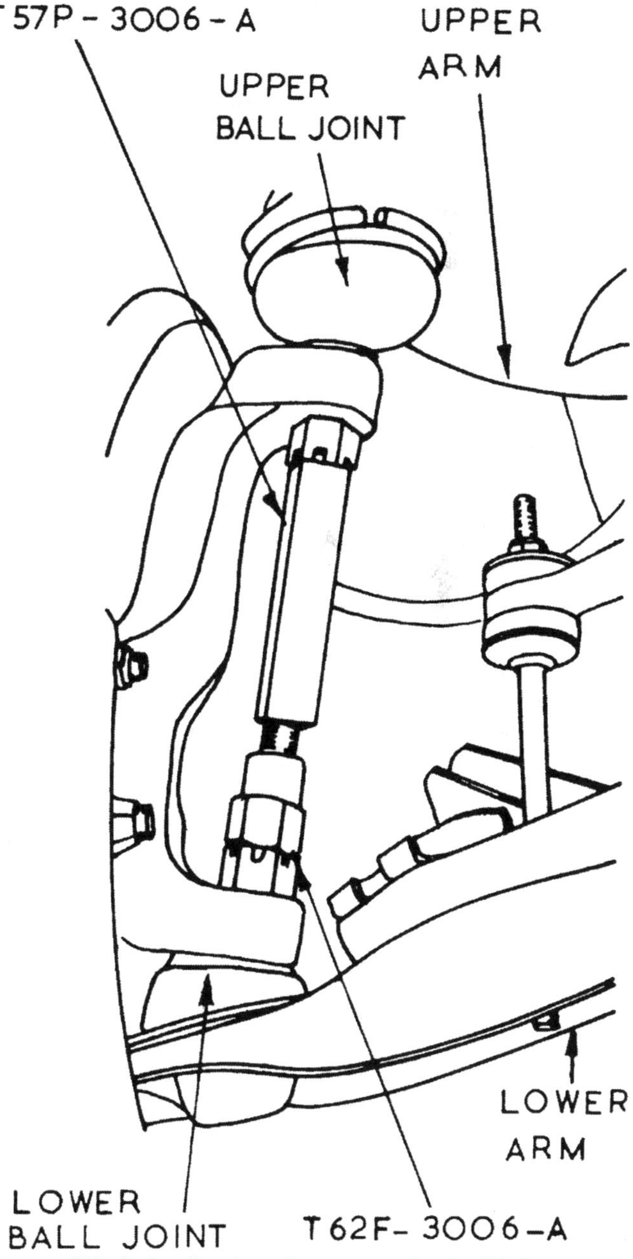

FIG 9:9 Freeing the suspension ball joints

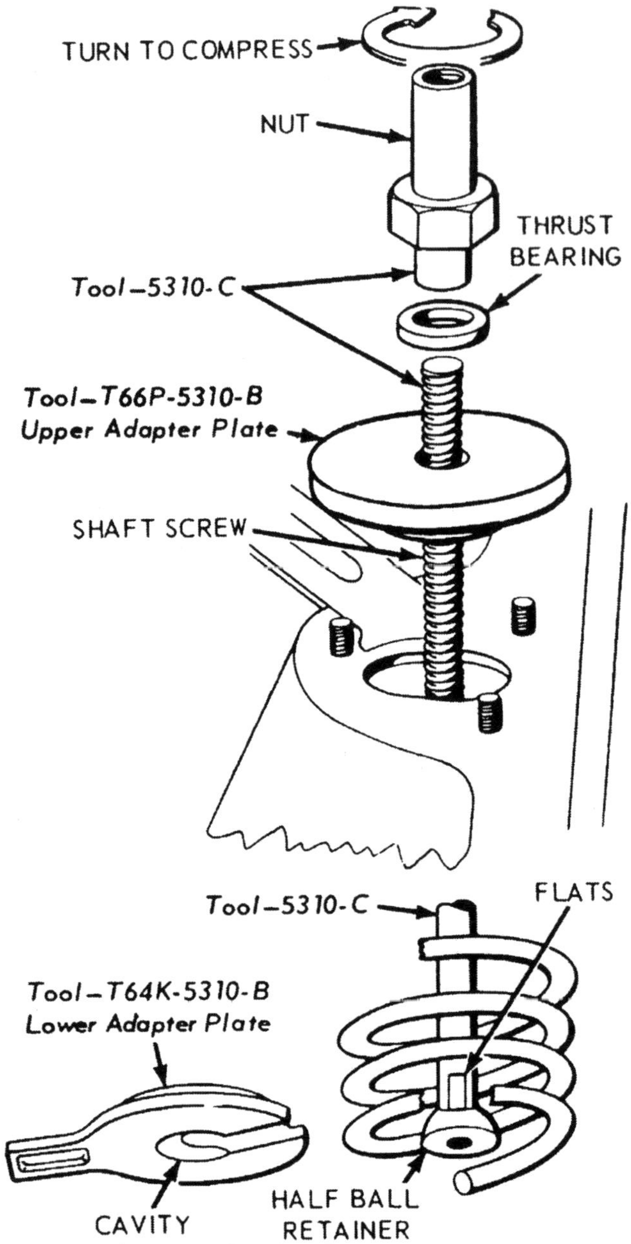

FIG 9:10 Road spring compressor parts

Remove the cotterpin from the ball joint nut and slacken the nut by several turns. Install the tool shown in **FIG 9:9**. The ball joint can be freed from the spindle by tightening the tool so that the ball joint pin is pressed out of the spindle. If the special tool is not available do not hit the end of the ball joint pin, as this will internally damage the ball joint if it is to be used again and will also rivet over the end of the pin. If no tool is available, slacken off the nut as before. Lever the arm firmly from the spindle and lay a block of metal beside the tapered eye on the spindle. Hit with a copper hammer on the opposite side of the tapered eye and the tapers will quickly free.

Partially drill through the rivets that secure the ball joint to the arm and then knock off their heads with a cold chisel. Punch out the stems so that the ball joint is freed from the upper arm.

Check the arm for cracks or distortion. Clean the holes and remove any burrs from them.

Secure the new ball joint in place using the special nuts and bolts supplied in the kit. **Do not use ordinary nuts and bolts and do not attempt to rivet the new ball joint into place.**

Install the spindle to the ball joint and secure it by tightening the nut to the correct torque. Lock the nut with a new cotterpin, tightening the nut further if necessary to align the cotterpin holes.

Lubricate the ball joint through a grease fitting until the dust cover just starts to swell. Install the road wheel and remove the support before lowering the car back to the ground.

Renewal of lower arm:

This is best left to a service garage, not because the work is beyond the power of the owner to carry out, but because the suspension geometry is controlled by the inboard attachments of the lower arm and by the positioning of the locating strut. If the owner does carry out the work, the vehicle should be taken to a service agent on completion so that the suspension geometry can be checked and adjusted.

The work is straightforward. Disconnect the lower ball joint, using the tool shown in **FIG 9:9**, after having supported the upper control arm as shown in **FIG 9:8**. Disconnect the stabilizer bar link and the strut from the lower control arm. Free the inboard attachments and remove the arm assembly.

The parts are refitted in the reverse order of removal. Try to set all parts as closely as possible to their original settings.

9:7 The road springs

Removal:

1 Remove the air cleaner from the engine to gain access room. Raise the front of the car and place it securely onto underbody stands. Remove the road wheel.
2 Remove the shock absorber as detailed in **Section 9:4**. Install the parts of the special tool shown in **FIG 9:10**. The lower plate is installed under the third coil of the spring from the bottom and is then locked into

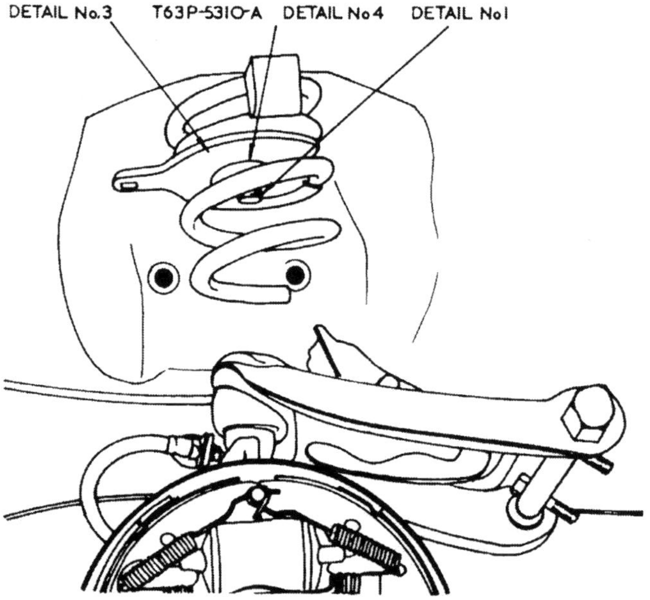

FIG 9:11 Removing the road spring

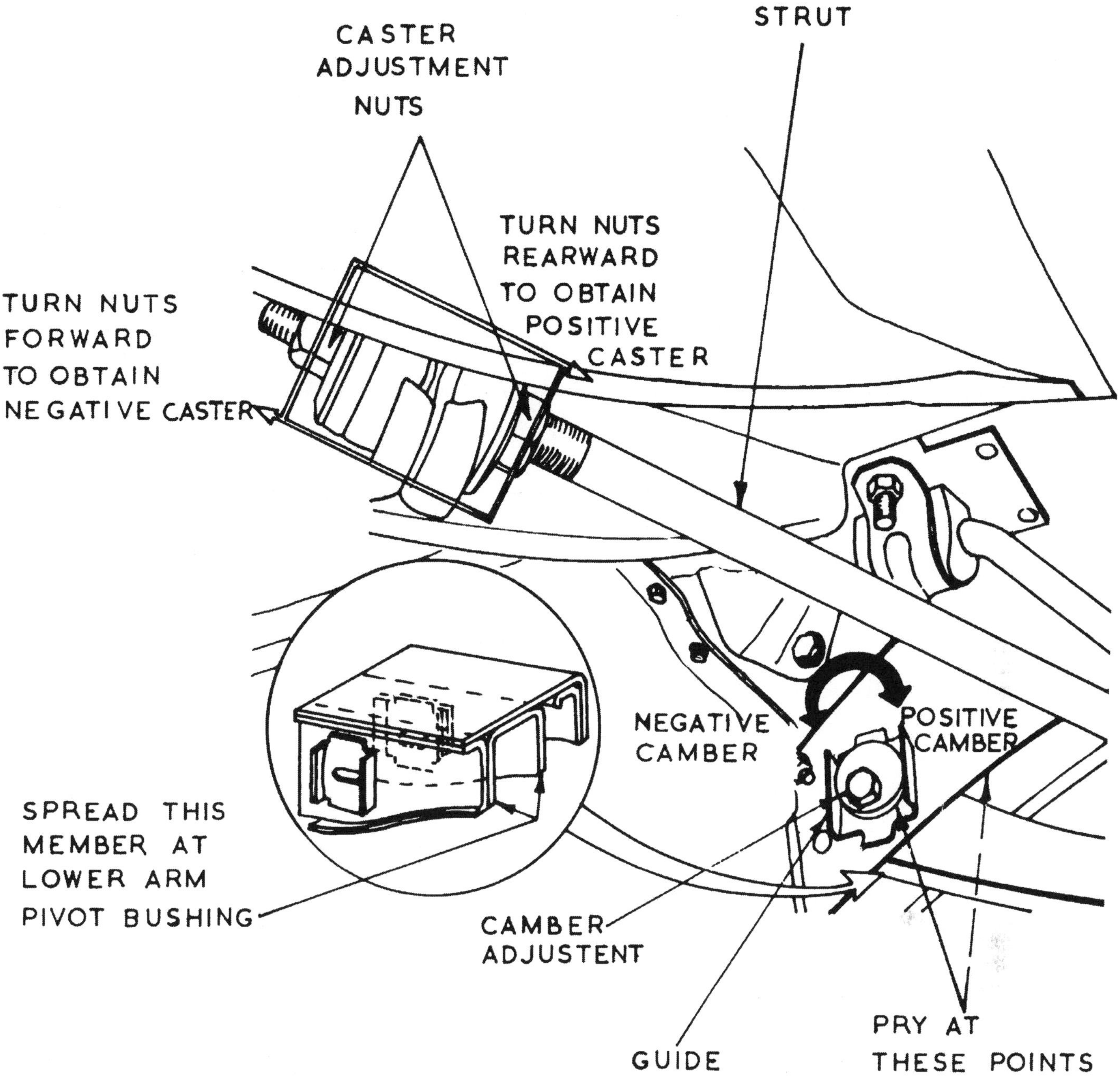

FIG 9:12 Suspension geometry adjustment points

the half ball retainer. The upper adaptor plate is held by the three studs of the shock absorber upper attachment. Use a suitable socket and ratchet drive to compress the spring until it is clear of the upper arm.

3 Remove the nuts that secure the upper arm to the frame, carefully collecting any shims fitted, and swivel the arm out of the way, as shown in **FIG 9:11**.

4 Carefully release the spring compressing tool. Once the pressure is released, remove the tool parts and take the spring out from the vehicle.
Check the spring for cracks, damage or weakening and renew it if it is defective.

The parts are installed in the reverse order of removal, using the special tools to compress the spring and hold it in position while refitting the upper arm. Make sure that the pivot for the upper arm is evenly spaced and that the correct keystone lockwashers are used as well as all shims fitted back into their original positions.

9:8 Suspension geometry

Accurate gauges and special tools are required to check and adjust the geometry. Incorrect geometry will cause poor road-holding and excessive tire wear. The owner is not capable of setting the suspension geometry to the accuracy required and therefore the task should be left to a service agent. The adjustment points for the suspension geometry are shown in **FIG 9:12**.

On the 1965 models the method is slightly different and the geometry is varied by adjusting the shims under the lower arm inboard pivot.

9:9 Trouble shooting guide

(a) Wheel wobble or tramp

1 Unbalanced front wheels and tires
2 Broken or weak road springs
3 Defective suspension arm inboard pivots
4 Worn or loose hub bearings
5 Worn or defective suspension ball joints
6 Loose road wheel attachments
7 Uneven tire wear

(b) Vehicle bottoms on suspension

1 Check 2 in (a)
2 Bumpers missing or worn
3 Defective shock absorbers

(c) Heavy steering

1 Neglected lubrication
2 Incorrect suspension geometry

(d) Excessive tire wear

1 Check 2, 3, 4 and 5 in (a) and 2 in (b)

(e) Rattles

1 Check 2, 3 and 5 in (a) and 3 in (b)
2 Stabilizer bar broken or bushings worn

(f) Excessive body roll

1 Check 2 in (a); 3 in (b) and 2 in (e)

CHAPTER 10

THE STEERING SYSTEM

10:1 Description

A recirculating-ball steering unit is fitted to all models covered by this manual and a sectional view of a typical unit is shown in **FIG 10:1.** The steering shaft is connected to the steering wheel, through the steering column, and therefore revolves as the steering wheel is turned. The ball nut moves up and down the spiral of the steering shaft as the shaft rotates, with the balls reducing friction, and as it moves it rotates the sector shaft by the sector gear teeth.

The sector shaft is linked by the Pitman arm to the steering linkage. A typical steering linkage is shown in **FIG 10:2**. An idler arm is mounted on the opposite side of the frame to the steering unit, and a rod operates between the steering unit Pitman arm and idle lever. Connecting rods then join the center rod to the steering levers on the wheel suspensions. When sector shaft and Pitman arm of the steering unit rotate, the center connecting rod moves from side to side and the spindle connecting rods and levers rotate the suspension spindle in its ball joints to give the steering action. Ball joints are used throughout the steering linkage to minimize friction.

Power-assisted steering can be fitted as an optional extra. This is either a Ford non-integral system or, more recently, a Saginaw integral system.

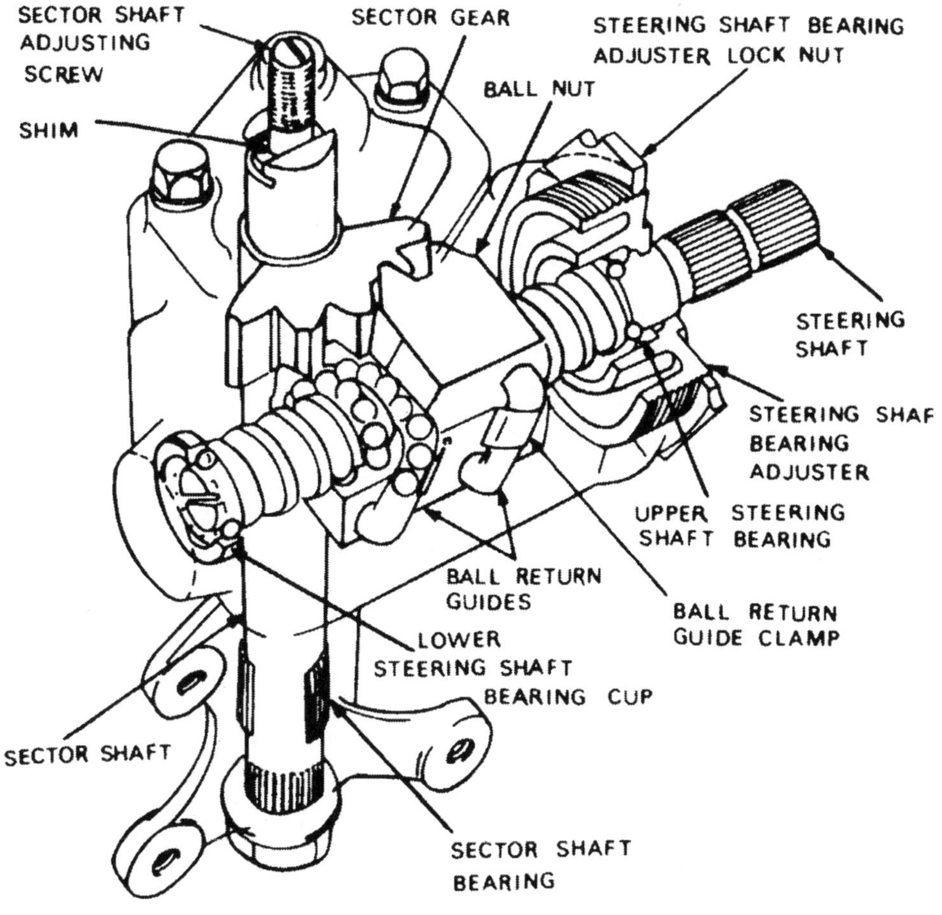

FIG 10:1 Sectional view of the steering unit

Either system derives its power from a hydraulic pump. The pump and fluid reservoir form a single assembly mounted on the engine and belt-driven from the crankshaft pulley.

The parts of the Ford system are fitted as an addition to the standard linkage as shown in **FIG 10:3**. The control valve is mounted on the steering linkage so that it moves with it to provide the required feedback of position. Under straight-ahead conditions, both ends of the power cylinder are fed with equal fluid pressure from the pump and the valve allows a continuous flow of fluid back to the reservoir. When the steering wheel is turned the internal parts of the valve are moved by the lever so that fluid pressure rises on one side of the piston. As a result, the cylinder moves and with it the steering linkage to which it is connected. The feedback action returns the valve to the neutral position when the steering has moved as far as required.

The Saginaw system replaces the standard steering box and contains an integral rotary valve and hydraulic slave unit. A one piece double-acting piston and rack is meshed with the sector shaft. It operates purely on fluid displacement when the valve is turned by the steering wheel. As the entire assembly is always full of fluid no periodic lubrication is needed. In addition, the fluid forms a cushion to absorb shock, some necessary road feel being transmitted to the driver through a torsion bar.

Two types of gearing are available, the conventional constant ratio gear or a new type of gear of variable ratio. The new type has a rack and sector tooth configuration providing a ratio of 16:1 at center progressing to 13:1 at either lock.

When the front wheels are clear of the ground, the steering must not be turned vigorously from lock to lock as the shock when the parts are halted may cause damage.

10:2 Maintenance

On all models check that the steering operates fully and freely without excessive play or stiffness. Check the dust covers on all the ball joints and renew them if they show incipient signs of failure. Lubricate the ball joints through a grease fitting, as for the suspension ball joints.

On models fitted with power steering, check the fluid level and belt tension, and grease the actuator ball stud.

Run the engine and turn the steering from full lock to lock a few times and then stop the engine. Withdraw the dipstick and check the fluid level. If necessary, top-up to the correct mark with automatic transmission fluid. If the level has fallen so low that air has been drawn into the system, the fluid will appear milky or frothy. Top up to the mark and then operate the steering from lock to lock with the engine running, adding more fluid to the reservoir as required and finally checking the level with the engine stopped.

The method of setting the drive belt tension is given in **Chapter 4, Section 4:4**. Note that the reservoir must not be levered and the pump should be moved with a wrench on the cast boss.

When lubricating the actuator ball stud, wipe the grease fitting clean before applying a grease gun. Inject grease until the dust cover just starts to swell.

10:3 The steering linkage

A typical steering linkage is shown in **FIG 10:2**. The spindle connecting rods are made up of ball joint ends and an internally threaded sleeve. One end of the sleeve

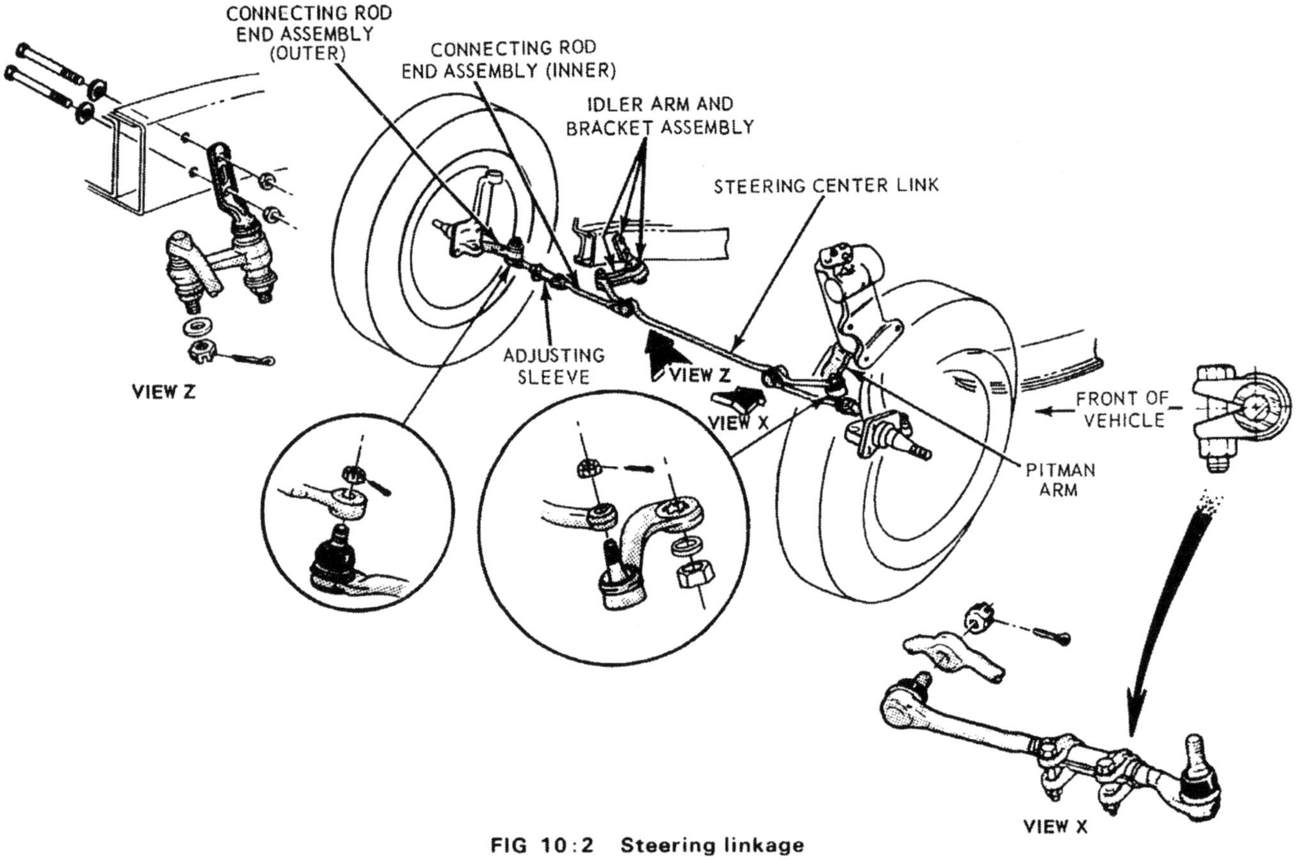

FIG 10:2 Steering linkage

and one ball joint have a lefthand thread while the other end is threaded with a righthand thread. The effective length of the connecting spindle can therefore be altered to set the front wheel alignment, or to adjust the position of the steering wheel, as described in **Section 10:7**. Clamps secure the threaded sleeve to the ball joint ends to prevent the adjustment from altering. The ball joints are non-adjustable and non-repairable so if one is found to be defective the ball joint and the end or lever to which it is attached must be renewed. Check for play by having an assistant gently turn the steering backwards and forwards, with the weight of the vehicle on the road wheels, and watching the linkage as it moves.

Freeing ball joints:

The safest and best way of disconnecting a ball joint is to use the special tool shown in **FIG 10:4**. Extract the cotterpin and remove the nut that secures the ball joint Fit the tool as shown and by tightening the bolt of the tool the pin of the ball joint will be pressed out.

If the special tool is not available the ball joint can still be freed. **Never hammer on the end of the threaded portion of the pin, as even with a slave nut fitted to protect the threads internal damage will be caused to the ball joint.** Undo the nut that secures the ball joint to the last few threads. Lay a block of metal on one side of the tapered eye and pull the end with the ball joint firmly away from the other member. Hit firmly with a copper hammer on the side of the tapered eye opposite to the block of metal, and the tapers will quickly free. The nut will prevent the members flying apart and causing possible injury or damage.

The ball joint is reconnected by pressing the pin into the tapered eye by hand, and tightening the nut to the correct torque before locking it with a new cotterpin.

If a new spindle connecting rod end is being fitted, the alignment of the front wheels and setting of the steering wheel will not be unduly altered if the number of threads required to unscrew the old end counted and the new end screwed back by the same number of turns. However, it is still advisable to have the front wheel alignment checked by a service agent.

The sleeve itself must be renewed if it is damaged or bent and under no circumstances should an attempt be made to straighten a bent sleeve or rod.

Pitman arm:

The ball joint on the Pitman arm is disconnected as described previously. On models where the ball joint is integral with the Pitman arm, a new complete arm must be fitted if the ball joint is defective or worn.

When removing the Pitman arm from the steering it is essentail to use the type of extractor shown in FIG 10:5. Hammering on the sector shaft or Pitman arm will transmit shocks up the sector shaft which will damage the ball nut and parts of the steering unit.

Before the Pitman arm is fully removed from the sector shaft, make light alignment marks across the arm end of the shaft to ensure that the arm will be installed in its original position.

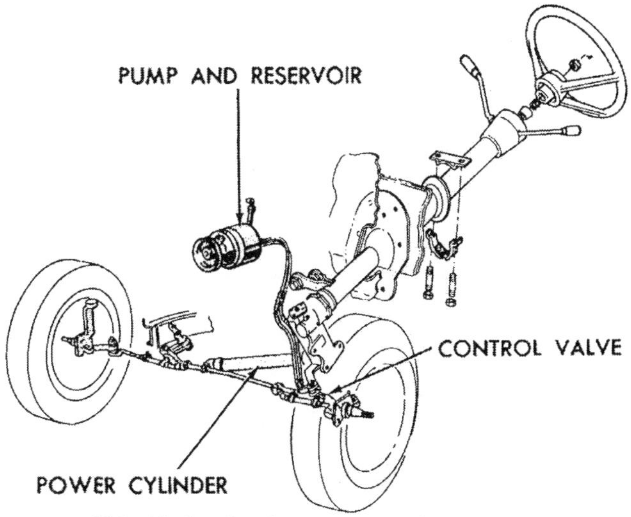

FIG 10:3 Ford power steering system

Idler arm:

If the idler arm ball joint or the pivot bush for the unit is worn a complete new assembly must be installed. Free the lever ball joint from the linkage and remove the bolts that secure the bracket to the frame. Install the new idler assembly in the reverse order of removal.

On 1965 and 1966 models the bushing that attaches the arm to the bracket is renewable but on all later models the complete assembly must be renewed.

10:4 Steering unit adjustments

There are only two adjustments that require to be made on the steering unit and they can be carried out, in the correct sequence, with the unit installed in the vehicle. The owner may find however that it is easier, and more accurate, to carry out the adjustments with the unit removed from the vehicle.

Lubrication:

This must be checked with the unit installed and the car on level ground. The adjustment points and filler plug are shown in **FIG 10:6**. Center the steering and remove the

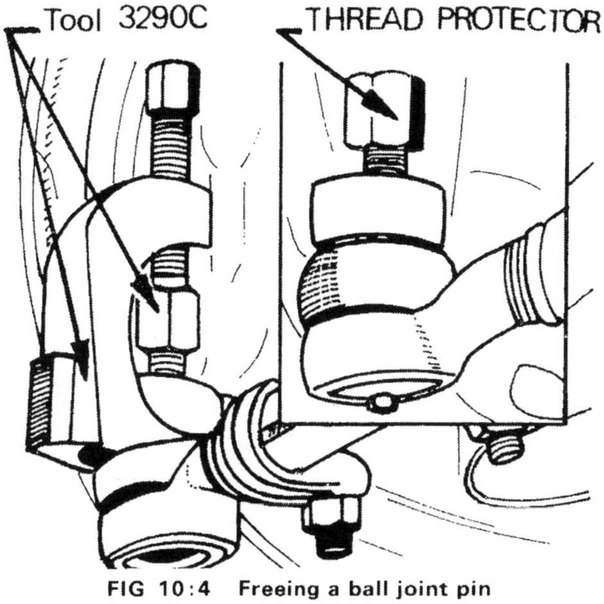

FIG 10:4 Freeing a ball joint pin

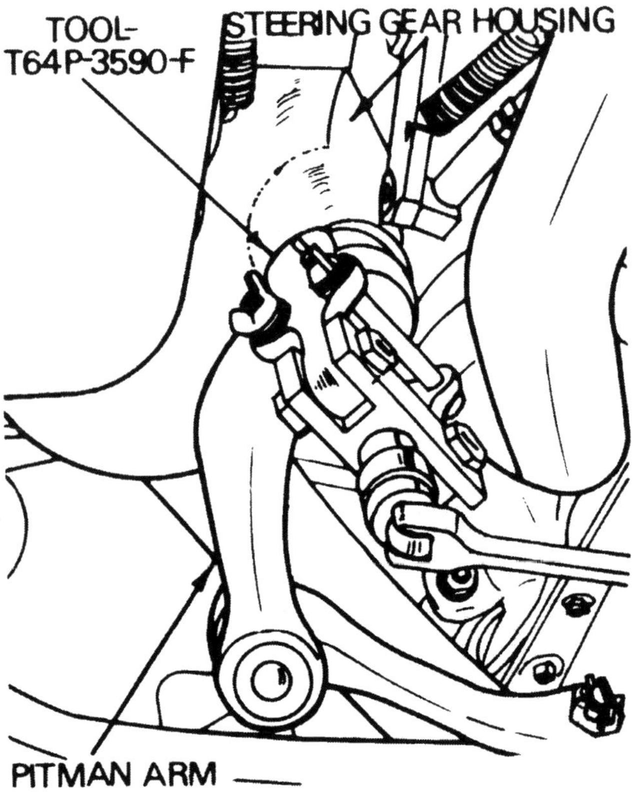

FIG 10:5 Freeing the Pitman arm from the steering unit

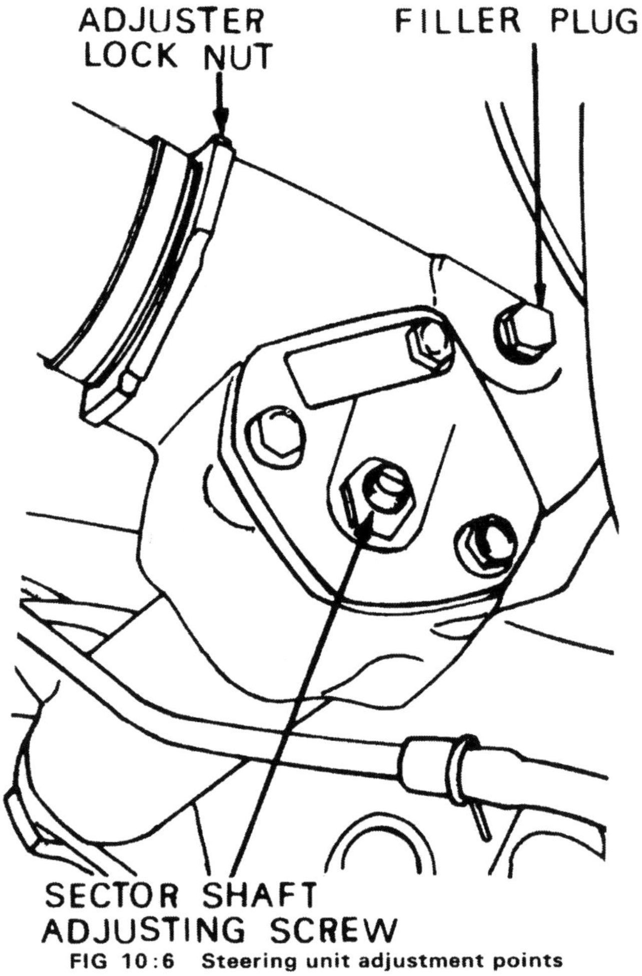

FIG 10:6 Steering unit adjustment points

filler plug as well as the lower housing-to-cover bolts. Use a clean piece of suitable rod or dowel to push all loose lubricant back into the unit so that the holes are clear. Slowly turn the steering to full left left lock and the lubricant should rise within the coverbolt hole. Slowly turn the steering to the opposite lock and the lubricant should rise in the filler hole. If the lubricant does not rise correctly in the holes, further lubricant should be added through the filler hole until the level is correct.

Steering column:

Remove the crash pad and sufficient hardware to expose the steering shaft nut that secures the steering wheel. Check the alignment of the shaft in the bearing. If the shaft is off-center, fit a spring scale to the end of the shaft and pull on the scale until the shaft is centered, noting the reading on the scale. If the pull required to center the shaft is less than 20 lb no action need be taken. If the pull is greater than 20 lb, the steering column must be realigned. A steering column out of alignment will cause the steering to be heavy.

Checking:

If the steering is stiff or has excessive play, the steering linkage, lubrication (of steering and suspension ball joints as well as the steering unit) and steering column should all be checked. If these are found to be satisfactory, adjust the steering unit. The steering shaft must be adjusted first and then the sector shaft.

Steering shaft:

Disconnect the Pitman arm from the sector shaft. Slacken the locknut on the sector adjusting screw and turn the adjusting screw anticlockwise by at least two full turns (see **FIG 10:6**).

Attach a torque wrench (measuring in lb inch) to the steering wheel attachment nut. Use the torque wrench to turn the steering approximately $1\frac{1}{2}$ turns either side of the center position. Slacken off the adjuster locknut for the steering shaft and turn the bearing adjuster until the correct torque load turns the steering (see **Technical Data**). Tightening the bearing adjuster will increase the torque required. When the bearing preload is correctly set, hold the adjuster and tighten the adjuster locknut. If the bearings feel rough or operate noisily, the steering unit must be dismantled for an examination of the parts.

Sector shaft adjustment:

This is only set after the steering shaft bearings have been adjusted. Turn the steering gently from full lock to lock while counting the turns required. Turn the steering back by half that number of turns so that it is centralized.

Use the torque spanner on the steering wheel nut to turn the steering. Turn the sector shaft adjustment screw clockwise until the correct torque is required to turn the steering (see **Technical Data**) through the center point. There should be no perceptible backlash with the steering 30 deg. either side of center. Hold the adjuster screw in position and tighten the locknut that secures it.

Install the hardware and crash pad to the steering wheel and install the Pitman arm in the steering unit.

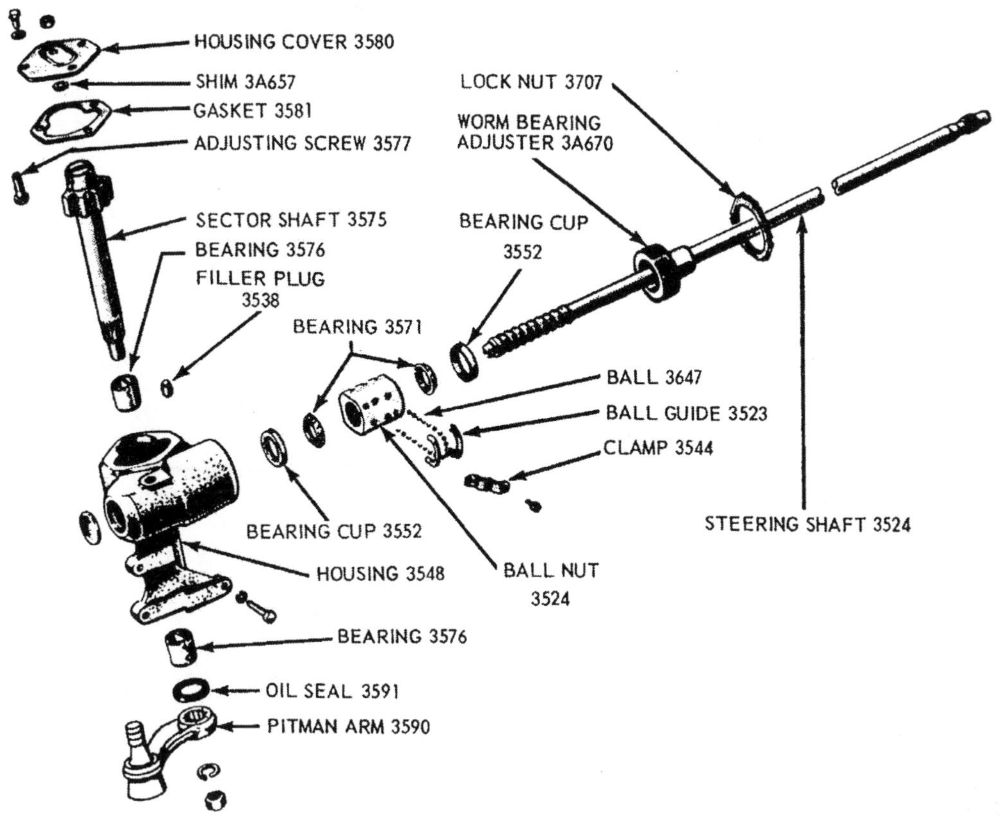

FIG 10:7 Steering unit components

10:5 Steering unit servicing

The components of the unit are shown in **FIG 10:7**.

Strict cleanliness is essential when reassembling the parts. All parts should be washed in suitable solvent and then lubricated before assembly.

1 Turn the sector shaft gently from stop to stop and count the turns required. Turn the sector shaft back by half that number of turns to set it in the central position. Remove the locknut for the sector shaft adjuster. Remove the three bolts that secure the cover to the casing. Free the cover by turning the adjuster screw clockwise until it winds out of the cover. Remove the adjuster screw from the sector shaft, collecting the shim that is with it. Withdraw the sector shaft from the housing.

2 Slacken the locknut for the bearing adjuster nut. Unscrew the adjuster nut with the locknut on it. Remove the steering shaft upper bearing assembly. Carefully withdraw the steering shaft assembly from the casing. **Do not allow the ball nut to run down the spiral as the ball guides and balls will be damaged when the nut reaches the end of the spiral.**

3 Clean the steering shaft and ballnut assembly by washing them in solvent. Rotate the shaft gently in the nut to wash the internal parts of the nut, taking care not to force it to the end of the spiral. Dry the parts with compressed air and lubricate them with thin oil. Rotate the shaft and check for stiffness or binding. If the shaft rotates freely there is no need to dismantle the parts. Only if there is stiffness or binding should the parts be dismantled.

4 If the sector shaft bearing and oil seal show signs of damage or wear they should be pressed out of the housing and new parts pressed back into place. If the bearing has rotated in the housing, a new housing must be used. If the bearing and oil seal are satisfactory, leave them in place but avoid washing the oil seal with any solvent.

Ball nut assembly:

If there is stiffness or binding in rotation, dismantle the parts and check them for wear or roughness. Hold the ball nut so that the ball guide tubes are uppermost. When dismantling the nut, hold the unit over a container with a pad of cloth in the bottom, as the pad will prevent the balls bouncing out as they fall. Remove the clamp and withdraw the two ball guides, taking care not to lose the balls from the guides. Turn the nut over and rotate the shaft in it until all the balls come out of the nut. The number of balls varies with type but there is always an even number.

Fit the nut back onto the shaft and install the ball guides, leaving the clamp off at this stage. Divide the number of balls into two even amounts, noting that each half of the set is fitted through one ball guide. Hold the nut with the ball guides uppermost and feed the balls back through the hole in each ball guide. Rotate the shaft backwards and forwards to distribute the balls in the circuit. When all the balls are in place, fit the clamp to secure the ball guides and cover the holes in the balls guides. Lubricate the nut with the correct lubricant (C3AZ.19578.A).

Assembly:

Check all the parts for wear or damage, renewing any that are defective. If there is difficulty in removing the cup for the lower bearing, tap the housing onto a clean piece

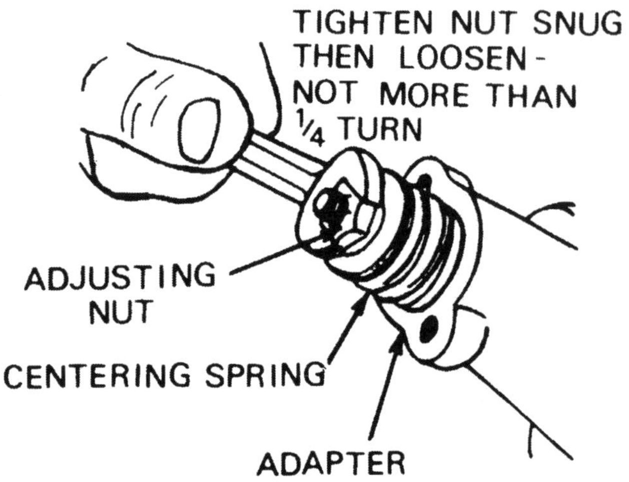

FIG 10:8 Centering the power steering valve spring

of wood to dislodge the cup. Renew both bearings if they are worn or show signs of fretting or breaking up. Lubricate the oil seal lightly. Install a new seal on the bearing adjuster nut if the old one is leaking or defective.

1 Install the lower bearing cup back into the housing. Lubricate all the bearings and install both bearings onto the steering shaft. Carefully slide the shaft and steering nut assembly back into the housing.

2 Lightly smear the threads of the bearing adjuster with sealing compound. Do not allow any sealing compound onto the female threads in the housing and do not allow any onto the bearings themselves. Clamp the housing gently in the jaws of a vice so that the sector shaft axis is horizontal and the splined end of the steering shaft is vertically upwards. Install the bearing cup, adjuster with oil seal and finally the locknut with the marking 'S' outwards. Leave the locknut slack and attach a torque wrench and adaptor to the splines of the steering shaft. Set the adjuster until the correct torque load is required to turn the shaft (see **Technical Data**). Hold the adjuster and tighten the locknut.

3 Fit the sector shaft adjuster and its shim to the sector shaft and use feeler gauges to check the clearance. If the clearance between the adjuster and the shaft with the shim in place exceeds .002 inch, then a new shim must be fitted to reduce the clearance below this limit.

4 Fill the unit with the correct amount of lubricant. Partially thread the adjuster back into the cover and then attach the sector shaft and adjusting shim to the adjuster. Rotate the steering shaft until the ball nut is in the correct position for meshing with the sector teeth. Lower the sector shaft, with cover and adjuster, down into the housing so that the teeth mesh. Lightly smear the threads of the attachment bolts with sealing compound and use them to lightly secure the cover, without tightening at this stage. Turn the sector adjuster anticlockwise until there is end clearance and then fully tighten the cover bolts. Set the mesh adjuster until the correct torque is required to turn the steering shaft, and then lock the adjuster with its locknut.

10:6 The power steering

Ford non-integral:

The valve is fitted with a centering spring which requires a force of approximately 4 lbs at the steering wheel to move. Once the spring moves, the valve spindle can also move and the power steering operate. This ensures that there is always some 'feel' in the steering. A check valve is fitted between the pressure and return lines so that when excessive pressure is reached, the valve opens and limits the pressure in the system.

Adjusting centering spring:

An incorrectly adjusted centering spring can cause stiff steering. The method of adjustment, after removal of the cover, is shown in **FIG 10:8**. When carrying out the adjustment do not overtighten the nut otherwise the stop pin may break at the travel regulator stop. Install the cover and apply a little silicone grease at the sealing areas.

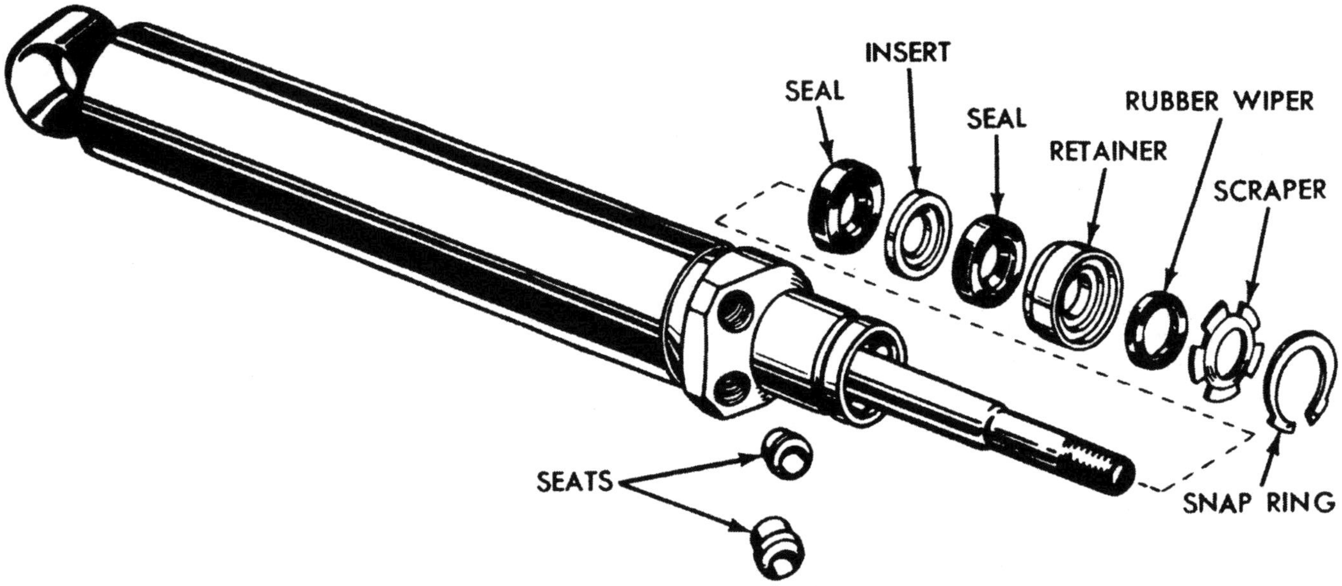

FIG 10:9 Ford power steering cylinder and seals

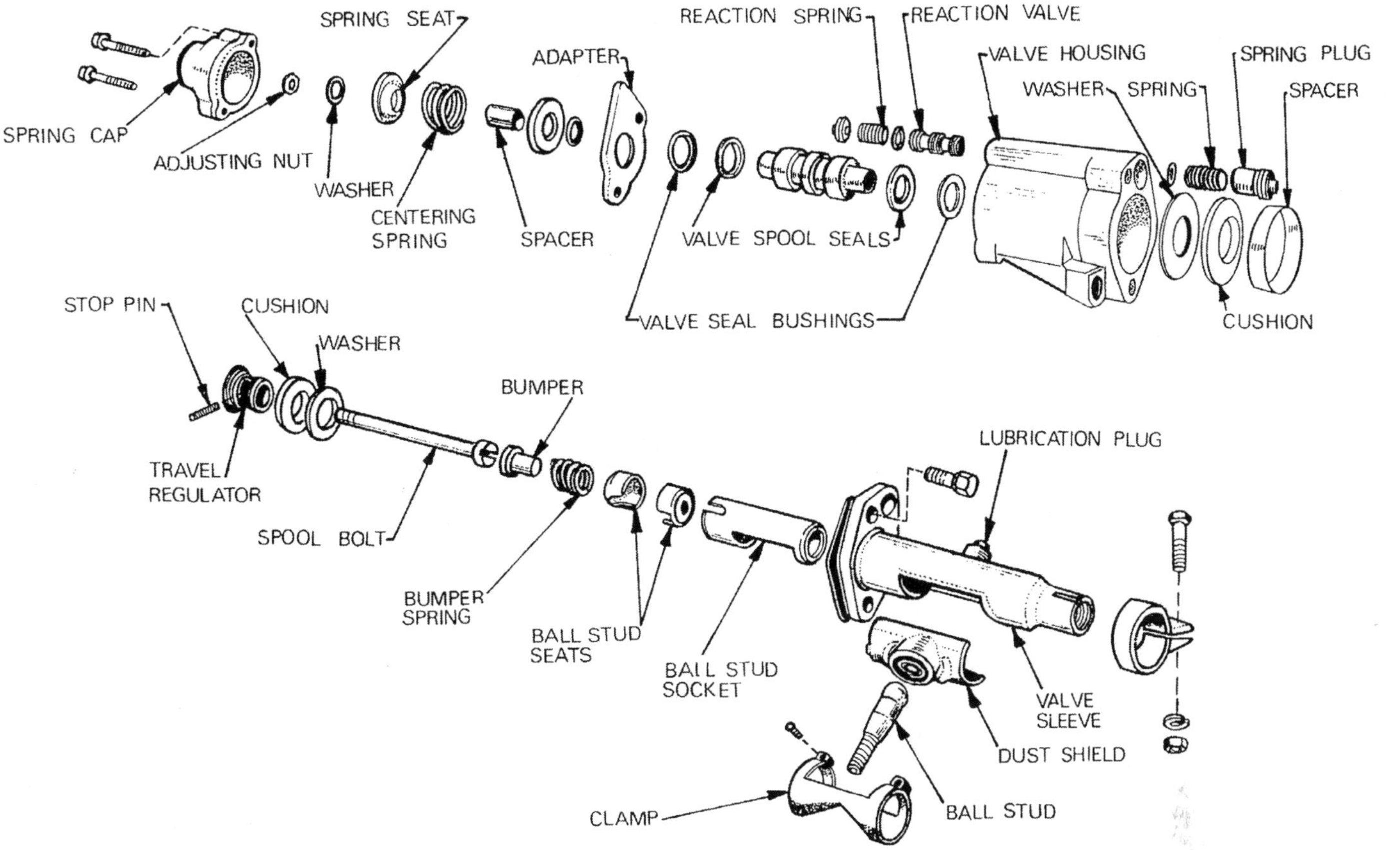

FIG 10:10 Ford power steering control valve

Pump:

A defective pump cannot be repaired and a new one must be fitted in its place.

A tachometer, pressure gauge, T-piece and shut-off valve are required to check the operation of the pump. Disconnect the pressure line from the pump and reconnect it with the T-piece and shut-off valve, with shut-off valve downstream of the T-piece and the pressure gauge on the third arm. Make sure that the shut-off valve is open. Start the engine and top-up to replace any fluid loss. Cycle the steering to expel any trapped air and then top-up to the correct level again. Leave the steering alone and the engine running at curb-idle. The pressure in the system should be approximately 50 lb/sq inch.

Speed the engine up to 1000 rev/min and close the shut-off valve. Note the maximum pressure and open the shut-off valve within **5 seconds. If the pump is left running longer with the shut-off valve closed the fluid in the pump will overheat and cause damage to the pump.** The pressure with the valve closed should be a minimum pf 750 lb/sq inch. A low pressure will be caused by a defective pump, though belt slip can also be a cause. If the pressure is above normal but the steering is stiff the fault is most likely to be in the valve rather than the pump.

Power cylinder:

The components of the power cylinder are shown in **FIG 10:9**. New seals are available in kits of spares. When installing new seals use a deep socket or suitable piece of tubing to press the seals back into position.

To remove the power cylinder, first disconnect the hydraulic lines. It is most advisable to mark the lines when disconnecting them as if they are incorrectly reconnected the steering wheel is liable to kick back with considerable force when it is turned. Catch the fluid as it drains in a suitable container. Remove the Pal nut and nut which secures the cylinder rod to the frame. Take off the washer and insulator to free the rod. Remove the castellated nut, after extracting the cotterpin, that secures the cylinder to the center link. Use the extractor No. 64P.3590.F to free the cylinder from the center link.

Install the cylinder in the reverse order of removal. Top-up the system and exercise it to bleed out air.

Control valve:

The components of the control valve are shown in **FIG 10:10**. The valve can be dismantled for renewal of the seals and deburring of the parts. If there is a major defect a new valve should be installed. If on disassembly the valve spool is found to be sticking, it can be smoothed down with Crocus paper, but take great care not to round off the sharp edges of the spool.

To remove the unit, disconnect all four hoses from it. Label the hoses to ensure that they will be correctly reconnected. Turn the steering gently fully left and right to force out the fluid. Slacken the nut and bolt on the clamp on the righthand side of the valve. Remove the rollpin that acts as a stop for the valve, as shown in **FIG 10:11**. Remove the nut and then free the ballpin as shown in **FIG 10:12**. Turn the steering fully to the left and unscrew the valve assembly.

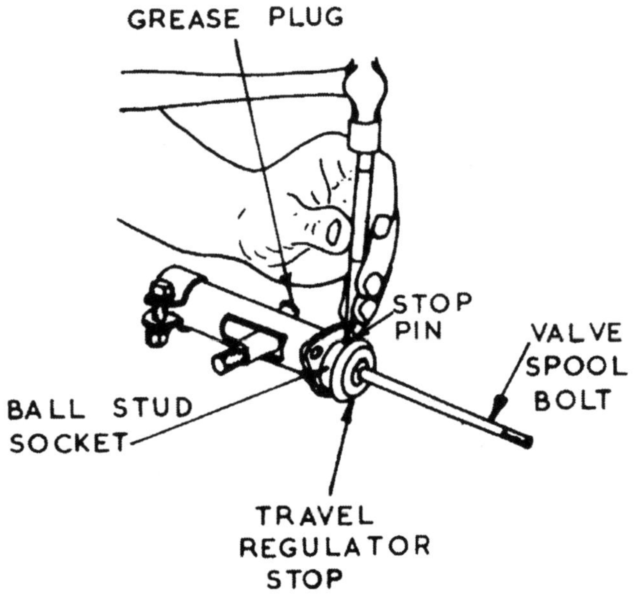

FIG 10:11 Removing the stop from the valve

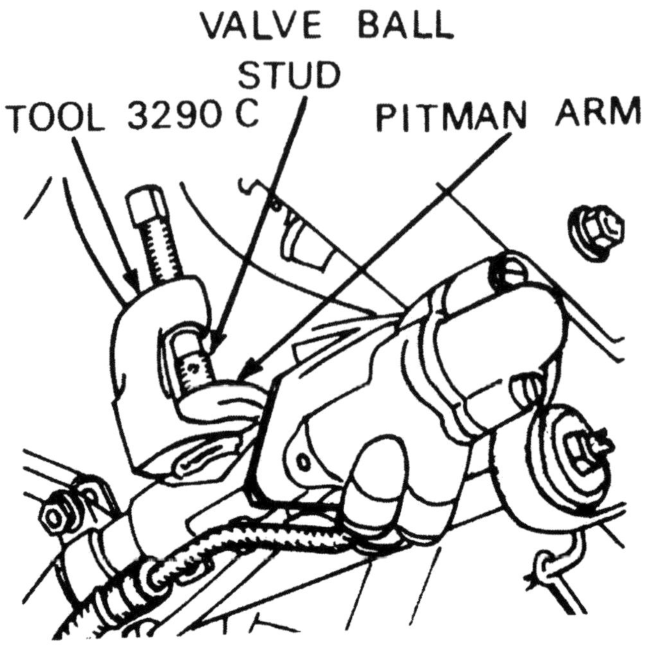

FIG 10:12 Removing the control valve ball stud

Fit the valve back into position and turn it until approximately 4 threads are left visible and then position the ball stud. Measure the distance between the center of the grease plug and the center of the stud, as shown in **FIG 10:13**. Screw the valve in or out until the dimension is correct at $5\frac{5}{8}$ inch (on 1965 models the dimensions should be set between $4\frac{1}{4}$ and $4\frac{3}{8}$ inch). When the dimension is correct, align the hole in the steering arm to the idler arm rod with the slot in the sleeve and install a new rollpin to lock the part. Tighten the clamp and install the remainder of the parts in the reverse order of removal. Fill the reservoir to the correct level and top up as required after cycling the system and checking for leaks.

Saginaw integral:

The only adjustment that should be performed on this system is to eliminate excessive lash between the sector shaft and rack teeth.

1 Disconnect the Pitman arm from the sector shaft and take off the steering wheel hub.
2 Uncouple the fluid return line at the reservoir, and cap the reservoir return line pipe. Place the end of the return line in a clean container and turn the steering wheel in both directions to discharge the fluid from the gear.
3 Turn the wheel, either way, a half turn off-center. Using a 24 lb inch torque wrench on the steering wheel nut, rotate the shaft slowly through a 20 deg. arc and check the torque. Make a note of the reading.
4 Turn the wheel back to the central position and check the torque again through a 20 deg. arc. The torque reading should be 6 lb inch greater than the previous reading. If not, refer to **FIG 10:14** and slacken off the locknut. With a $\frac{7}{32}$ inch Allen wrench, turn the screw inwards and recheck the torque in the off-center and central positions. Continue to tighten the screw until the central position is equal to 6 lb inch in excess of the off-set position.
5 Reconnect the Pitman arm and replace the steering wheel hub. Connect the fluid return line to the reservoir and fill the reservoir with C1AZ.19582.A fluid to the proper level.

Removing the power unit:

1 Disconnect the pressure and return lines from the steering gear. Plug the pipelines to prevent the entry of dirt. Remove the two bolts which secure the flexible coupling to the steering column.
2 Raise the vehicle and unscrew and remove the Pitman arm retaining nut, collect the lockwasher. To withdraw the Pitman arm from the sector shaft obtain a suitable puller or use puller T64P.3590.F. Do not hammer on the end of the puller in an attempt to help the removal of the Pitman arm as damage may occur to the gear or the cover.
3 If the car is fitted with manual transmission remove the clutch release lever retracting spring to provide clearance for the removal of the steering gear.
4 Support the steering gear and remove the three steering gear attaching bolts. Work the assembly to free the flexible coupling off the column and withdraw the steering gear assembly.

Refitting:

5 Fit the flexible coupling in position on the steering shaft. Turn the steering wheel to bring the spokes in a perfect horizontal position. Center the steering gear input shaft.
6 Slide the input shaft into the flexible coupling and locate the assembly in position in the frame side rail. Fit the attaching bolts and tighten them to the correct torque.
7 Ensure that the road wheels are in the straightahead, position then fit the Pitman arm on the sector shaft, tighten the retaining nut to the correct torque.

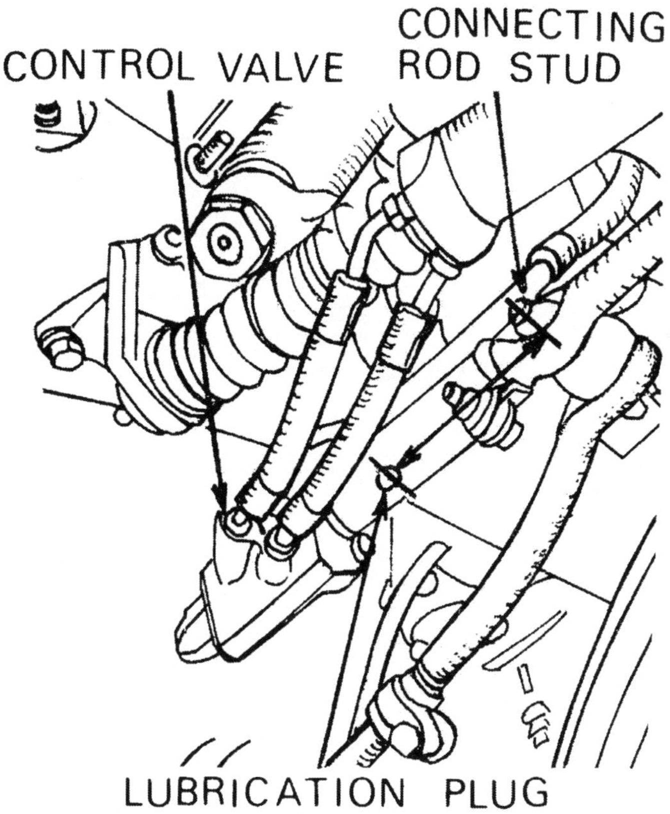

FIG 10:13 Installing the control valve

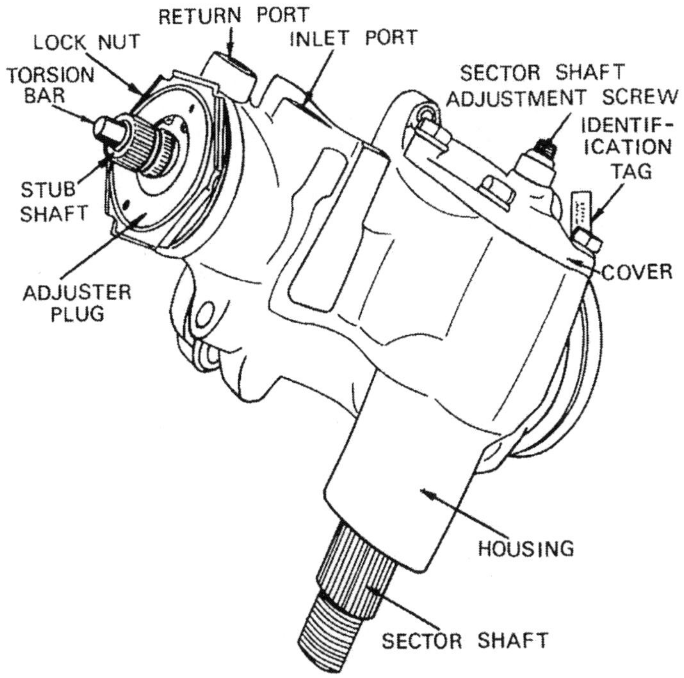

FIG 10:14 Saginaw power steering unit

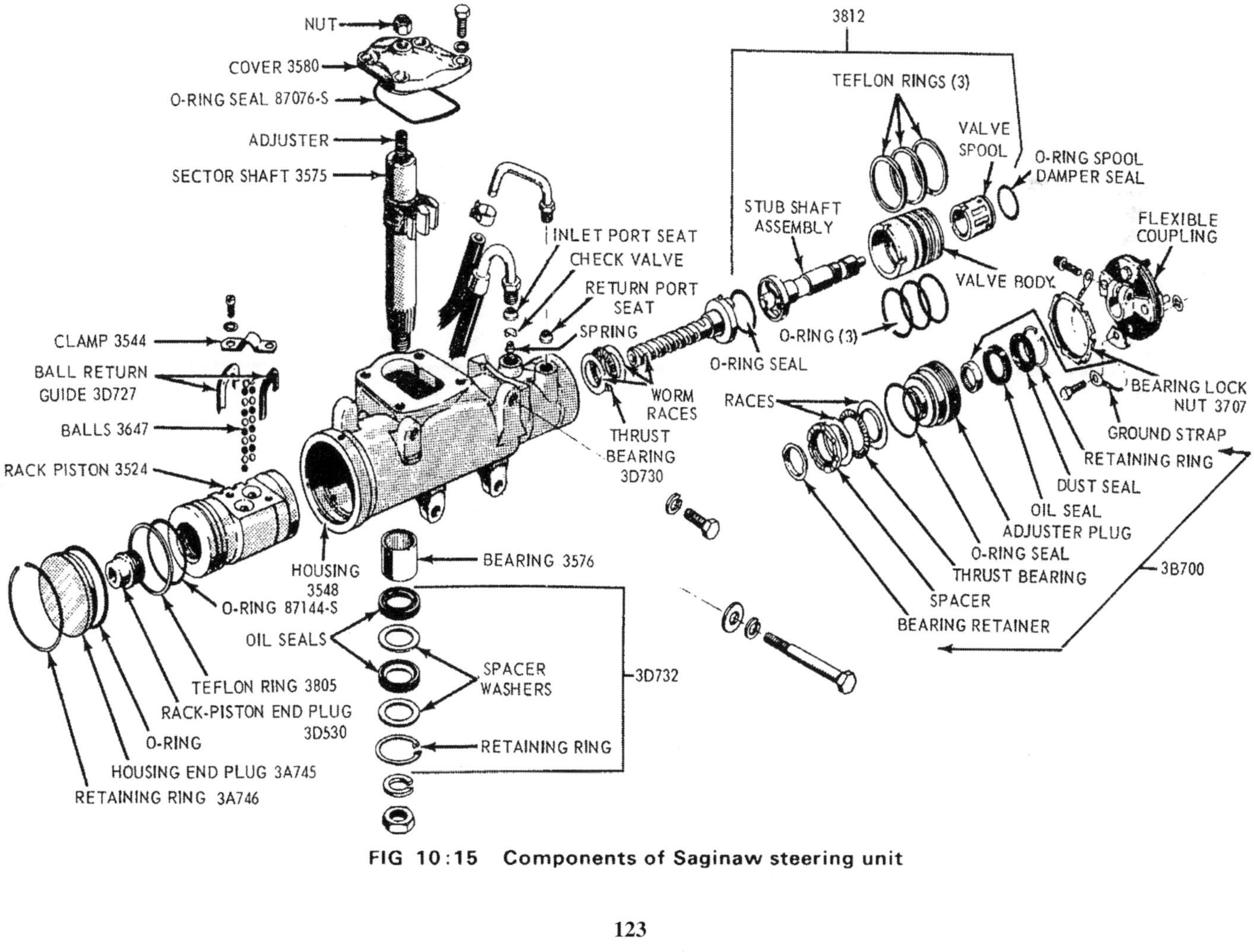

FIG 10:15 Components of Saginaw steering unit

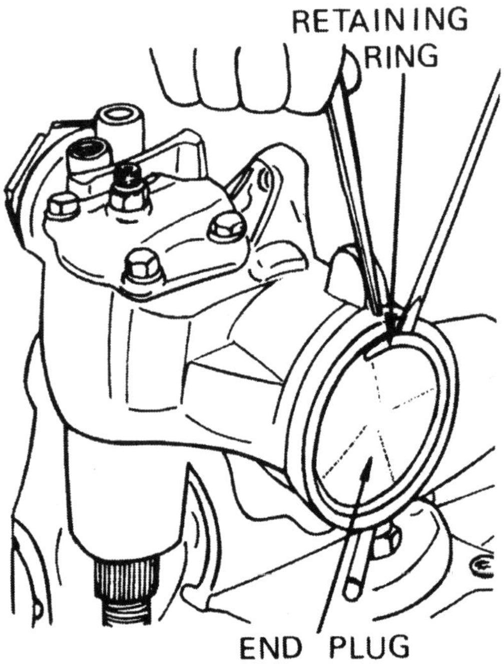

FIG 10:16 **Removing the end plug retaining ring**

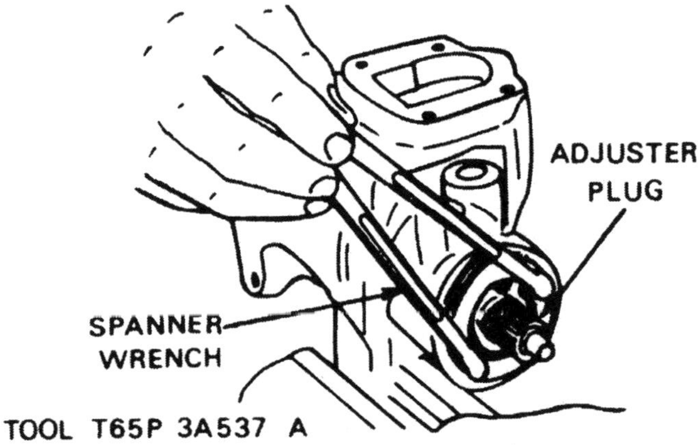

FIG 10:17 **Unscrewing adjuster plug with special tool**

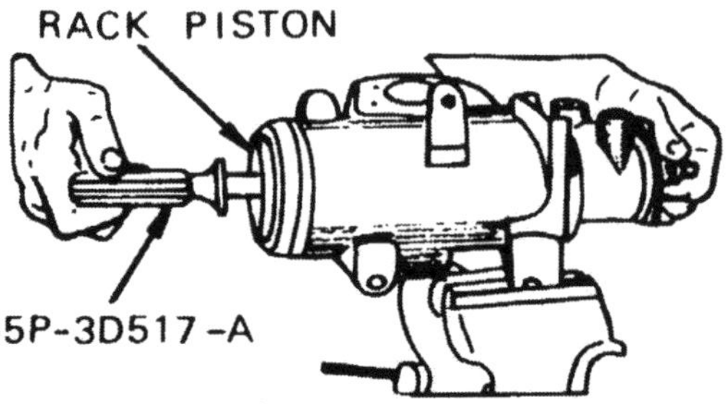

FIG 10:18 **Removing rack/piston**

8 Secure the flexible coupling to the input shaft and the steering column. Reconnect the fluid pipes to the steering gear.

9 Disconnect the ignition coil wire to prevent the engine from starting and fill the pump reservoir with C1.19582.A fluid. Crank the engine with the starter and continue adding fluid until the level remains constant. Whilst cranking the engine, turn the steering wheel approximately 30 deg. each side of the center position.

10 Recheck the fluid level and top-up as required. Reconnect the coil wire, start the engine and allow it to run for several minutes. Stop the engine and recheck the level of fluid. Top-up as required.

Servicing the steering gear (see FIG 10:15):

1 Remove the steering gear as previously described and mount it in a vise. Refer to **FIG 10:16** and rotate the end plug retaining ring to bring one end opposite the probe hole in the housing. With a probe through the hole, unseat the ring and work it out of its groove with a screwdriver.

2 Rotate the input shaft in an anticlockwise direction, using a $\frac{3}{4}$ inch 12-point socket wrench, to force the end plug out of the housing. Do not rotate the shaft more than is necessary to effect the removal of the plug, as further rotation will allow the balls to fall out and the rack/piston to disengage itself from the sector shaft.

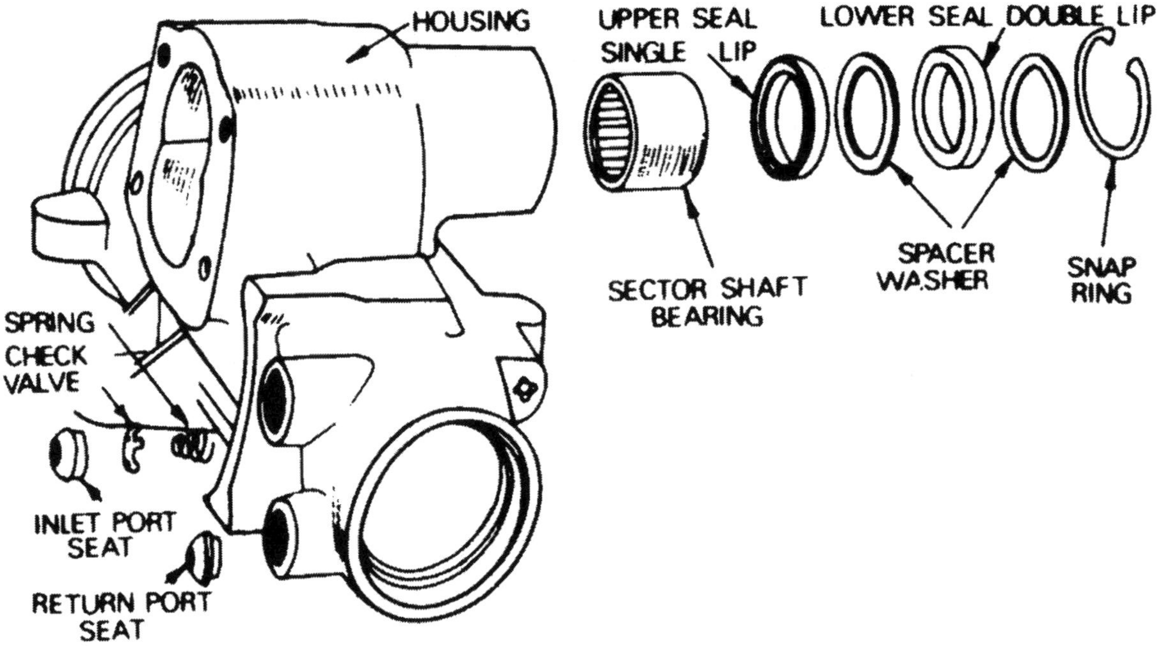

FIG 10:19 **Sector shaft bearing and seals**

Remove the O-ring from the housing and discard it.

3 Draw the piston inward, approximately a half turn on the input shaft, and turn the end plug anticlockwise out of the piston using a $\frac{1}{2}$ inch drive extension.

4 Unscrew the locknut from the sector shaft adjustment screw and discard it. Remove the four retaining set-screws from the cover and rotate the adjusting screw with an Allen wrench until the cover is free from the housing. Lift off the cover and its seal. Discard the seal.

5 Using a soft-faced hammer, tap the end of the sector shaft to remove it from the housing, make sure the sector shaft gearteeth are central to the cover aperture and do not bind on the housing.

6 Refer to **FIG 10:17**, remove the adjuster plug locknut and with the special wrench T65P.3A537.A remove the adjuster plug.

7 Insert tool T65P.3D517.A into the end of the rack/piston, as shown in **FIG 10:18**, until it contacts the wormshaft. With the other hand rotate the stub shaft in an anticlockwise direction until the worm is free of the rack/piston. Carefully withdraw the rack/piston from the housing with the tool, make sure the tool is held all the way in the piston to prevent the balls from dropping out. Do not dismantle the rack/piston at this stage.

8 Withdraw the stub shaft and valve assembly from the housing. Extract the worm gear, lower thrust bearing and the races from the housing.

Servicing the housing (see FIG 10:15):

9 Refer to **FIG 10:19** and remove the snap ring from the housing and the lower spacer washer. Extract the lower seal, the upper spacer washer and the upper seal with tools T58L.101.A and T59L.100.B as shown in **FIG 10:20**.

10 Check the sector shaft bearing for wear or damage. If it is in good condition do not remove it, if not, press it out of the housing and discard it. To fit a new bearing, work from the upper end of the housing and press it into position until the upper end of the bearing is .030 inch below the housing bore, the identifying number of the bearing should face outwards.

11 Obtain new sector shaft seals and lubricate them with clean hydraulic fluid (C1AZ.19582.A). Install the seal with the single lip first followed by the upper spacer washer; using the special tool T65P.3D642.A as shown in **FIG 10:21** drive the seal and washer into the bore. Do not allow the seal to bottom in the bore but drive it in far enough to permit the fitting of the lower seal, spacer washer and snap ring. Install the double lip seal and the lower spacer washer and drive them into the bore only as far as is necessary to be able to fit the snap ring. Fit the snap ring and ensure that it is seating correctly.

12 Still referring to **FIG 10:19**, to remove a damaged port seat, obtain a $\frac{5}{16}$ inch 18 US thread course tap and coat it with petrolatum; this to prevent the chips from entering the port passage. Tap through the seat.

13 Obtain a small piece of tubing that has an inside diameter the same size or slightly larger than the seat orifice and approximately $\frac{1}{2}$ inch long. Obtain a bolt that has the same thread as the tapped hole, preferably one with a long thread, fit a nut onto it and a $\frac{5}{16}$ inch flat washer which has an overall diameter large enough to cover the open end of the tube. Thread the bolt into the seal, far enough for the bolt to have a firm grip, screw the nut and washer down until it abuts the tube and make sure the tube is central and there is sufficient clearance for the seal to be pulled through it. Tighten the nut until the seal is pulled out of its seating by the bolt (see **FIG 10:22**). Extract both seats by the same method and remove the check valve and spring from the inlet port.

To fit new seats, drive them into the bores as shown in **FIG 10:23** and do not forget to fit the check valve and spring into the inlet port.

14 A stop ring is fitted into the piston bore to limit the piston travel, do not attempt to remove this ring, if the ring is damaged the whole unit must be renewed.

Servicing the adjuster plug (see FIG 10:15):

15 Lift the adjuster plug off the stub shaft assembly and prise the thrust bearing retainer off the plug as shown in **FIG 10:24**. Take care not to score the bearing bore. Discard the retainer.

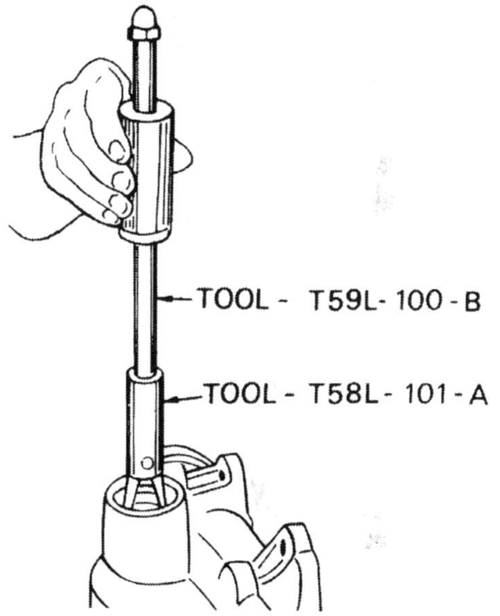

FIG 10:20 Removing sector shaft seals

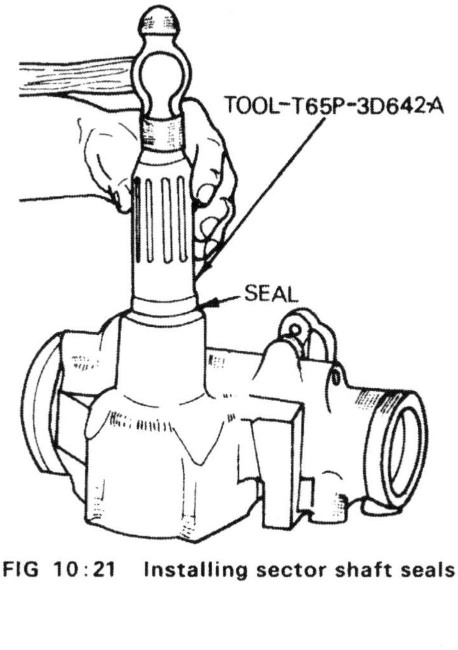

FIG 10:21 Installing sector shaft seals

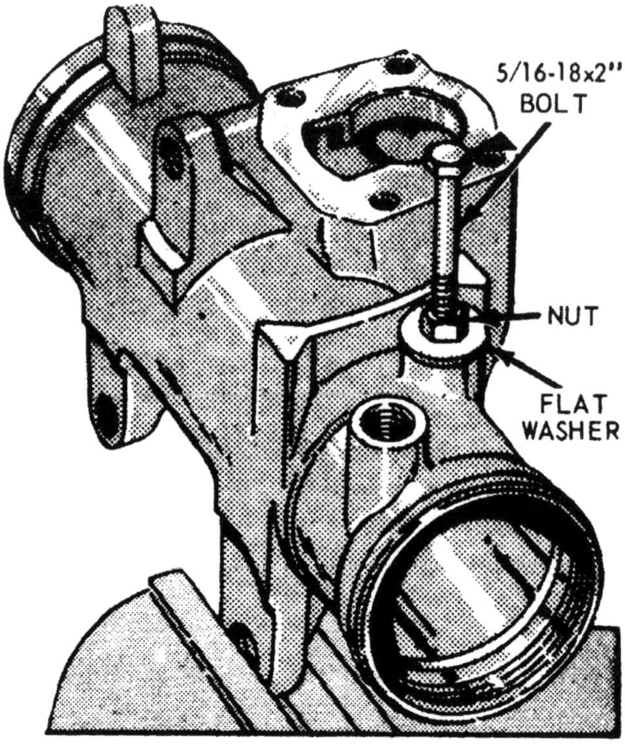

FIG 10:22 Removing port seat

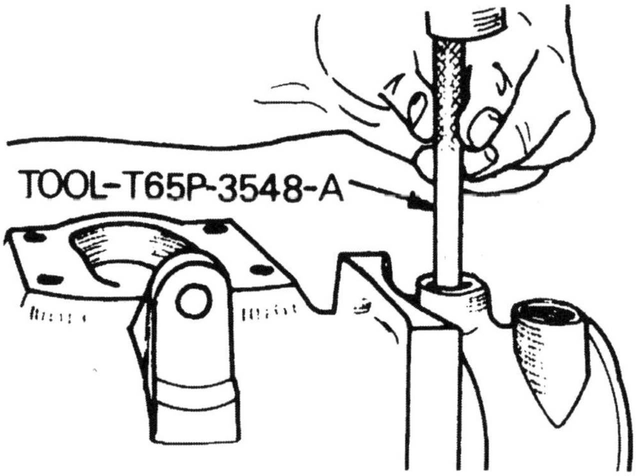

FIG 10:23 Installing port seat

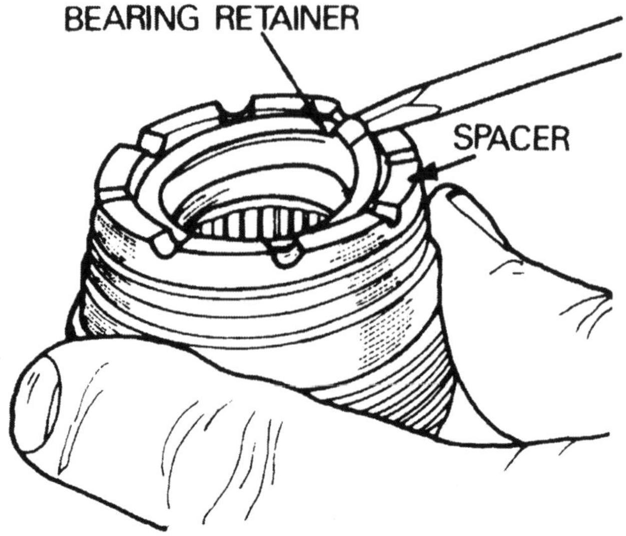

FIG 10:24 Removing thrust bearing retainer

16 Lift the spacer, thrust bearing, the races and the O-ring from the plug. Discard the O-ring (see **FIG 10:25**). Take out the retaining ring and extract the dust seal and oil seal. Discard both seals.

17 Examine the needle bearing for damage and wear, if it does not have any of these faults do not remove it. If it is unserviceable, remove it by drifting it out of the adjuster as shown in **FIG 10:26**. Fit the new bearing with the same tool and locate it with the manufacturers identification marks towards the tool. Drive the bearing into the bore until it is flush with the bottom surface of the stub shaft seal bore.

18 Lubricate a new oil seal with the correct hydraulic fluid and press it into position allowing sufficient clearance for the dust seal and snap ring. Lubricate the dust seal and fit it with the lips inwards, secure the seal with the snap ring and make sure it is seated correctly.

19 Lubricate the new O-ring with vaseline and install it on the adjuster plug, assemble the thrust bearing outer race, thrust bearing, inner race and spacer. Fit the new retainer, drifting it into position with a small punch.

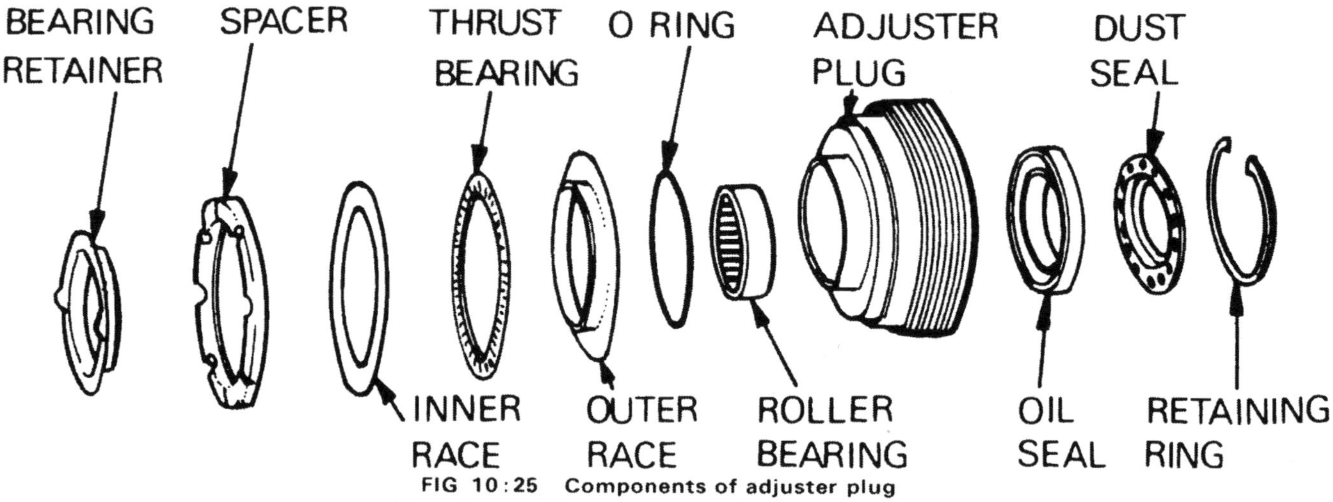

FIG 10:25 Components of adjuster plug

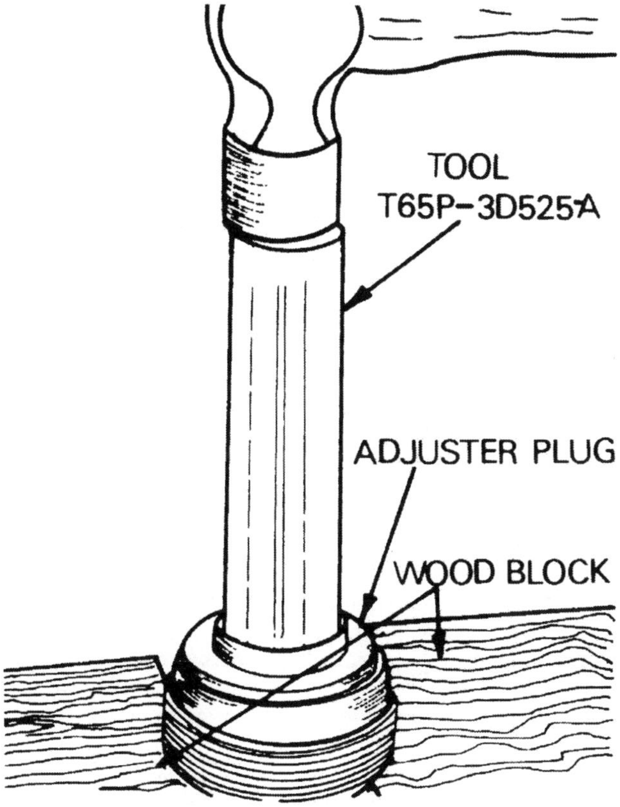

FIG 10:26 Removing or installing adjuster plug bearing

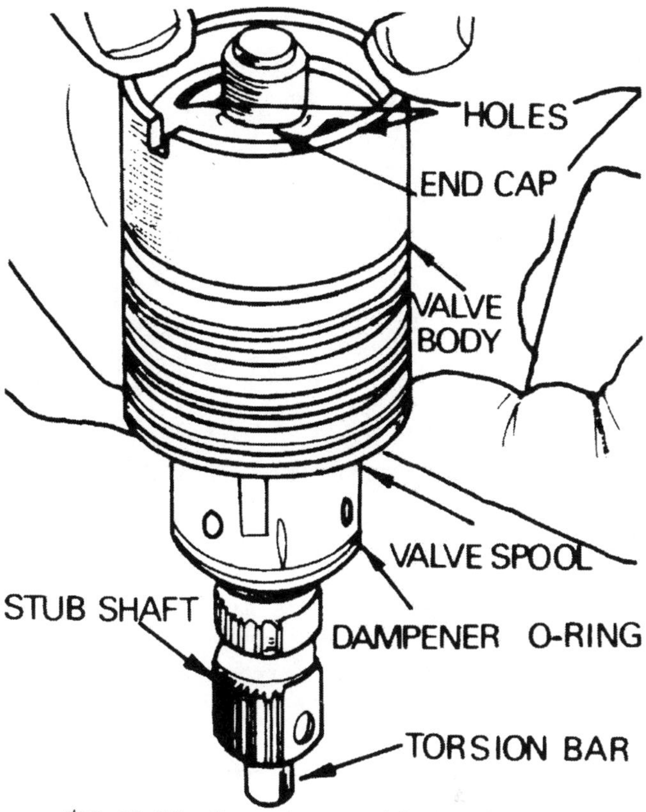

FIG 10:28 Removing spool from valve body

Servicing the stub shaft and valve assembly (see FIG 10:15):

20 The control valve and spool is a select fit and cannot be serviced except to remove the spool for the purpose of renewing the dampener O-ring, the teflon seals or the O-rings under the teflon seals.

21 Refer to **FIG 10:27** and remove the O-ring that is adjacent to the end cap. Discard the O-ring.

22 **Extreme care must be used when removing the valve spool**. The clearance between the valve spool and the valve body bore may be as low as .0004 inch and the slightest tilting of the spool may jam it in the valve body. To remove the spool, hold the valve assembly as shown in **FIG 10:28**, using a small rod through the openings of the end cap, push lightly on the spool until it is protruding from the valve and can be withdrawn by hand. Use a steady oscillating pull to prevent jamming. If slight sticking does occur, gently reverse the direction of the spool to free it and then try again. If serious jamming occurs, do not use force in the attempt to free the spool. First, partially remove the stub shaft assembly by the method shown in **FIG 10:28**, then tap the torsion bar lightly on the workbench to dislodge the cap from the valve body. The stub shaft assembly can now be removed from the body. By examining the spool in the valve body it

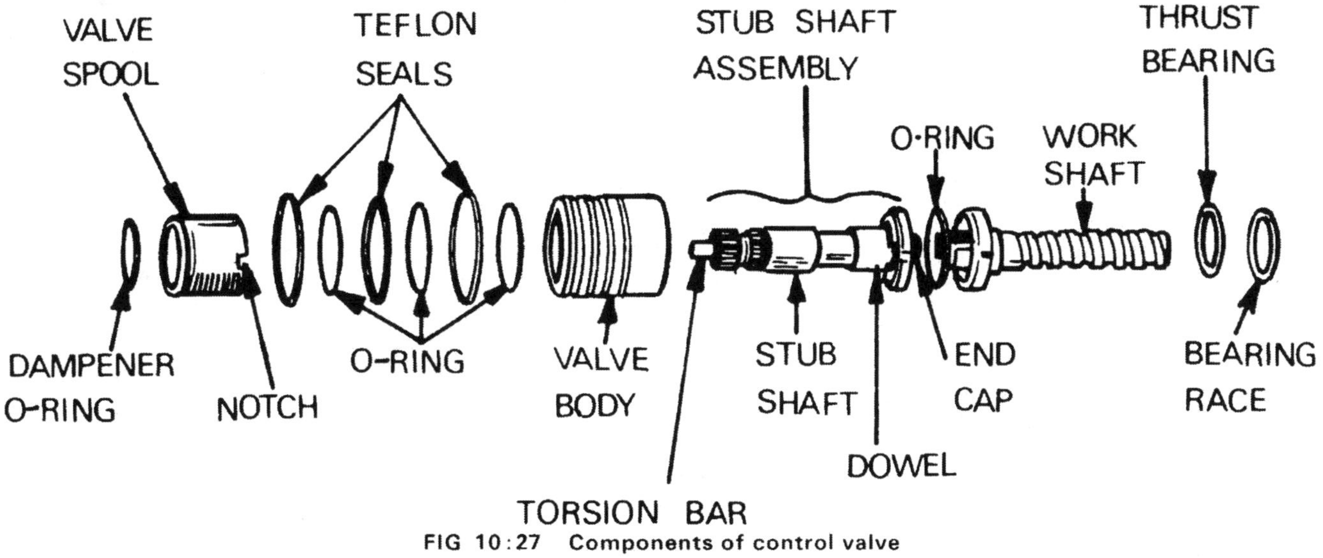

FIG 10:27 Components of control valve

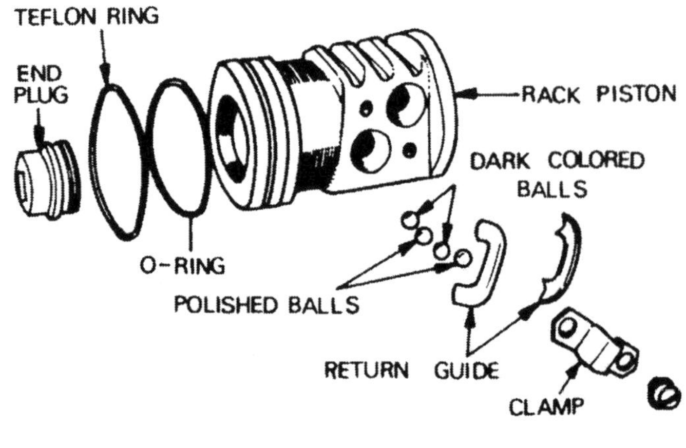

FIG 10:29 Components of rack/piston

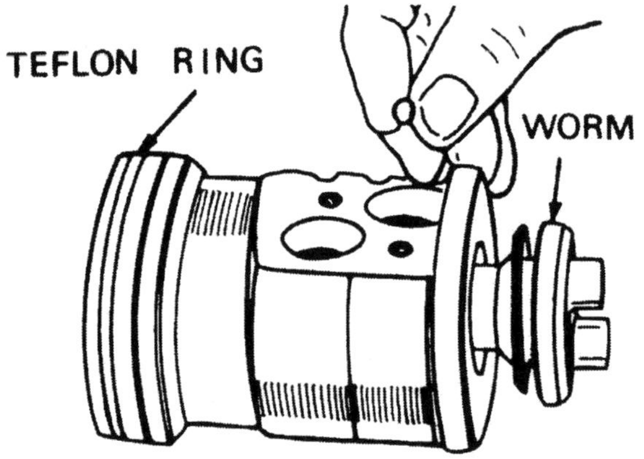

FIG 10:30 Assembling the worm balls into the rack/piston

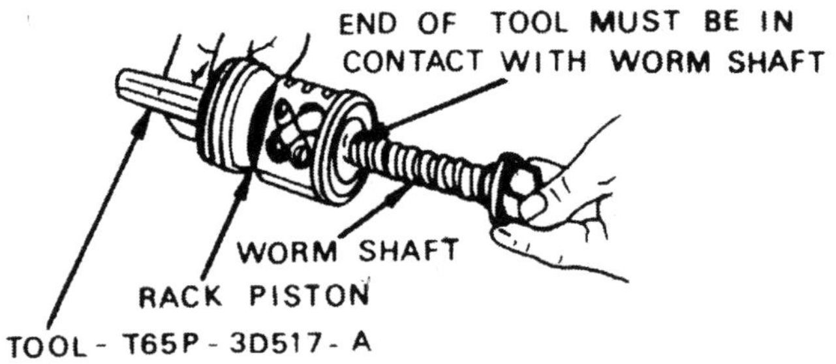

FIG 10:31 Fitting special tool into rack/piston

can be determined which way it is tilted. Align the spool with a few taps from a plastic or hide mallet, do not tap with anything metallic. When the spool can be rotated in the bore it can be removed. Remove the O-ring dampener seal from the spool and discard it.

23 The valve rings are made from teflon and it is most unusual if they require renewal, but if excessive wear is evident it will be necessary to cut the rings and the O-ring seals to remove them.

24 Lubricate new O-rings and teflon rings in the correct hydraulic fluid (C1AZ.19582.A) and fit them on the valve body, take care not to twist them and, when sliding the teflon rings over the O-rings, do not stretch them more than is necessary. Lubricate a new dampener O-ring and fit it into the groove on the spool.

25 Install the stub shaft in the valve body so that the groove in the cap engages with the pin in the valve body. Tap the cap lightly with a soft faced hammer to seat it correctly.

26 Dip the valve spool in clean hydraulic fluid and insert the notched end of the spool into the valve body and over the shaft, rotate it slightly to make the entry easier and ensure the dampener O-ring is distributed evenly. Guide the spool into position to engage the notch with the locating dowl in the body. **During the operation do not allow the stub shaft assembly to slip out of its engagement with the valve body pin, otherwise the spool will be permitted to enter the valve body too far and the dampener O-ring on the spool will engage in the valve body fluid grooves, which will prevent the removal of the spool.**

27 If the O-ring does get jammed in the fluid grooves, first try and withdraw it, by pulling hard with a lot of rotary motion. If this method fails place the valve body on a flat surface with the notched end upwards, tap the spool with a wood or plastic rod so that the O-ring is cut and the spool can be removed. Fit a new O-ring and proceed again with operation 26.

Servicing the rack/piston (see FIG 10:15):

28 Refer to **FIG 10:29** and cut the O-ring and teflon ring from the piston. Place the piston assembly on a piece of clean lint-free cloth and take out the two screws which secure the clamp to the ball return

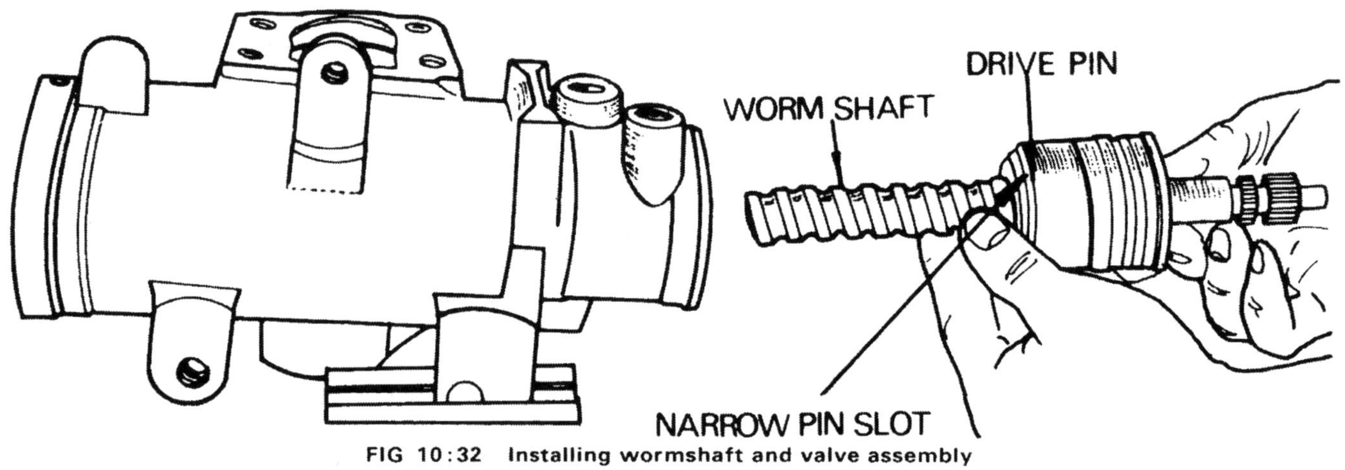

FIG 10:32 Installing wormshaft and valve assembly

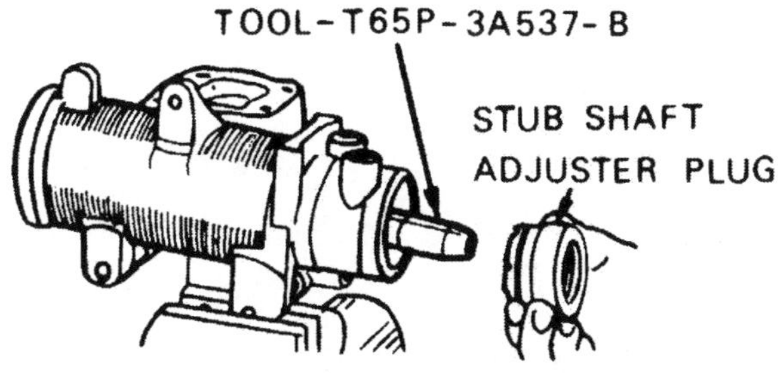

FIG 10:33 Fitting the adjuster plug

guides. Remove the special tool T65P.3D517.A (see operation 7) and lift the ball guides off the piston. Check that the ballbearings are complete, a constant ratio gear has 22 balls and a variable ratio has 24 balls.

29 Thoroughly clean all the parts.

30 Lubricate a new O-ring and teflon ring in the correct hydraulic fluid, slide the rings into position and avoid stretching them more than is necessary.

31 Insert the worm gear into the rack/piston, refer to **FIG 10:30** and load eight dark colored balls and eight bright balls **alternately** into the guide hole nearest the teflon ring. Rotate the worm gear anticlockwise whilst loading to feed the balls through the circuit.

32 Liberally coat one of the ball guides with petrolatum to secure the balls and place the remaining balls **alternately** into the guide ensuring that the end ball in the guide, is an alternative color to the adjacent one in the piston. Place the other half guide in position and insert them both into the holes in the piston. Fit the clamp and lock the screws, with lockwashers, to the correct torque.

33 The worm gear groove is ground with a high point in the center of the shaft and when the rack/piston passes over this high point there should be a preload of 1 to 4 lb/sq inch. To check the preload, position the rack/piston in a soft jawed vise with the worm shaft upwards. Do not tighten the vise too tightly otherwise there will be some distortion. Fit the valve and stub shaft assembly on the worm and engage the worm drive pins. Rotate the wormshaft until it extends from the rack/pinion to a dimension of $1\frac{1}{4}$ inch, measured from the wormshaft thrust bearing face. This is the center position. Place a $\frac{3}{4}$ inch 12-point socket over the stub

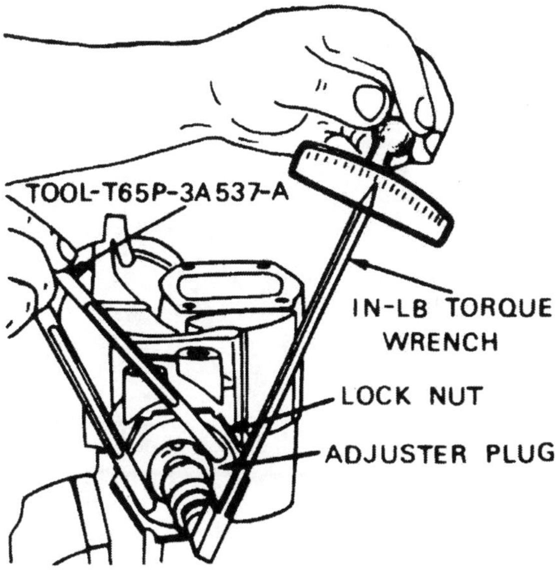

FIG 10:34 Adjusting the thrust preload

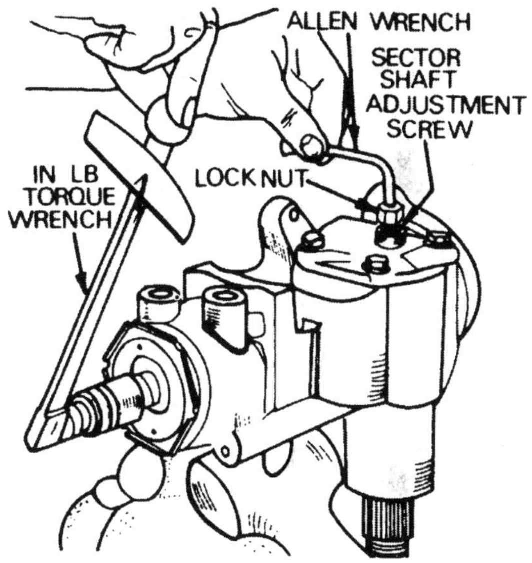

FIG 10:35 Adjusting sector shaft mesh load

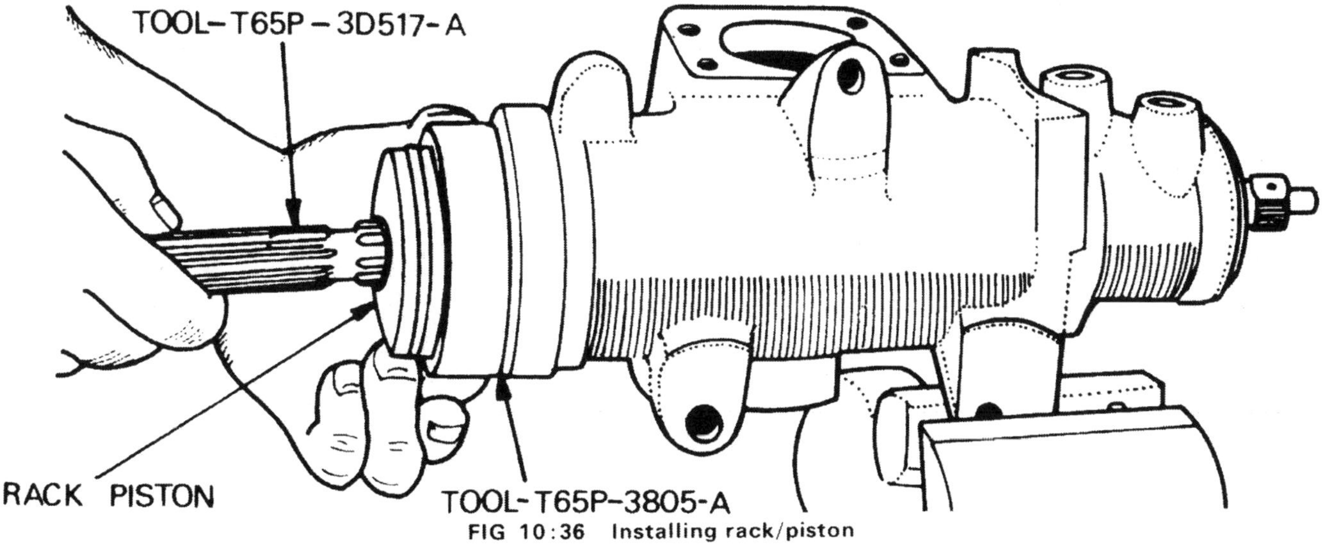

FIG 10:36 Installing rack/piston

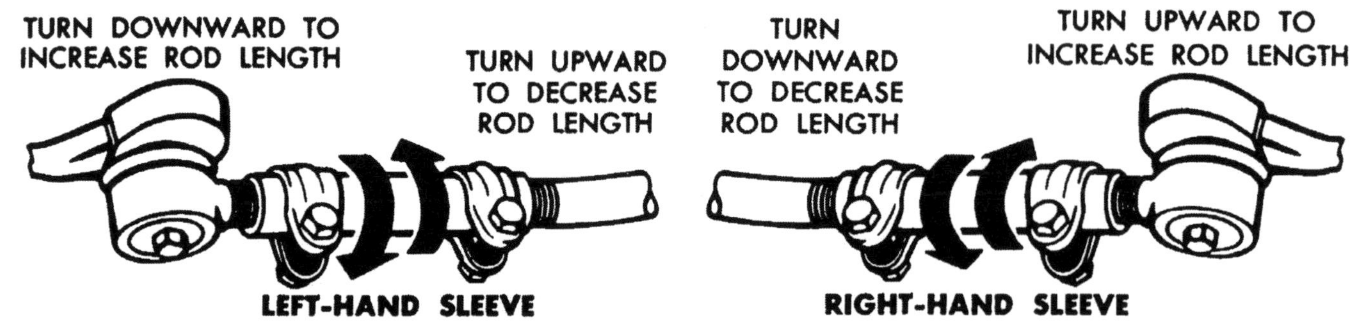

FIG 10:37 Adjusting the spindle connecting rods

shaft splines, fit the lb/sq inch torque wrench and rotate the wrench through an arc of approximately 60 deg. in both directions. Work the wrench several times before taking the highest average reading, which should be between 1 to 4 lb inch. If the reading is too high or too low a new set of balls must be fitted. The code number of the fitted balls is stamped on the rack/piston, the sizes of replacement balls and their code number is shown in **Technical Data**. Note the code number on the rack/pinion, refer to **Technical Data** and bear in mind that to increase the load a size larger balls must be fitted. If there is no code number on the rack/piston, the original code size was No. 7. A size larger ball will increase the preload by approximately 1 lb/sq inch.

34 When the preload is correct remove the valve assembly from the worm and the rack/piston from the vice. Insert tool T65P.3D517.A into the plug end of the rack/piston as shown in **FIG 10:31** ensuring that it makes contact with the wormshaft. Apply pressure on the tool whilst the wormshaft is rotated out of the piston/rack. Leave the tool in place until the piston is installed in the housing.

Assembling steering gear (see FIG 10:15):

35 Secure the steering gear housing in a vise, lubricate the wormshaft, the lower thrust bearing and the races with steering gear fluid and assemble them together in the correct order. Align the drive pin on the wormshaft head with the narrow pin slot in the valve body, but before locating the wormshaft head, fit the O-ring. Lubricate the valve teflon rings with petrolatum.

36 Position the valve assembly and wormshaft in the housing bore as a unit (see **FIG 10:32**). Install it by

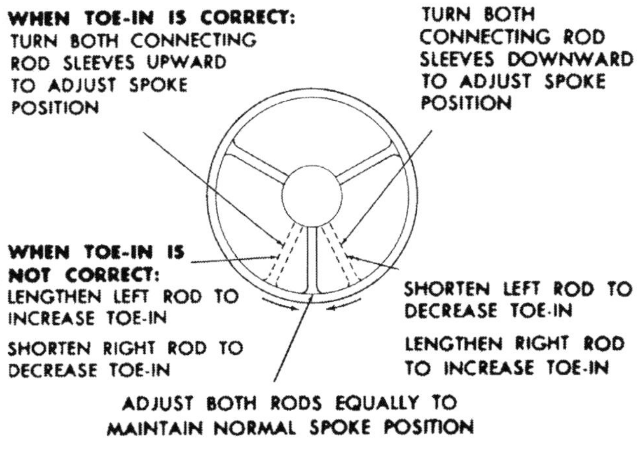

FIG 10:38 Setting the steering wheel position

pushing on the outer diameter of the valve body with the finger tips. Do not push on the stub shaft otherwise the adverse condition of spool jamming will occur as described in operation 26. Ensure that the teflon rings are not binding inside the bore. When the fluid return hole in the gear housing is fully visible the valve assembly is correctly seated.

37 Fit special tool T65P.3A537.B over the end of the stub shaft as shown in **FIG 10:33**. Lubricate the fitted adjuster plug O-ring with petrolatum. Install the adjuster plug over the tool and stub shaft, tightening it just sufficiently to ensure the internal parts are properly seated. Remove the special tool. Fit the adjuster plug locknut loosely on the plug.

38 Tighten the adjuster plug with special wrench T65P.3A537.A as shown in **FIG 10:34** and at the same time rotate the stub shaft. As soon as additional drag is felt on the stub shaft, back off the adjuster plug one eighth of a turn. Fit a $\frac{3}{4}$ inch 12-point socket and a torque wrench to the stub shaft and measure the drag. Make a note of the reading. Refer to **FIG 10:35** and tighten the adjuster plug again to obtain a 1 to 3 lb inch preload in excess of the previous reading. Tighten the locknut and recheck the thrust bearing preload. Total preload must not exceed 8 lb inch.

39 Refer to **FIG 10:36** and fit special tool T65P.3805.A in the housing. Pick up the rack/piston and the inserted special tool T65P.3D17.A, lubricate the teflon ring and O-ring with petrolatum and slide the rack/piston into the bore until it contacts the wormshaft. Maintain pressure on tool T65P.3D517.A, turn the stub shaft clockwise to engage the wormshaft gear in the rack/piston and continue turning until the center rack groove in the rack/piston is aligned with the sector shaft roller bearing. Remove the special tools.

40 Lubricate and fit a new cover seal to the cover, pick up the sector shaft and thread the cover onto the adjuster screw until it bottoms, thread it off one and a half turns. Install the sector shaft so that the center gear tooth meshes with the center rack groove of the rack/piston. Ensure the cover seal is in place, position the cover and secure it with the four screws. Tighten them to a torque of 30 to 35 lb ft. Loosley fit the adjuster locknut.

41 Fit the end plug into the rack/piston and tighten it to a torque of 50 to 100 lb ft. Lubricate the new O-ring and fit it into the housing, fit the end cap making sure it is seated against the O-ring and, if necessary, tap it lightly to accomplish this. Fit the retaining ring and make sure it is seated correctly in the housing.

42 Turn the stub shaft to the full left or full right position and then turn it back one and threequarter turns to

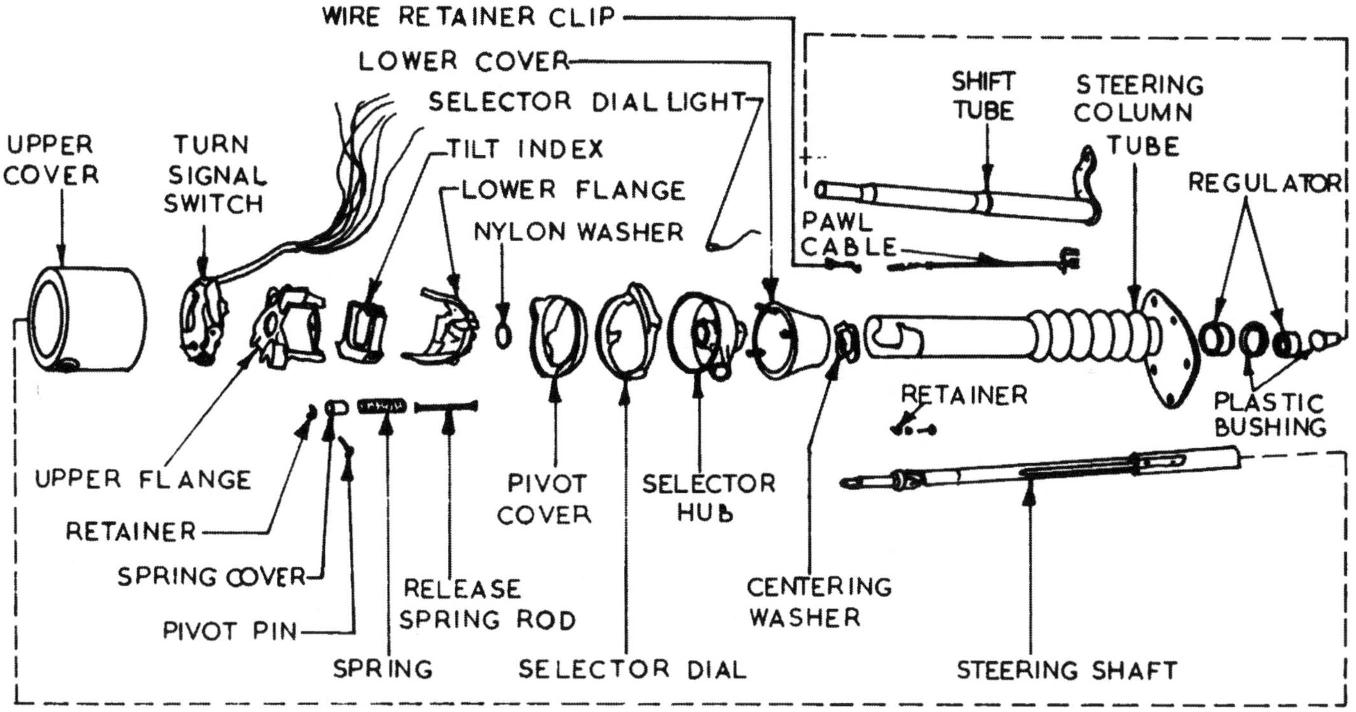

FIG 10:39 Typical steering column components

centralize it. Refer to **FIG 10:35** and fit a torque wrench to the stub shaft, measure the preload whilst tightening the adjusting screw with an Allen key. The preload to obtain must be 4 to 8 lb inch in excess of the total preload and drag that was obtained in operation 38. Readings are to be made through an arc not exceeding 20 deg. with the gear on center. Tighten the locknut to a torque of 30 to 35 lb ft.

43 Refit the steering gear assembly to the car as previously described.

10:7 Steering geometry

The front wheel alignment should be checked by a service agent using accurate gauges. Incorrect alignment will cause excessive tire wear, leaving the tread of the tire worn with a characteristic feathered edge on one side.

The toe-in of the wheels, and the setting of the steering wheel, is carried out by altering the effective lengths of the spindle connecting rods, as shown in **FIG 10:37**. The setting of the steering wheel, without altering the toe-in, is carried out as shown in **FIG 10:38**.

10:8 The steering column

On the earlier models the steering shaft is integral with the steering unit and the column and steering unit are removed from the engine compartment. On all later models a safety collapsible column is fitted and it is connected to the steering unit by a coupling. Under heavy impact, the column collapses in a controlled manner and prevents injury to the driver. Care must be taken with this type of column as if the collapsible portion is damaged no repairs are possible and a new column must be installed.

When removing or installing the steering wheel on a later column, force must not be used and hammering must be strictly avoided. Use an extractor to remove the steering wheel, after its attachment nut has been removed, as hammering can cause damage to the collapsible portion. A typical later column diassembled is shown in **FIG 10:39** and the attachments in **FIG 10:40**.

A tilt-type column can also be fitted. The column is operated by a vacuum motor and allows the driving position to be set by four up and four down from the central position. When the ignition is switched off and the left door opened, the vacuum release valve is energized (but only when both operations are carried out) and the vacuum operates the vacuum motor to release the parking pawl. The spring pressure moves the steering wheel upwards and to the right to the tilt-away position, where it remains until pulled down again by the driver. The vacuum and electrical systems of the column are shown in **FIG 10:41** and the components in **FIG 10:42**.

If the collapsible portion of the column is damaged, a new column must be installed as it is not possible to carry out repairs.

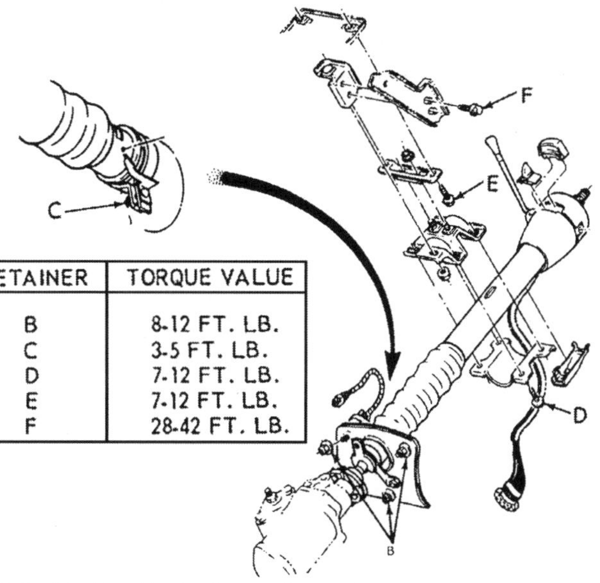

RETAINER	TORQUE VALUE
B	8-12 FT. LB.
C	3-5 FT. LB.
D	7-12 FT. LB.
E	7-12 FT. LB.
F	28-42 FT. LB.

FIG 10:40 Typical steering column attachments

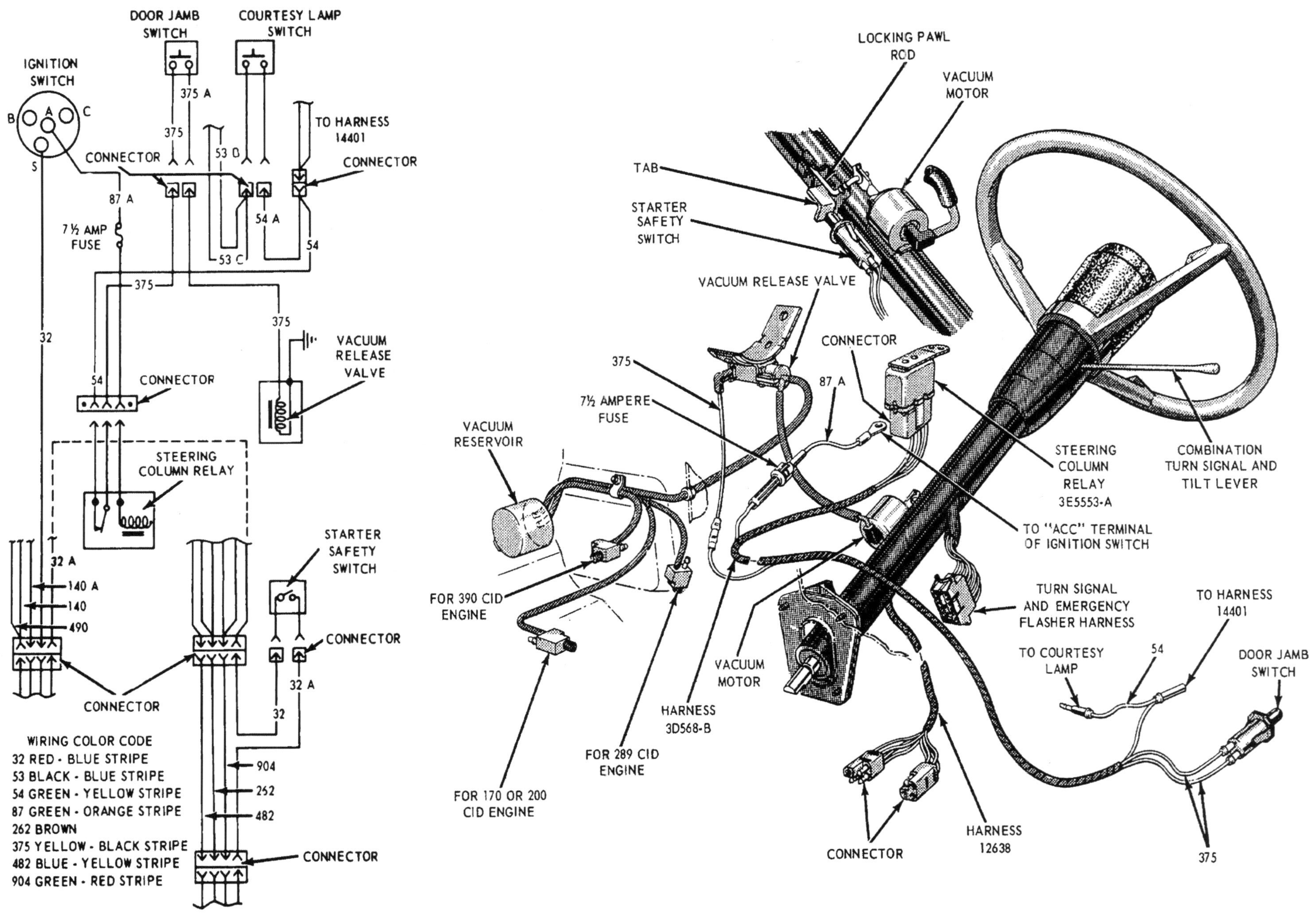

FIG 10:41 Tilt column vacuum and electrical systems

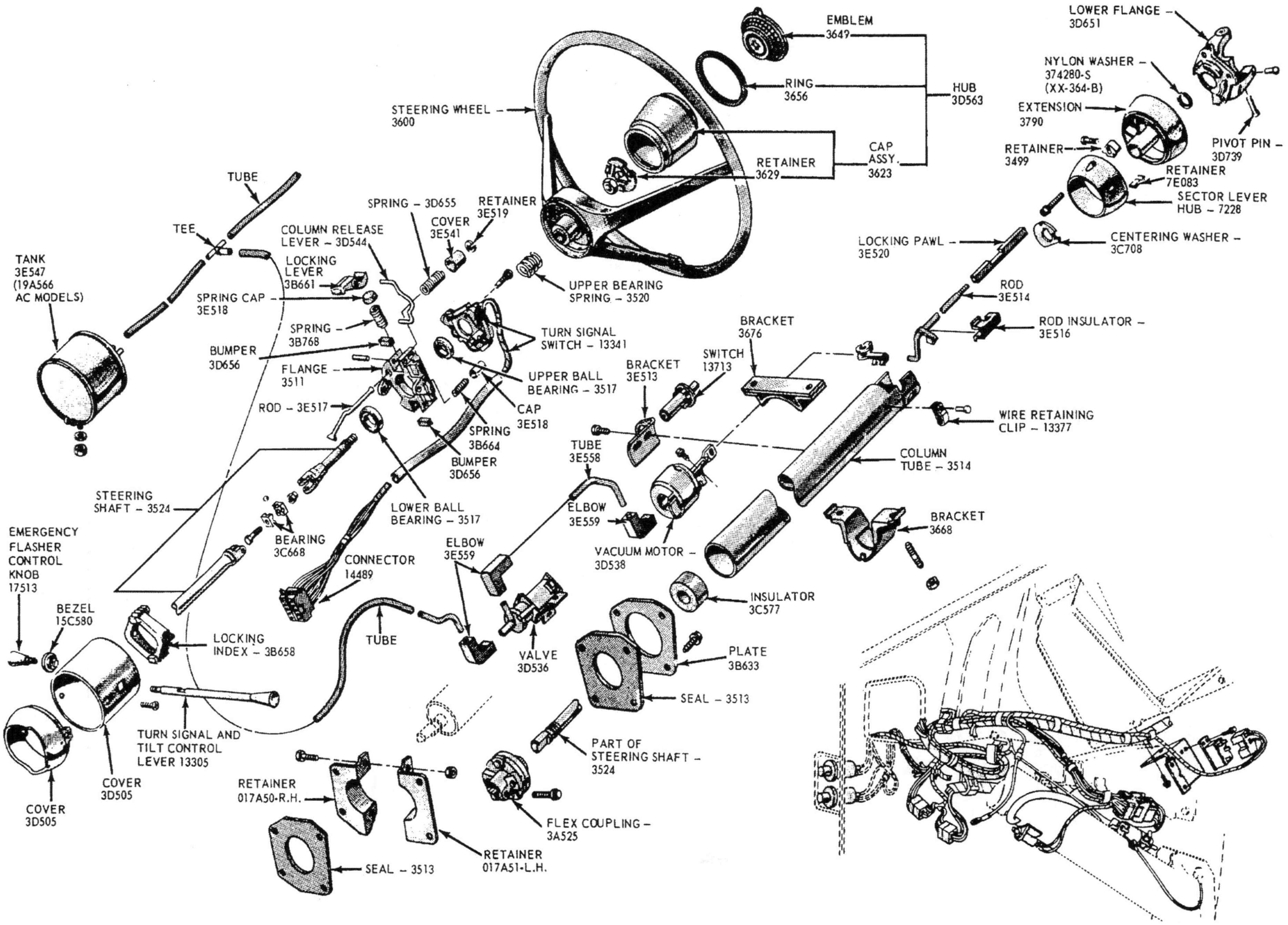

FIG 10:42 Tilt column components

10:9 Trouble shooting guide

(a) Wheel tramp or shimmy

1 Worn hub bearings
2 Defective or worn suspension
3 Unbalanced wheels and tires
4 Slack or worn steering connections
5 Incorrect suspension geometry

(b) Wander

1 Check 2, 4 and 5 in (a)
2 Front and rear suspensions out of alignment (frame damage)
3 Uneven tire pressure or wear

(c) Hard steering

1 Check 5 in (a)
2 Neglected lubrication
3 Steering unit incorrectly adjusted
4 Steering column bent or misaligned
5 Defective bearings in steering column
*6 Defective pump
*7 Defective valve assembly
*8 Slipping drive belt
*9 Control valve centering spring not accurately set
*10 Low fluid level

(d) Lost motion

1 Check 1 and 4 in (a); 3 and 9 in (c)
2 Defective column coupling
3 Loose steering wheel or badly worn splines

*(e) Noisy operation

1 Check 6, 8 and 10 in (c)
2 Hoses incorrectly fitted

*Power steering only

CHAPTER 11

THE BRAKING SYSTEM

11 :1 Description

Hydraulically operated brakes are fitted to all four wheels. On standard models all four wheels are fitted with drum brakes but disc brakes can be fitted as an optional extra to the front wheels. The master cylinder generates the hydraulic pressure and the pressure is led to the brakes by a system of pipes and flexible hoses. A typical hydraulic system is shown in **FIG 11 :1**.

On the earliest models a single master cylinder is fitted but on all later models a tandem master cylinder is used. On the later system the front brakes are hydraulically independent from the rear brakes and a failure in one half will still allow the other half to operate. A pressure differential valve is fitted to warn of failure in one half of the system.

A power unit can be fitted as an optional extra. The unit operates on vacuum from the intake manifold and is installed between the brake pedal and master cylinder, to increase the pressure on the master cylinder piston. If the unit fails, direct pressure can be transmitted to the master cylinder, though heavier pedal pressure will be required for a given retardation.

The rear drum brakes operate by a mechanical cable system, shown in **FIG 11 :2**, for parking the car. Later models have a foot-operated parking brake system (see **FIGS 11 :22** and **11 :23**).

11 :2 Maintenance

Fluid level:

Wipe the top of the master cylinder clean and slide the retainer to one side. On simple master cylinders unscrew the cap. On tandem master cylinders remove the filler cap complete with the rubber diaphragm. Fill the reservoir to within $\frac{1}{4}$ inch from the top with fresh Ford Brake Fluid (Extra Heavy Duty Part No. C6AZ.19542.A) or equivalent (meeting at least specification SAE.70.R3). Correctly install the filler cap.

On models fitted with disc brakes there will be a slow steady drop in level as the pads wear. If there is an excessive regular need for topping-up or the level falls suddenly between checks there is a leak in the system. **The leak must be found and rectified otherwise the brakes may fail.**

Adjustments:

All brakes are self-adjusting. The disc brakes will automatically adjust if the service brake pedal is pumped hard several times. The drum brakes automatically adjust when the vehicle is backed and the brakes firmly applied. If the car is used for much freeway motoring with little backing it is advisable to back the car several times and apply the service brake pedal firmly. Between each application backwards, drive forwards and apply the brakes.

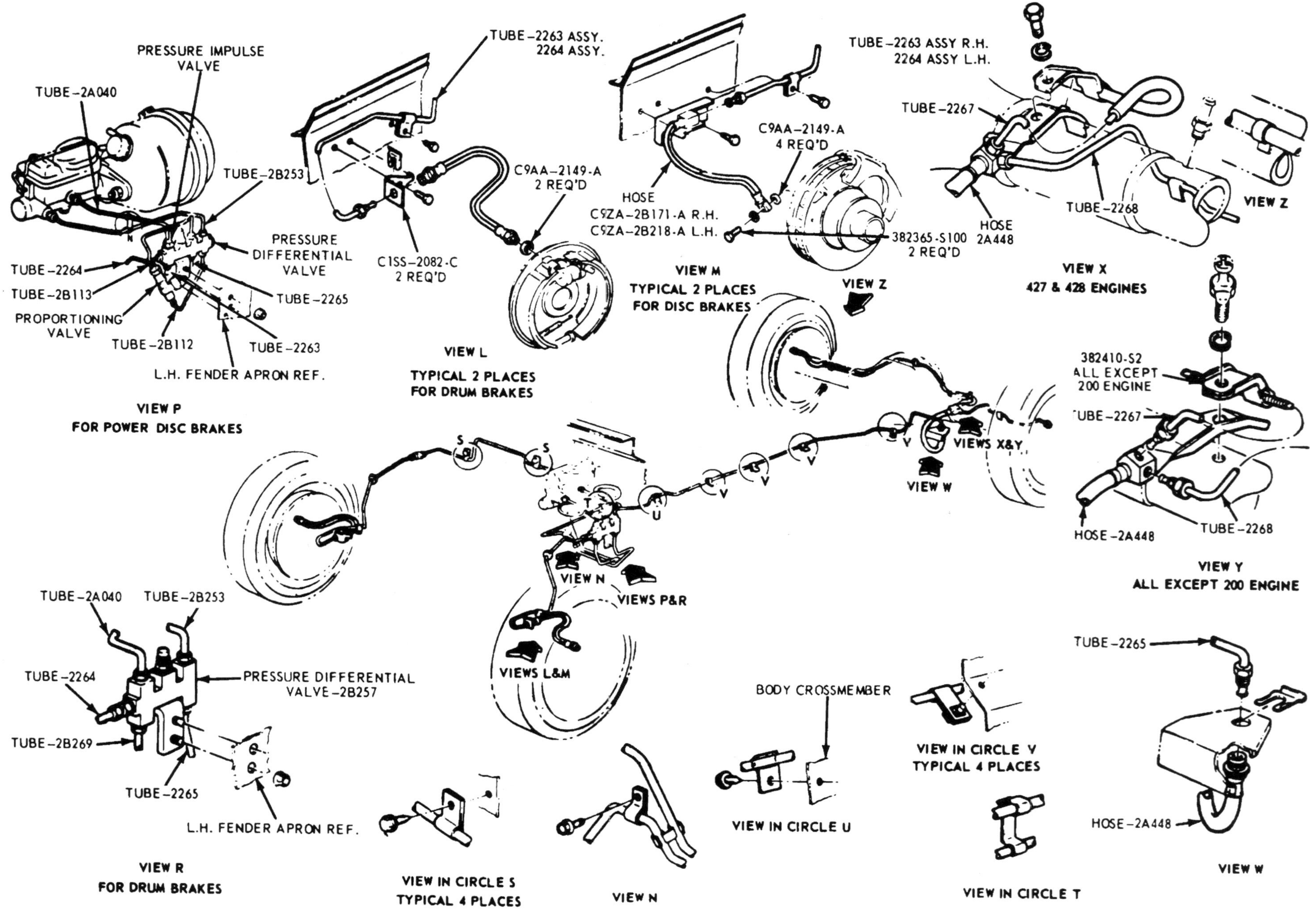

FIG 11:1 The brake hydraulic system

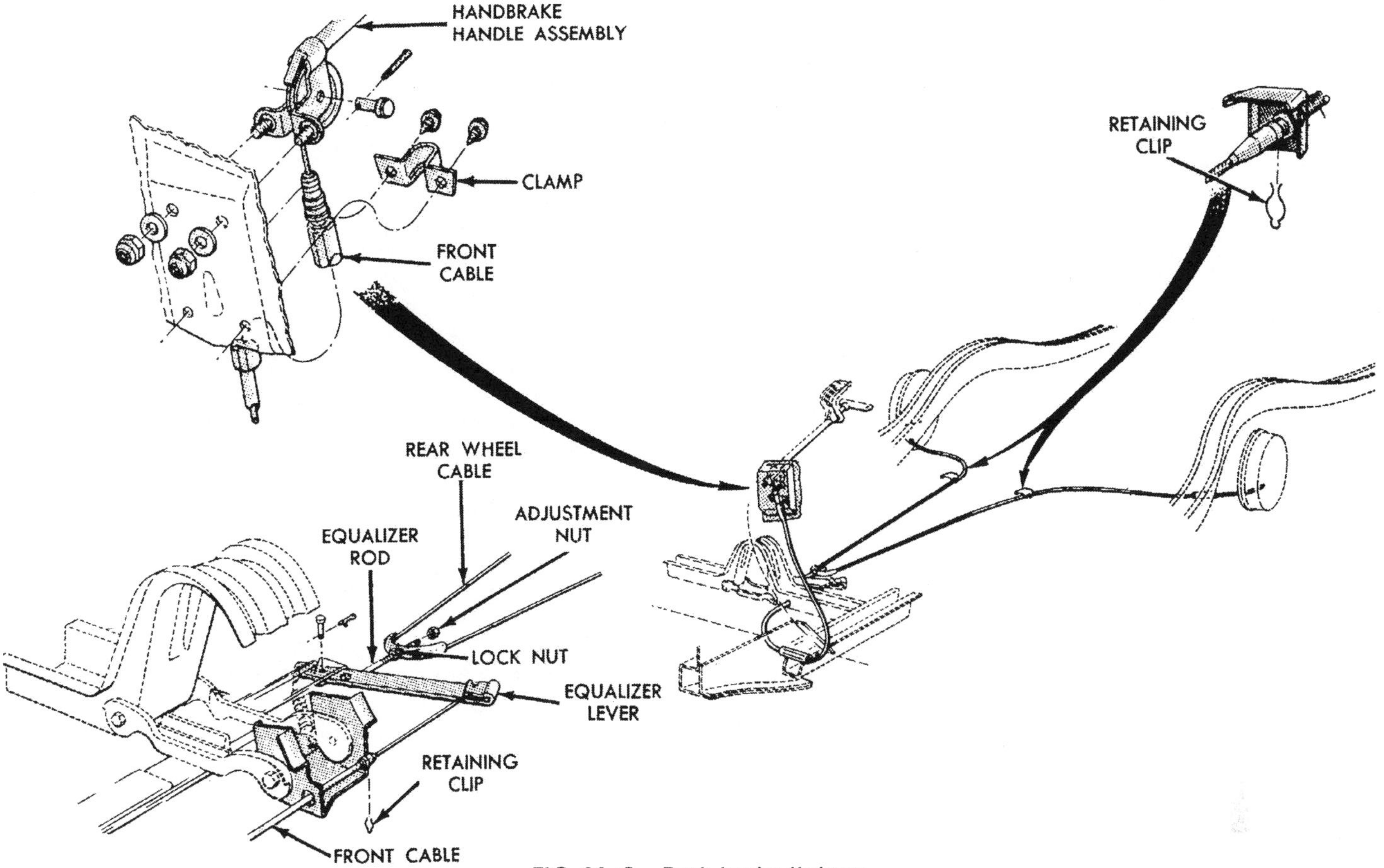

FIG 11 : 2 Park brake linkage

At regular intervals remove the brake drums, and check that the automatic adjusters are operating correctly and that the linings are in good condition. New shoes (in complete axle sets) must be installed if the linings are worn down to within $\frac{1}{32}$ inch of the rivet heads or if they are contaminated with grease or oil.

At the same time check the thickness of the linings on the pads of the disc brakes. New pads must be installed if the lining material has worn down to .030 inch at the front edge or if the linings are contaminated with grease or oil.

Preventative maintenance

In addition to visually checking the brake linings at regular intervals also check the brake lines and hoses. Lines which are corroded or dented must be renewed as the damage will cause a weak spot which may fail under the stress of heavy braking. Similarly renew flexible hoses if they show signs of perishing or chafing.

Brake fluid accumulates moisture from the air and also becomes overheated in the brakes, so it is advisable to renew the fluid at intervals of 18 months. Either the system can be drained through the bleeder screws, and then refilled and bled with fresh fluid, or the system can be bled until fresh fluid comes out of each bleeder screw.

At intervals of 3 years, drain the system completely and flush through with denatured alcohol. Remove and dismantle all the components, discarding all old seals. Check all components for wear or defects, renewing parts as required, and assemble them with new seals. Fill and bleed the system with fresh brake fluid.

11 : 3 Flexible hoses

When removing or installing a flexible hose take great care not to twist or strain the flexible portion. When the hose has been installed the line on it must be straight within narrow limits. The hose attachments are shown in **FIG 11 : 1**.

If a hose is blocked and cannot be cleared by blowing compressed air through it then the hose must be discarded and a new one installed in its place. Never attempt to clear hoses by poking through them as this will damage the internal surface.

Removal:

Hold the adjacent hexagons of the hose with a wrench to prevent the hose rotating. Undo the union nut that secures the brake line to the hose, using another wrench. Take care that the union nut does not twist the metal line as the nut rotates. Gentle cleaning with a brass-wire brush and a few drops of penetrating oil or old hydraulic fluid will help to prevent this occurring. Remove the clip or locknut that secures the end of the hose to the frame bracket, and withdraw the hose from the bracket. The other end of the hose can now be undone using a wrench on the hexagons and allowing the flexible portion of the hose to rotate freely.

Install the hose in the reverse order of removal. If seals are fitted to the end of the hose, these should be renewed. Check that the hose is free from chafing on obstructions on the frame or muffler after it has been installed, particularly allowing for movement of the suspension or steering.

FIG 11:3 Backing off the brake adjusters

11:4 The brake drums

Slacken the road wheel attachments slightly and raise the appropriate side of the car until the wheel is clear of the ground. Remove the road wheel. On the front wheels remove the complete hub assembly (see **Chapter 9, Section 9:3**). On the rear wheels, remove the three Timmerman nuts and draw off the drum.

If the drum is held by the brake shoes, slacken the brakes off. Knock out the access plug at the back of the brake carrier plate and insert a thin screwdriver through the slot to disengage the adjusting lever from the adjusting screw, as shown in **FIG 11:3**. Use the brake adjusting tool to back off the adjusting screw as shown. Take great care not to damage or burr the teeth on the adjuster gear otherwise the automatic adjuster may fail to operate.

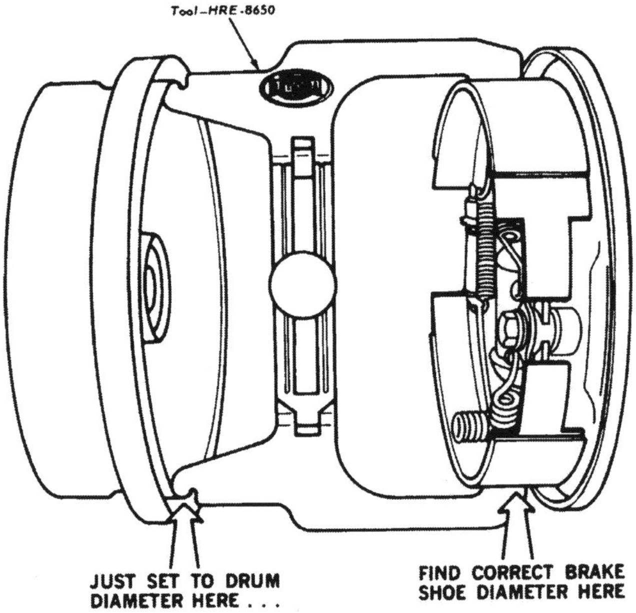

FIG 11:4 Setting the brake shoe adjustment

Examination:

Blow and brush away all loose dirt and dust from the brakes and drum and then wipe down the brake carrier plate with a cloth moistened in denatured alcohol. Do not use any other solvents as they will attack the material of the seals. Do not allow any solvent onto the brake linings themselves. Take the opportunity to examine the wheel cylinders for leaks and the linings for thickness.

Wash the drum operating face in any suitable solvent that evaporates completely. When the surface is clean take care not to allow greasy fingers onto it, unless they are washed off again, and do not allow any grease or oil onto the brake linings.

Examine the surface of the drum for scores. Light damage may be machined off, with the drum mounted on an arbor and not in a chuck, provided that .060 inch is the maximum amount of stock removed. If the drum is machined to less than .030 inch oversize standard linings may be used, but if the drum is more than .030 inch oversize then oversized linings must also be used. If the scoring or damage is too deep to be machined off a new drum must be installed.

If it is found that a drum is scored but the linings are only partially worn and fit for further service, no action should be taken until the linings are worn to the limit ($\frac{1}{32}$ inch above rivet heads) at the thinnest point. New shoes should then be installed and the drum machined. The old shoes will have bedded in with the drum and will give effective retardation. Very poor braking will result from new linings and scored drums, and similarly with a new drum and scored shoes.

Check the drum for cracks. Hang the drum on a wooden handle and tap it lightly with a metal object. If the drum rings true it is satisfactory while a crack will cause it to ring flat. Discard the drum if it is cracked. If the shoes are also scored, new shoes must be installed at the same time.

Installation:

Do not apply the service brake pedal while the drum is off otherwise the hydraulic pressure will shoot the pistons out of the wheel cylinder. As a safety precaution, wrap the wheel cylinders with strong elastic bands to hold the pistons in place.

Measure the internal diameter of the drum with Rotunda calipers, as shown in **FIG 11:4**. Clamp the calipers so that they do not alter from the setting, and use them to set the brake shoes. Rotate the calipers around the shoes so that several points are checked. The brake shoes are adjusted by holding the adjusting lever off the teeth and rotating them as shown in **FIG 11:3**. Check that the adjuster rotates freely and if necessary lightly lubricate the threads with a thin film of C1AZ.19590.B grease. Apply very small quantities of high-melting point grease to the points where the brake shoes contact the carrier plate and anchor pin. **Avoid getting lubricant onto the linings and avoid touching them with greasy or dirty hands.**

Install the drum in the reverse order of removal. On the front wheels set the hub clearance correctly (see **Chapter 9, Section 9:3**).

When all the parts are installed, roll the vehicle forwards to check that the brakes are not binding. Finally adjust the brakes by backing and firmly applying the brakes, with a forward roll and brake application between each backing application.

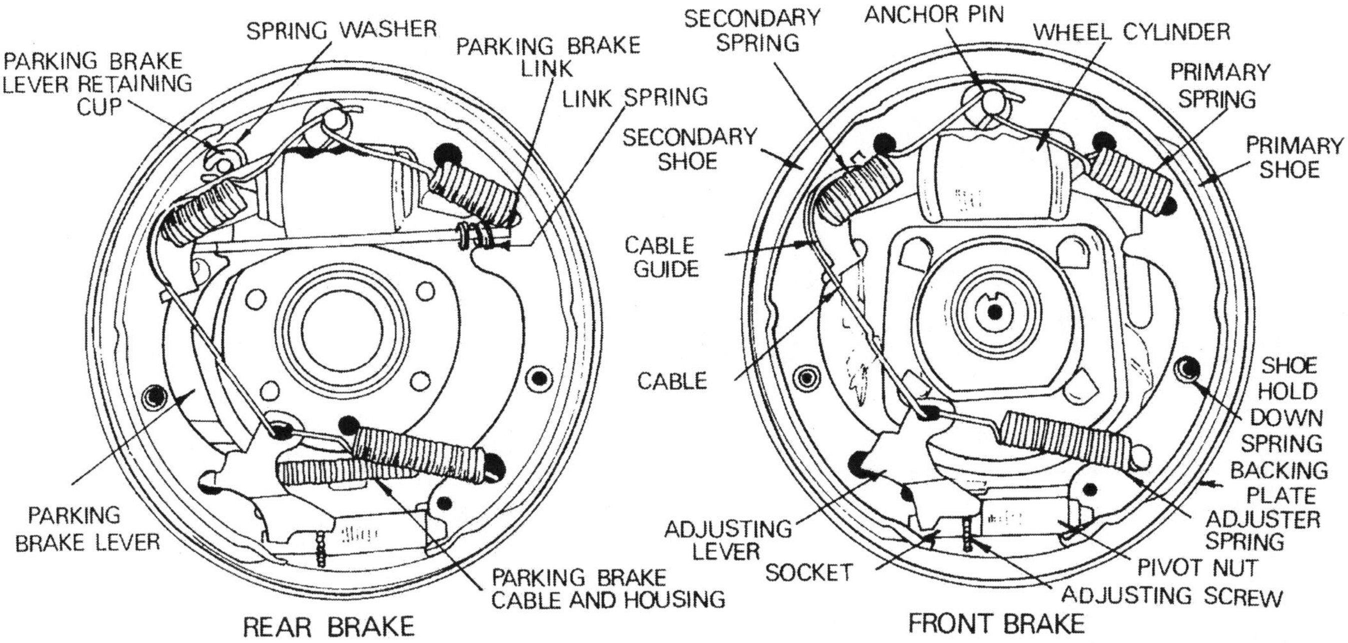

FIG 11:5 Drum brake assemblies

11:5 Brake shoes and adjusting screws

Typical front and rear brake assemblies, with the drums removed, are shown in **FIG 11:5**.

As the car backs and the brakes are applied, the movement of the shoes causes the anchor pin to pull the cable which acts on the adjusting lever. If there is sufficient play the adjusting lever will move the adjuster by one tooth to take up that play, but when the brakes are correctly adjusted there is insufficient movement to allow the adjusting lever to grip and turn the tooth.

The parts of the adjuster are marked to ensure that they will be installed on the correct side. If the incorrect parts are installed the adjuster will not operate. The markings are shown in **FIG 11:6** for the righthand side parts. The nut for the lefthand adjuster is identified by a single groove.

Removing parts:

1 Remove the brake drum as described in the previous section. Remove the brake shoe hold down springs and parts. Hold the pin in the carrier plate with a finger behind the plate and grip the washer with a pair of pliers. Press the washer in against the spring and turn the washer through 90 deg. so that the head of the pin passes through the washer and the washer and spring can be removed. Withdraw the pin from the back of the carrier plate.

2 Clamp the pistons into the wheel cylinder to prevent them being accidentally blown up. Free the return springs from the anchor pin, using the tool shown in **FIG 11:7**. If the special tool is not available use a strong length of cord through the eye of the spring to pull it off the anchor pin, while extending the spring.

3 Remove the shoes and the adjuster as an assembly and then dismantle the parts on the bench. On the rear brakes, remove the parking brake link and spring, and also disconnect the parking brake cable from the parking brake lever. After the shoes have been removed from the brake, remove the retaining clip and spring washers to free the parking brake lever from the secondary shoe.

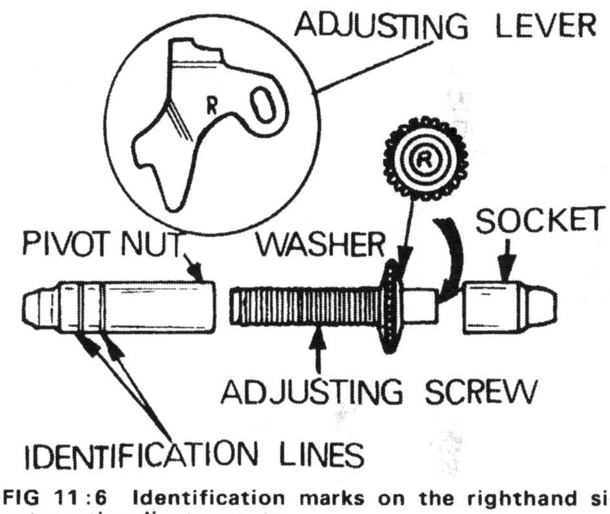

FIG 11:6 Identification marks on the righthand side automatic adjuster parts

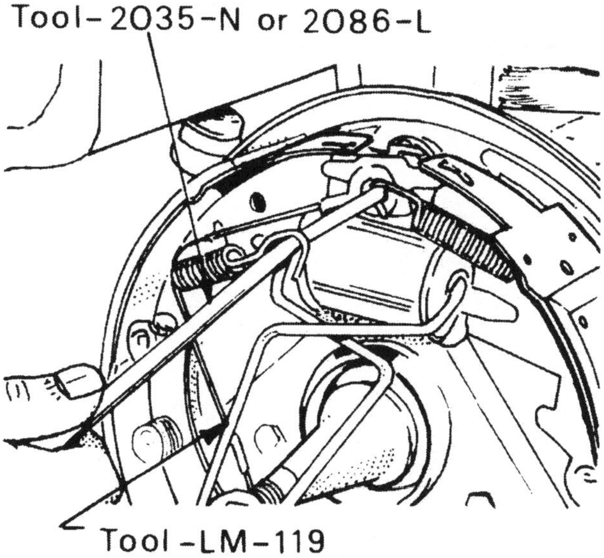

FIG 11:7 Freeing the shoe return springs

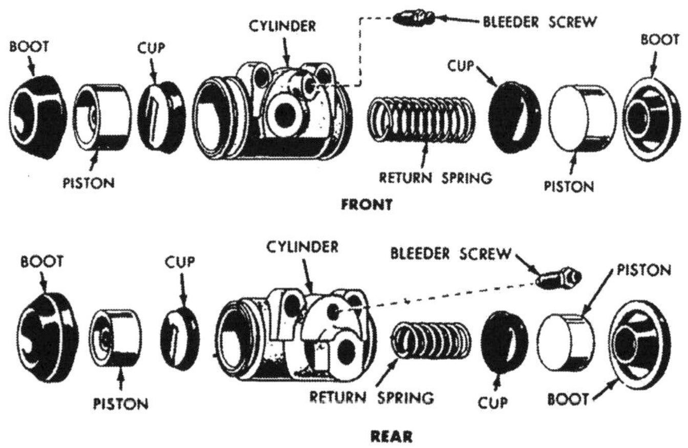

FIG 11:8 Typical wheel cylinder components

Inspection:

Disassemble the components of the adjusting screw. Check the linings for wear.

New linings must be fitted in axle sets if the old ones are worn down to within $\frac{1}{32}$ inch of the rivet heads or if they are contaminated with grease or oil. **Do not attempt to remove contaminants from the shoes by baking or washing as any treatment drastic enough to completely clean the shoes will also ruin them.** New linings ground to shape are available, but if the owner has not had experience of riveting it is better to fit factory reconditioned exchange shoes complete.

At the same time as renewing the linings or shoes, install new shoe return springs. The springs receive heat from the brakes and after long use they lose their temper and are no longer fully effective.

Check the adjusting cable for damage or weakening. The cable must completely fill or extend beyond the crimped portion. If the cable appears weak or damaged renew it, as a poor cable is one of the commonest causes of incorrect automatic adjustment. Check the cable guide for damage and make sure that the groove for the cable lies parallel to the shoe web, with the body of the guide flat against the web.

Check the pivot hook on the adjusting lever. The hook surfaces should be square with the body of the lever.

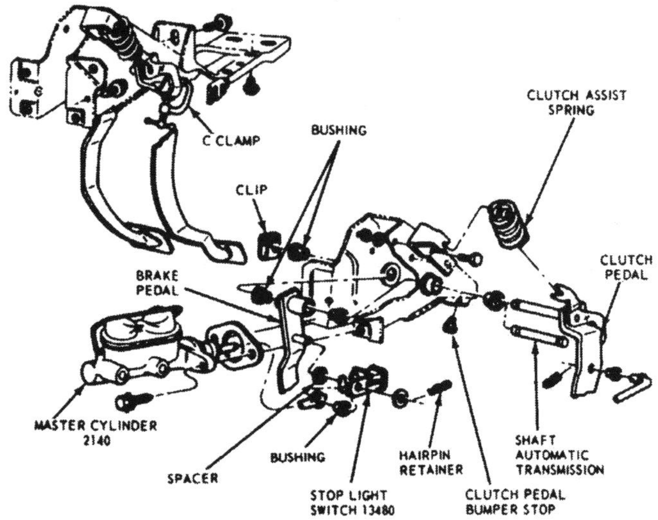

FIG 11:9 Master cylinder installation, without power unit

Renew any defective or damaged parts. Clean by wiping with a piece of cloth moistened with denatured alcohol but do not allow any solvent onto the linings themselves.

Installation:

The parts are installed in the reverse order of removal. Lubricate pivot points lightly with high temperature grease. The cable must be installed with the crimped side of the eye towards the carrier plate and the cable must be routed in the guide and not between guide and plate. Be sure that all the parts lie flat on the anchor pin.

When the parts have been installed, the adjusting lever should contact the teeth of the adjuster $\frac{3}{16} \pm \frac{1}{32}$ inch above the center line of the adjusting screw.

Check the operation of the adjuster. Pull the adjuster cable, on the portion between the guide and anchor pin, toward the secondary shoe web far enough to lift the adjusting lever past a tooth on the wheel. The lever should spring into position behind the tooth and when the cable is released the lever should move the adjuster by one tooth.

Adjust the brakes to accept the drums and finally adjust the brakes by applications in reverse.

11:6 Wheel cylinders

These are accessible after the drums and brake shoes have been removed (see previous two sections). The components are shown in **FIG 11:8**.

Disassembly and removal:

The unit can be disassembled while still attached to the carrier plate and only if the bore is found to be scored need the cylinder itself be removed from the carrier plate.

Remove the links and the rubber boots then extract the internal parts from the cylinder.

If the cylinder requires removal, take off the bleed screw. On front wheels, disconnect the flexible hose at its bracket and then free it from the cylinder. On rear wheels, free the metal line from the cylinder by undoing the union nut but do not pull the pipe away, as this will damage the pipe. Remove the wheel cylinder bolts and take off the cylinder. On the rear wheels it will free leaving the pipe in place.

Cleaning and inspection:

Discard the old cup seals and boots. Wash the parts in denatured alcohol. If the bore of the cylinder is worn or scored discard the unit and install a complete new one.

Assembly and installation:

Wet the internal parts with clean hydraulic fluid. Fit one cup seal back into the bore, **taking great care not to bend back or damage the lips of the seal as it is inserted.** From the open end, install the return spring and other cup seal. Be sure that the spring is correctly located between the seals and then insert the pistons, flat faces leading. Secure the parts in place with the new rubber boots.

The wheel cylinder is installed in the reverse order of removal. Once all the parts, including brake drums, are in place fill the master cylinder reservoir and bleed the braking system.

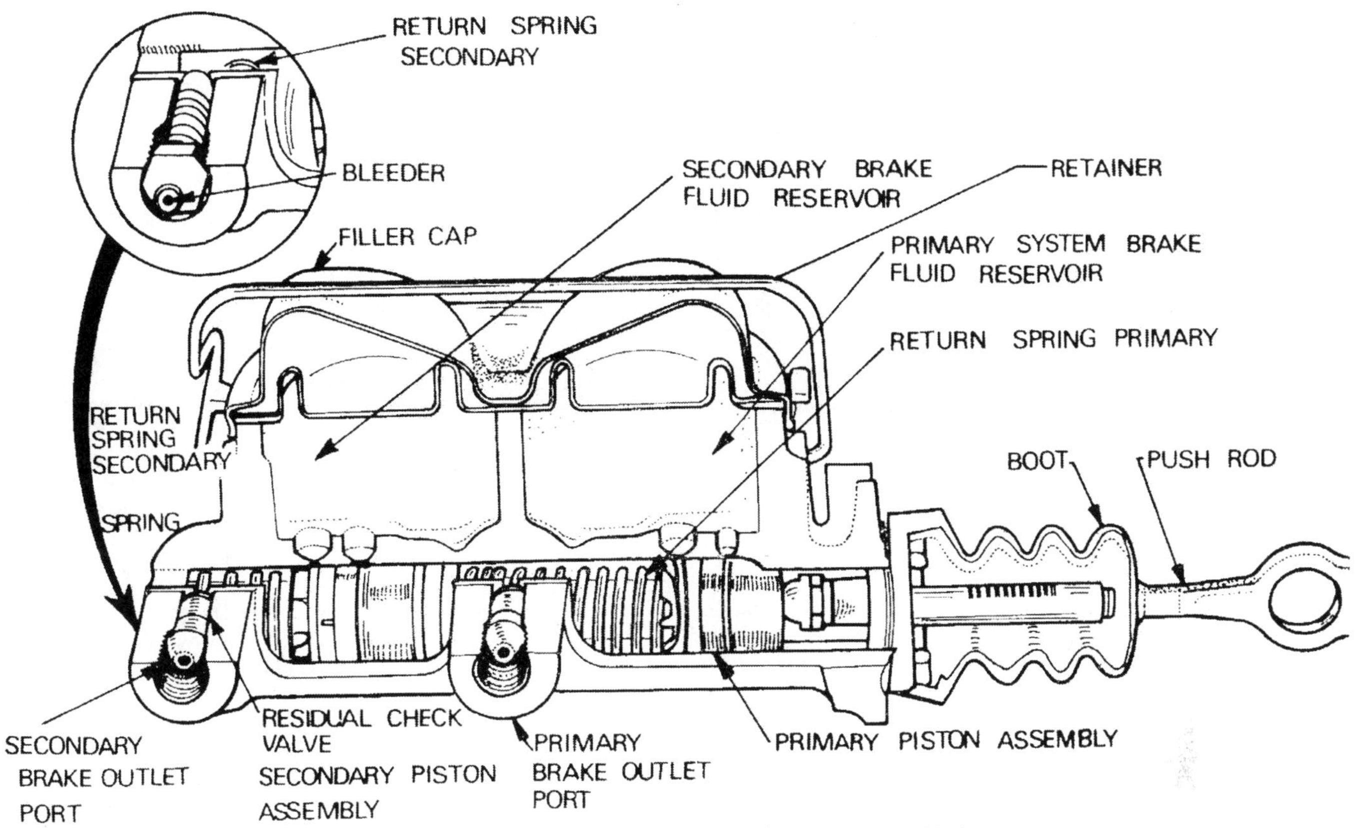

FIG 11:10 Sectional view of tandem master cylinder

11:7 The master cylinder

A typical installation is shown in **FIG 11:9**.

Removal:

Siphon out the fluid from the reservoir with a syringe. Disconnect the lines to the master cylinder, using rags to catch any spillage of fluid. **Hydraulic fluid rapidly softens and removes paint so spillage should be avoided.** From inside the car disconnect the master cylinder pushrod from the brake pedal. Remove the two attachment nuts and withdraw the master cylinder.

Install the master cylinder in the reverse order of removal. Once the unit is in place, fill the reservoir with fresh fluid and bleed both sets of brakes.

Servicing:

Pour out any remaining fluid from the reservoir and wash the unit externally with denatured alcohol. A sectional view of the master cylinder is shown in **FIG 11:10** and the components in **FIG 11:11**.

Pull back the rubber boot and remove it with the pushrod. Gently press the primary piston assembly down the bore and remove the snap ring with a special pair of snap ring removers. Still applying pressure to the primary piston, remove the stop bolt and its O-ring which secure the secondary piston. The internal parts can now be removed from the cylinder. Tap the open end against the palm of the hand to dislodge them or apply gentle air pressure through a port if the internal parts stick in place.

Discard the old seals, fitting new ones from a service kit on reassembly. Wash the parts in denatured alcohol and lay them out in order on clean paper. Inspect the bore. If there is only light etching wear or pitting, the bore may be honed out by a specialist firm but if the damage is deep a new master cylinder must be installed.

Install the new seals using only the fingers, after wetting them with clean hydraulic fluid. Make sure that the seals face in the correct direction. Wet the internal

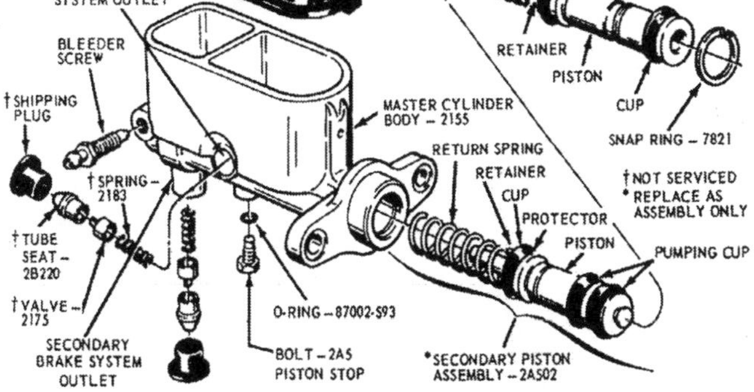

FIG 11:11 Tandem master cylinder components

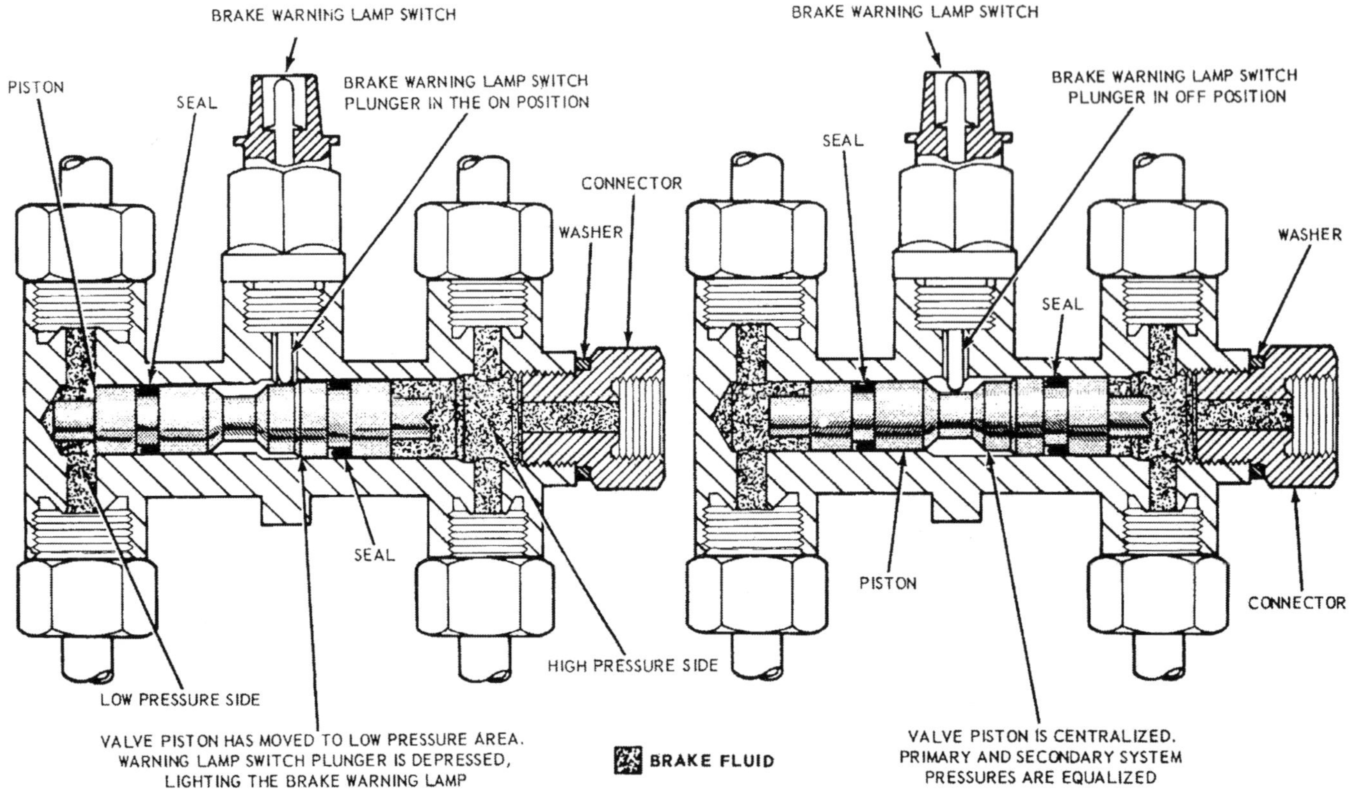

FIG 11:12 Sectional view of the pressure differential valve

parts with hydraulic fluid and carefully press them back down the bore. **Take great care not to bend back or damage the lips of the seals as they enter,** and insert them with a slight twisting motion. Hold the pistons down and install the stop bolt and snap ring. Secure the pushrod in place with a new rubber boot.

Pressure differential valve:

This is fitted to give warning or failure in one half of the braking system. A sectional view of a typical unit is shown in **FIG 11:12**. When there is a pressure differential, the valve piston is forced across to the low pressure side and operates the plunger so the electrical switch which in turn lights the warning lamp.

The unit cannot be serviced and if it is defective a new valve must be installed. The method of removal is straightforward, after disconnecting the electrical lead and four hydraulic lines. When the valve has been installed, both brake systems must be bled.

The method of centralizing the valve piston after it has moved is described in **Section 11:8**.

Single master cylinder:

This is only fitted to earlier models. The components of the unit are shown in **FIG 11:13** and the method of removal, servicing and installation is very similar to that for the tandem unit, except that there is only one piston. No pressure differential valve is fitted with single master cylinders.

11:8 Bleeding the brakes

This is not routine maintenance but it is required whenever air has entered the hydraulic system. Air can enter if the fluid level in the reservoir is allowed to drop so far that air is drawn into the master cylinder and it will also enter when parts are dismantled for servicing. If air has entered through a leak, the leak must be repaired.

If only one half of the system, with tandem master cylinders, has air in it then only that half of the system requires bleeding, as the two halves are hydraulically independent.

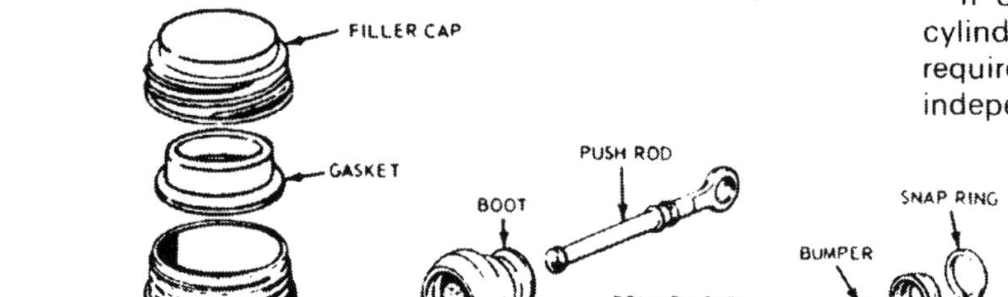

FIG 11:13 Early master cylinder components

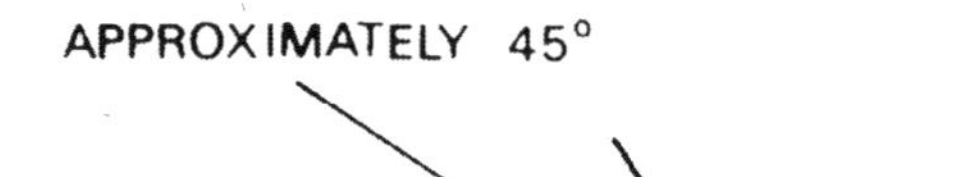

FIG 11:14 Modified wrench for brake bleeding

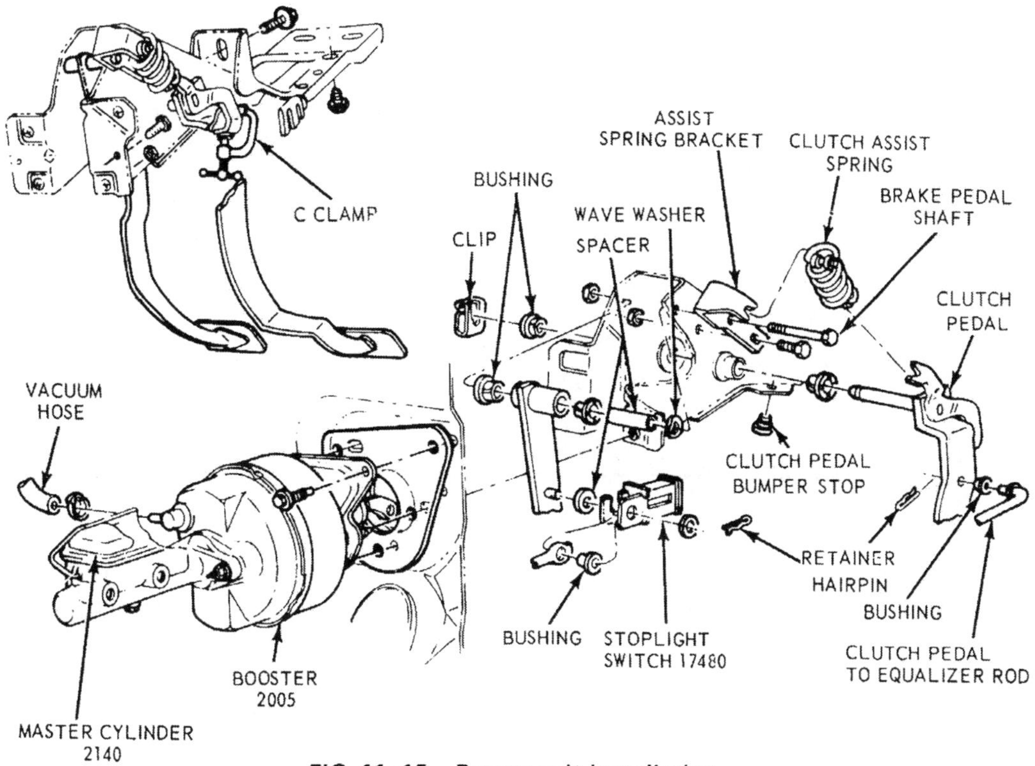

FIG 11 :15 Power unit installation

Fluid that has passed through the system should be discarded. Only if it is perfectly clean may it be used again but even then it should be stored for at least 24 hours in a clean sealed container to allow all the air to disperse. **Never return fluid directly from bleeding to reservoir. Never use fluid that has been stored in an open or dirty container.**

Before commencing the operation, remove the filler cap from the reservoir and top it as far as possible without spillage. Keep an eye on the level as the operation progresses and top up the reservoir when the level has fallen approximately to the half-way mark. If the level is allowed to drop too far air will again be drawn into the system.

Bleed the brakes in the order of decreasing pipe runs, starting at the brake with the longest run. If the master cylinder is fitted with a bleeder screw, bleed at this point first. Do not attempt to bleed through the hole for the secondary piston stop bolt as damage will be caused if the secondary piston moves back beyond the bolt position. Remove the road wheels if this makes access to the bleeder screws on the brakes easier. The bleeder screws will be easier to operate if a $\frac{3}{8}$ inch wrench is bent as shown in **FIG 11 :14**.

Attach a length of plastic or rubber tube to the bleeder screw of the brake to the bled. Be sure that the tube is a snug fit on the bleeder screw. Dip the free end of the tube into a little fluid in a clean glass container.

Open the bleeder screw by $\frac{3}{4}$ turn. Have an assistant press the foot brake pedal right down through its full range of movement. When the pedal has reached the end of its travel, close the bleeder screw and have the pedal released. Repeat this sequence until the fluid coming from the bleeder tube is an unbroken jet with no air bubbles in it. Close the bleed screw and remove the tube before releasing the pedal. Transfer the tube to the next bleeder screw of the system.

If the system has been fully dismantled it may be difficult to remove all the air. Bleed the brakes until a firm safe pedal pressure is obtained and the brakes are safe to use. Use the car for a day or so and then bleed the system again. In the meantime all the minute bubbles will have coalesced into larger ones which are easier to bleed out. In extreme cases, take the vehicle to a service agent where the system will be bled using pressure equipment.

Centralizing the pressure differential valve:

If the valve has been operated accidentally during the bleeding operation it can be recentralized. The valve can similarly be recentralized if it has operated because of a fault in the system but the fault must first be rectified.

A reverse pressure will be required to move the piston back into its correct position. If the piston has moved because of excess pedal pressure during the bleeding operation, complete the bleeding on that part of the system. Switch on the ignition to ACC but do not start the engine. The warning lamp will now be alight. Open a bleeder screw on the opposite system and apply steadily increasing pressure to the brake pedal until there is a click from the valve and the lamp goes out. Close the bleeder screw. Do not depress the pedal further. If excessive pressure is applied the piston will move over to the opposite side and it will have to be moved back to center by reversing the pressure again.

Before using the vehicle, be sure that a firm pedal pressure is obtained.

11 :9 The power unit

The attachment of the power unit is shown in **FIG 11 :15**. Removal of the unit is very similar to removing the master cylinder alone (see **Section 11 :7**). The unit is removed with the master cylinder attached, or the master cylinder can be removed separately.

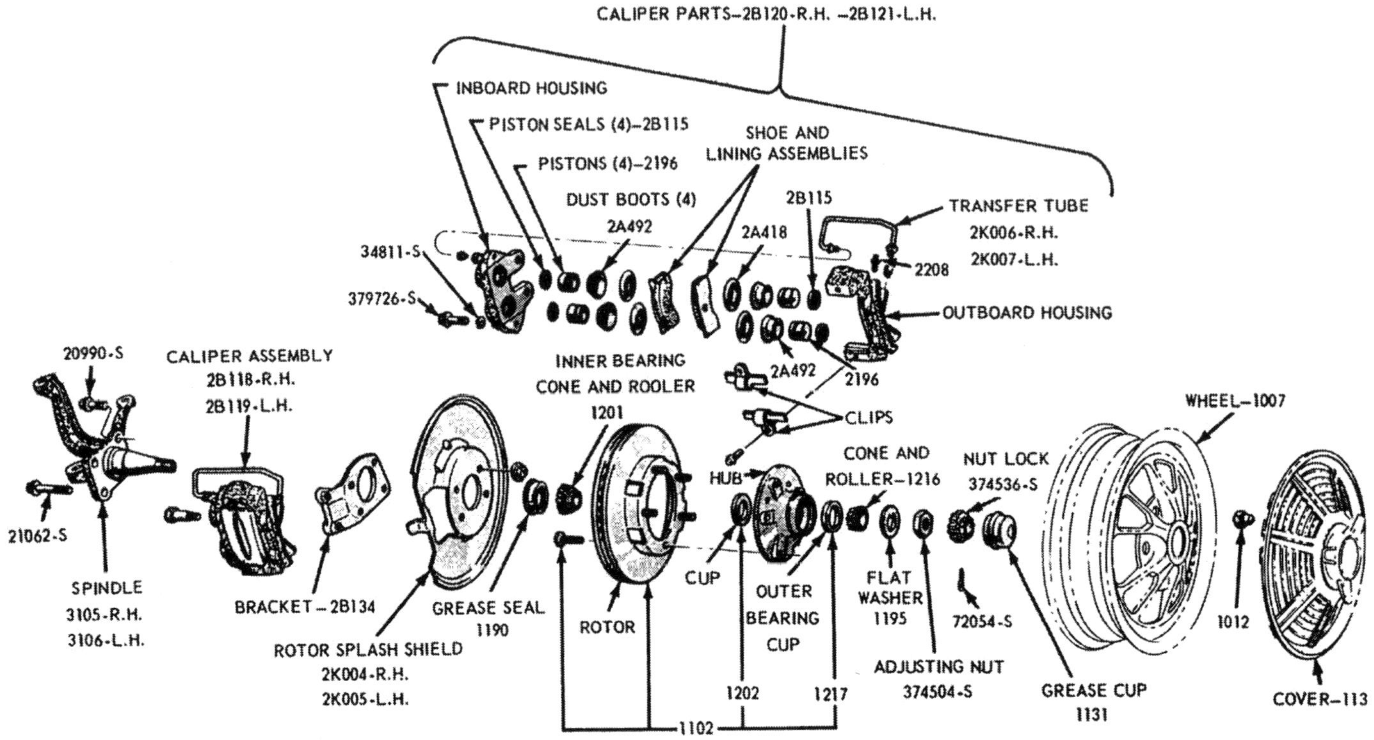

FIG 11:16 Early disc brake components

Testing:

Apply the park brake and pump the service brake several times to use up any vacuum in the reservoir. Apply the service brake and note the pressure required. Still holding the service brake pedal down, start the engine. If the unit is functioning satisfactorily, the pedal will gradually sink down as the vacuum builds up and the pressure required to hold the pedal will reduce.

If the unit does not operate, check the vacuum supply and the non-return valve on the unit. Disconnect the vacuum hose and remove the valve from the body of the unit. Renew the valve if it is defective. Check the hoses for blockage or kinks.

On earlier models the unit can be dismantled and reassembled with new seals, available in service kits, but on later models the unit must be renewed if it is defective.

The only adjustment possible on later models is setting the protrusion of the pushrod.

On the models where a renewable air filter is fitted to the unit, the air filter should be renewed at regular but long intervals. This will entail removal of the unit from the car and peeling back the dust cover over the pushrod to the brake pedal so that the old filter can be prised out and a new one installed.

11:10 The disc brakes

The earlier type of disc brakes is shown in **FIG 11:16**. Though the figure shows the two halves of the caliper separated, this is for illustrative purposes only and the owner must under no circumstances separate them.

In 1968 the type of caliper was changed and the later type is shown in **FIG 11:17**.

Renewing pads on earlier brakes:

Exchange shoes are supplied with linings ready bonded onto them. The owner should not attempt to bond new linings onto the old shoes.

New shoes must be fitted if the old linings are worn down to less than .030 inch thickness or if they are contaminated with oil or grease. All fouled shoes should be renewed at the same time to preserve the braking balance.

1 Jack up the front of the car and remove the front road wheels.
2 Siphon out sufficient fluid from the reservoir to lower the level to at least the half-way mark. If this precaution is not observed, fluid will overflow as the pistons are pressed back into the calipers. Discard the old fluid siphoned out and top up to the correct level ($\frac{1}{4}$ inch from top) with fresh fluid after the new pads have been installed and the brakes adjusted.
3 Remove the bolts that secure the shoes in the calipers. Press the shoes firmly but carefully towards the calipers so that the pistons are pressed back into their bores. The pistons will only move slowly and pressure with have to be maintained for at least a minute per shoes. Take great care not to score or damage the caliper and rotor.
4 Grip the lugs on each shoe with two pairs of pliers, are shown in **FIG 11:18**, and withdraw each shoe in turn.
5 Brush and blow out loose dust and dirt from inside the caliper. **Only denatured alcohol, special cleansing solvent, or hydraulic fluid may be used for cleaning the parts.**
6 Install the new shoes in the reverse order of removal being sure that they are correctly seated, and secure them in place. **Do not allow grease, solvent or oil onto the friction linings.**

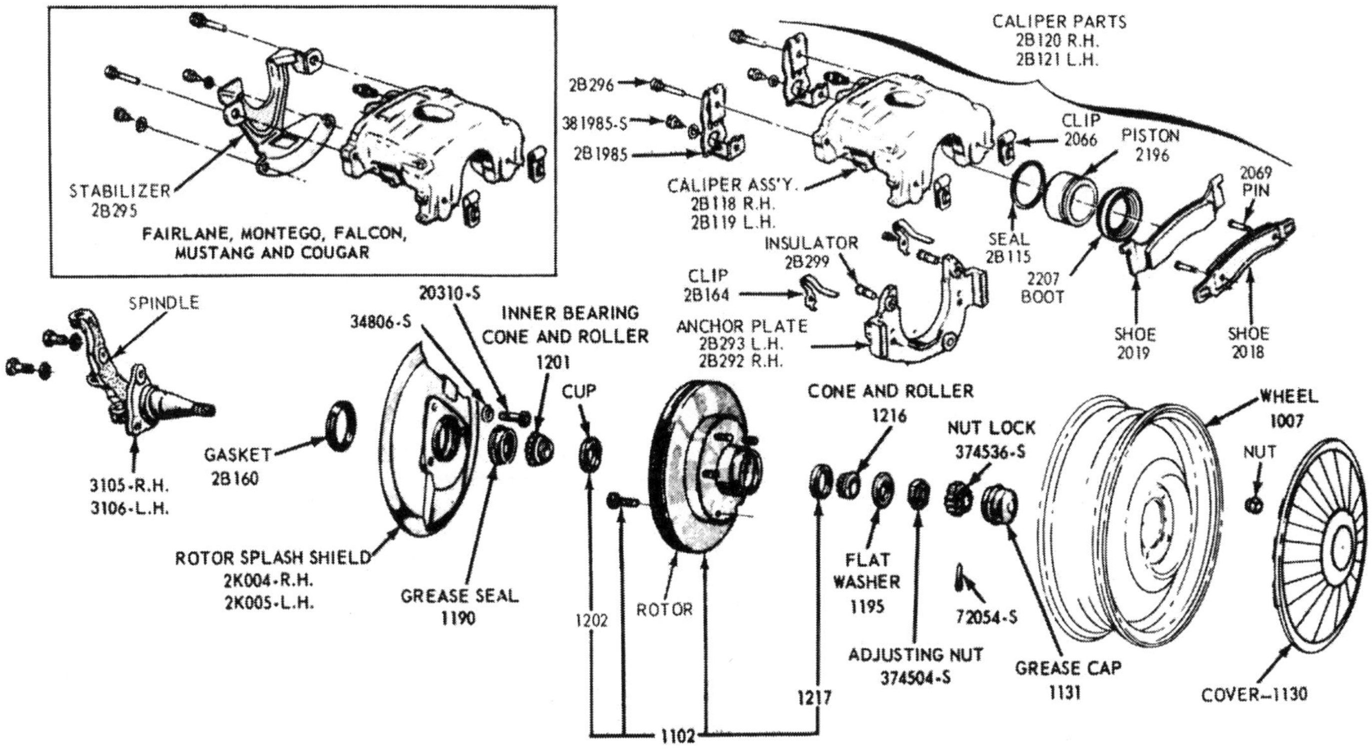

FIG 11:17 Later disc brake components

7 Install the road wheel. Pump the service brake pedal hard several times to take up the adjustment in the brakes and top-up the reservoir to the correct level with fresh fluid.

Renewing pads on later brakes:

1 Jack up the front of the car and remove the road wheel. Disconnect the brake line from the caliper and plug it to prevent dirt entering or fluid leaking out. Cut the safety wire and take out the two bolts that secure the caliper. Lift the caliper assembly off the brake rotor. It may be necessary to press the piston back into its bore to free the caliper. Press firmly for at least a minute against the inner shoe to move the piston back.

2 Refer to **FIG 11:17**. Slide off the outer two shoe retaining clips. Remove the two retaining pins from the outer shoe and then remove the shoe from the stationary caliper. Slide the inner brake shoe outward until it is free of the hold-down springs and then remove the shoe.

3 Remove the caliper locating pins and the stabilizer attachment bolts, then remove and discard the stabilizer. Remove the locating pin insulators from the anchor plate.

4 Press the piston fully to the bottom of its bore. Install new insulators in the anchor plate and position the caliper in the anchor plate.

5 Position the new stabilizer and install the caliper locating pins so that they engage by approximately four threads. New caliper locating pins must be installed if the old ones are corroded. **Water or isopropyl alcohol may be used to lubricate the pins but never ordinary grease or oil.**

6 Position the outer shoe on the caliper and install the retaining pins and clips. Install the inner shoe so that the ears on the shoes are on top of the anchor plate bosses and under the shoe hold-down springs. Position the shoes ready for sliding over the rotor, and rotate a hammer handle between the linings to position the stabilizer.

7 Fit the caliper back onto the suspension and secure it with the two bolts, tightening the upper bolt first. Lock the bolts with a new piece of safety wire and tuck the ends of the wire out of the way. Bleed the brakes.

8 Have an assistant apply moderate pressure to the service brake pedal and tighten the stabilizer attachment screws and caliper locating pins. Pump the brakes hard several times to adjust them.

Caliper removal:

The removal of the later type of caliper is dealt with in renewal of the pads.

On the earlier models, jack up the front of the car and remove the road wheel. Disconnect the steel transfer line from the caliper, leaving the other end attached to the flexible hose. Take out the two bolts that secure the caliper assembly to the suspension, **taking care not to slacken the bridge bolts which secure the caliper halves.** Slide the caliper off from the rotor.

The caliper is installed in the reverse order of removal. Check that the rotor runs centrally between the caliper halves with an equal gap of .090 to .012 inch on either side. Fill and bleed the brake system then pump the service brake pedal hard several times to adjust the disc brakes.

The rotor:

This can be freed from the hub, after the hub has been removed from the spindle (see **Chapter 9, Section 9:3**). Mark the hub and rotor so that they will be reassembled in the same relative positions.

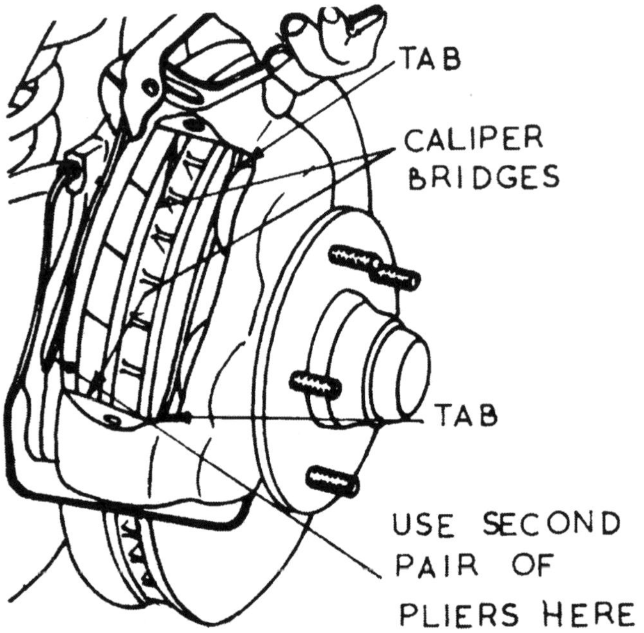

FIG 11:18 Removing the pads from an early disc brake

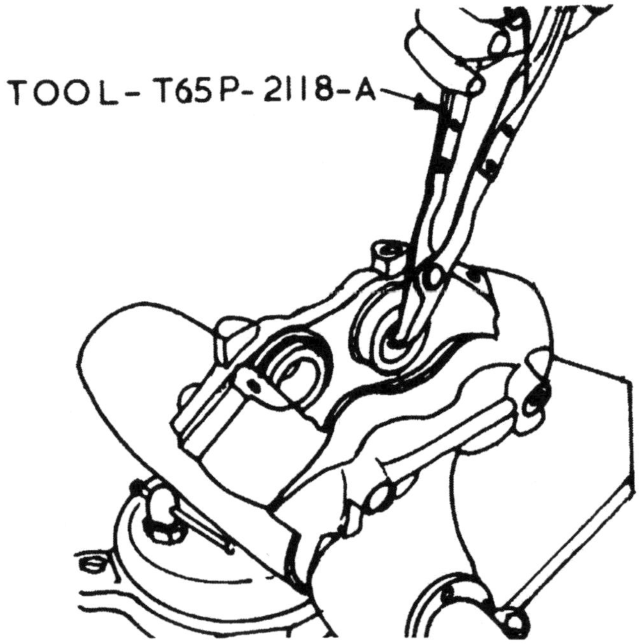

FIG 11:19 Removing the pistons from an early caliper

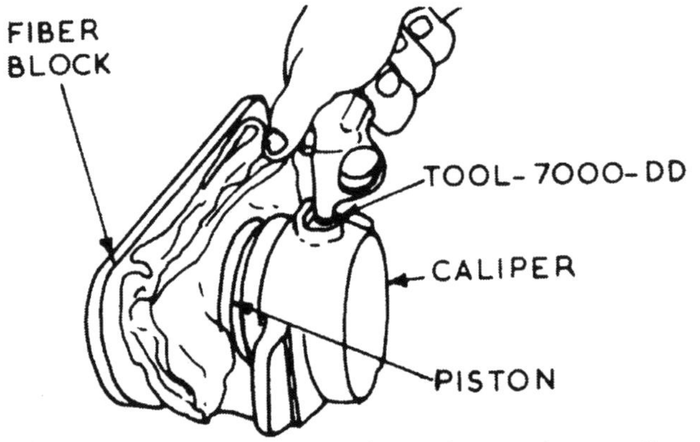

FIG 11:20 Removing the piston from a later caliper

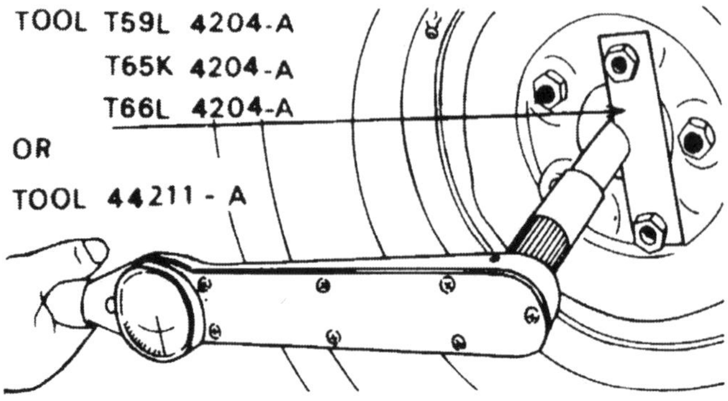

FIG 11:21 Checking the park brake torque

If the rotor is excessively scored or has excessive runout, the rotor may be machined down on a Rotunda machine at the service agents. Only a limited amount of stock can be removed but the agent will check.

Dirt and deposits, which affect braking efficiency, can be removed by soaking the rotor with trichlorethylene and gently scrubbing with old emerycloth.

Before installing the rotor to the hub, be sure that the mating faces are scrupulously clean and free from particles of dirt. Dirt will cause the rotor to have excessive runout.

When the assembly has been installed on the spindle and the wheel bearings correctly set, use the dial indicator to check the runout of the rotor near to the outer edge. If the runout is excessive, try rotating the rotor with respect to the hub to replace the runout but if this fails the rotor must either be renewed or machined.

Servicing a caliper:

Remove the caliper and take out the shoes as instructed earlier. On the earlier models, remove the dust boots and then withdraw the pistons as shown in **FIG 11:19**. On the later models, use air pressure to remove the piston as shown in **FIG 11:20**. If the piston sticks, tap gently around the caliper to free it.

Remove and discard the piston seals and rubber dust boots. Wash all the parts in denatured alcohol or special cleansing fluid. **Do not use any other form of solvent on the parts.** Examine the bores and renew the caliper complete if they are worn or scored in any way. Check the piston(s) for damage or wear and renew them if required.

Insert the new piston seals by hand, after wetting them with hydraulic fluid. Fit the dust boots to the pistons and press the pistons back into place. Take great care not to cock the pistons as they are inserted. The remainder of the parts are then installed in the reverse order of removal.

Proportional valve:

This is fitted to limit the pressure to the rear brakes. The valve cannot be serviced and it must be renewed if defective. Both brake systems must be bled after installing the valve.

11:11 The park brake

The linkage for the system is shown in **FIG 11:2**, and it is adjusted at the adjusting nut shown. Under no circumstances must the linkage be used to adjust the rear brakes, and the rear brakes must be in adjustment before adjusting the linkage.

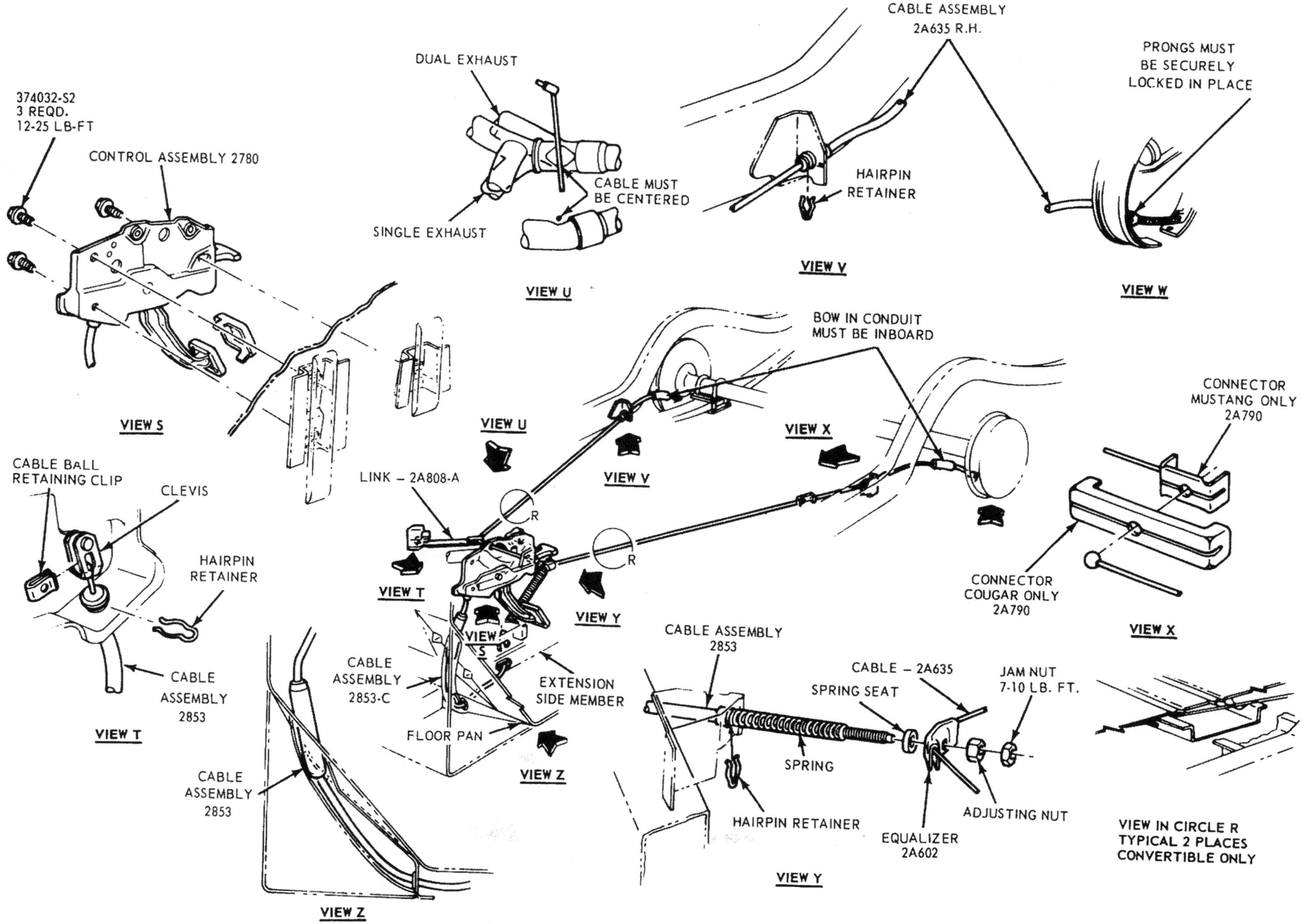

FIG 11:22 Foot operated park brake

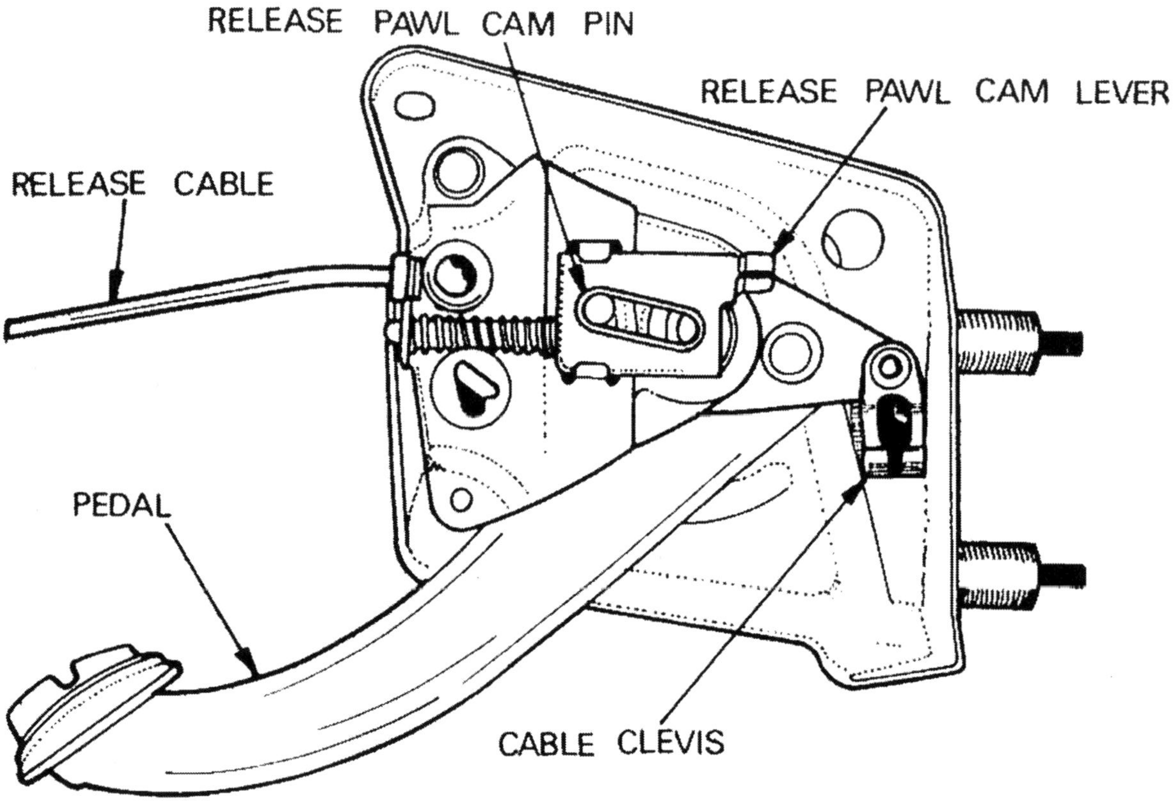

FIG 11:23 Parking control

Pull the park brake handle out to the third notch and raise the rear wheels clear off the ground. Adjust the nut until a torque of 100 lb ft is required to run the road wheel in a forward direction. The method of checking the torque required to turn the rear wheel is shown in **FIG 11:21**. Release the park brake and check that the wheels rotate freely. Remove the torque wrench and adaptor and lower the car back to the ground.

From 1971 an independent foot-operated parking brake control has been incorporated which actuates the rear wheel brakes shoes through a cable linkage. The system is shown in **FIG 11:22** and the pedal operating system in **FIG 11:23**.

To adjust the parking brake linkage, first ensure that it is fully released. Place the transmission in neutral and raise the vehicle. Refer to **FIG 11:22** and tighten the adjusting nut against the cable equalizer enough to cause rear wheel brake drag, then loosen the adjusting nut until the rear brakes are fully released. There must be no brake drag whatsoever. Lower the vehicle and check the action of the parking brake.

Removing foot-operated control (see FIG 11:22):

1 Ensure the brake is fully released and remove all tension from the rear cables by backing off the jam nut and the adjusting nut from the equalizer (View Y).

2 If fitted, disconnect the brake warning light switch lead from the switch.

3 Remove the front cable ball retaining clip from the clevis (View T). If fitted, remove the brake warning light switch.

4 Disconnect the cable ball from the notch in the brake clevis and remove the three screws that secure the pedal bracket assembly to the left cowl inner side panel. Pull the control away from the cowl panel, remove the hairpin retainer which secures the front cable assembly to the control assembly.

5 Remove the central assembly.

Refitting:

6 Replace the parts in the reverse order, adjust the brake as previously described and check the parking action.

11:12 Trouble shooting guide

(a) Spongy pedal

1 Air in the hydraulic system
2 Fluid leak in the hydraulic system
3 Gap between linings and shoes
4 Excessive hub play with disc brakes

(b) Excessive pedal movement

1 Check 1, 2 and 4 in (a)
2 Excessively worn friction linings
3 Defective automatic adjusters
4 Very low fluid level in reservoir

(c) Brakes grab or pull to one side

1 Wet or oily friction linings
2 Distorted, cracked or damaged drum or rotor
3 Seized piston in wheel cylinder or caliper
4 Seized parking brake cable
5 Mixed linings of different grades
6 Uneven tire pressures or wear
7 Defective suspension or steering

CHAPTER 12

THE ELECTRICAL SYSTEM

12:1 Description

A 12-volt electrical system is fitted as standard. The correct polarity must be observed when connecting components, particularly items such as the alternator or radio which contain polarity sensitive components. Irreparable damage can be caused to such components if they are incorrectly connected.

Power for the system is produced by an alternator driven by a belt from the engine, and a lead/acid battery acts as a storage unit. The output of the alternator is controlled to the demands of the system and state of charge of the battery by a regulator unit. The battery supplies the very heavy current required by the starter motor for starting the engine, supplies the current for the accessories when the engine is not running, and also acts as a stabilizer for the system voltage.

The charging circuit must not be disconnected when the engine is running. To prevent induced voltages from damaging the alternator or its regulator the charging circuit must be disconnected when boost charging the battery or carrying out arc-welding repairs to the bodywork.

A test lamp and leads, or any 0-20 voltmeter, may be used to test the continuity of a circuit and to trace wiring faults but cheap instruments must not be used to test or adjust components. Cheap instruments are incapable of measuring to the accuracy required and for such tests high-grade, preferably moving-coil, instruments must be used.

Typical schematic wiring diagrams are given in the **Appendix** at the end of the manual to enable those with electrical experience to trace and rectify wiring faults.

Instructions for servicing electrical components are given in this chapter, but it must be pointed out that it is a waste of time and money to attempt to repair items which are seriously defective, either electrically or mechanically. Such units should either be exchanged for new or reconditioned units, or they should be discarded and new units installed in their place.

12:2 The battery

Regular maintenance is essential to ensure long battery life and easy starting. If the battery is neglected it will rapidly fail as well as producing starting difficulties, and the performance of the whole electrical system will suffer.

In cold weather the battery loses efficiency and at the same time more demands are made on it, as the engine is stiffer to turn and the accessories are used more frequently.

A battery that has limped through summer will give trouble in the fall and fail in the winter. If the battery is in good condition, efficiency can be restored in extreme cold by removing the battery from the vehicle and storing it in a warm room overnight.

The alternator produces a high charging current down to low engine speeds but if the vehicle is used for short-mileage journeys and city driving a charge-up on a trickle charger at regular intervals will ensure that the battery is at full charge at all times, making starting easier.

Maintenance:

1 Always keep the top of the battery clean and dry. Wipe away spillage and dirt. If spillage has caused corrosion to form on the battery surrounds, remove the battery and clean the surround. Wash with a dilute solution of ammonia or baking powder disolved in warm water and then flush liberally with clean water. Prevent further corrosion from forming by painting with anti-sulphuric paint when the parts are dry.

2 Keep the battery terminal connectors clean and tight, high resistance at this point being a common cause of starting difficulties. Preferably use special wire brushes to clean the terminal posts and connectors. Corrosion may be washed off with dilute ammonia followed by plenty of clean water. Smear the connectors and posts with non-metallic grease (such as petroleum jelly) to prevent further corrosion.

3 At regular intervals, approximately monthly, check the electrolyte level in the cells. The check should be carried out more frequently in the summer and in hot climates. Remove the cell vent caps and top with pure distilled water until the level is just over the ring in the filler well. A single cell that uses more water than the remainder is suspect. **Do not inspect the level with a naked light as the gases given off are explosive. Use nothing but pure distilled water and avoid additives which are said to charge the battery or increase battery life. Ready-mixed electrolyte should only be added to replace spillage or leakage.**

Testing:

A hydrometer provides the easiest method available to the owner of checking the battery. If it is suspected that the battery is defective or the starter circuit poor, take the car to a service garage where the battery can be checked on a heavy-duty discharge tester and the voltage drop measured in the starter heavy-duty leads.

Draw sufficient electrolyte up into the instrument to ensure that the float is free from the sides and bottom ensure that the float is free from the sides and bottom. of the electrolyte. A dirty appearance or specks floating in it indicate that this cell is defective and this will be confirmed if the reading of the cell varies by more than .050 from the other cells.

If the level is so low that electrolyte can not be drawn up into the instrument, top up and then charge the battery or use the car for a ½ hour to mix the electrolyte and water.

The specific gravity readings give the following indications:

For climates below 32°C (90°F)

1.270 to 1.290	Cell fully charged
1.190 to 1.210	Cell half-charged
1.110 to 1.130	Cell discharged

Replace spillage with electrolyte of 1.270 specific gravity

For climates above 32°C (90°F)

1.120 to 1.230	Cell fully charged
1.130 to 1.150	Cell half-charged
1.050 to 1.070	Cell discharged

Replace spillage with electrolyte of 1.120 specific gravity

The figures are given assuming a standard electrolyte temperature to 16°C (60°F). To convert the actual reading to standard, add .002 for every 3°C (5°F) increment above standard temperature and subtract the same for every increment below.

Charging:

The maximum charging current and period of charge for the various battery capacities and charge states are given in the table.

The rates given are the maximum and no harm will be done if the battery is charged at a lower rate for a longer time.

A battery is fully charged when it is gassing freely and the specific gravity does not rise in three consecutive readings taken at hourly intervals. **The gases given off are explosive, so do not bring naked lights near the battery and avoid making sparks when connecting the battery.**

A poor battery may sometimes be partially reclaimed by giving it three or four cycles of discharge and charge. Discharge the battery down to a cell voltage of 1.8 volts by connecting it to a lamp bank taking approximately 5-amps (approximately 60 watts). Recharge the battery at a 5-amp rate and then repeat the cycle. At the end of a few cycles, charge the battery up ready for use.

Specific gravity reading	*Charge rate (Amperes)*	*Battery capacity—amp/hr*			
		45	55	70	80
1.125 to 1.150*	35	65 min.	80 min.	100 min.	115 min.
1.150 to 1.175	35	50 min.	65 min.	80 min.	95 min.
1.175 to 1.200	35	40 min.	50 min.	60 min.	70 min.
1.200 to 1.225	35	30 min.	35 min.	45 min.	50 min.
Above 1.225	5	**	**	**	**

*If the specific gravity is below 1.125, use the indicated high rate of charge for the 1.125 specific gravity, then charge at 5 amperes until the specific gravity reaches 1.250 at 80°F.

Charge at 5 ampere rate **only until the specific gravity reaches 1.250 at 80°F.

Storage:

Short-term storage of a well maintained battery presents no problems. If the battery is to be stored for long periods, be sure that it is fully charged, with the top dry and all exposed metal well covered with non-metallic grease. Store the battery in a cool room well away from extremes of temperature or frost.

At monthly intervals give the battery a freshening-up charge and at three-month intervals discharge the battery and fully charge it again. If this precaution is not taken the battery will slowly self-discharge and the plates sulphate up.

12:3 Servicing electric motors

All electric motors for automobiles follow the same general design principles and therefore there are many similar constructional features. To save repetition, general instructions for the servicing of electric motors are gathered into this section. It should be noted that some motors are sealed on manufacture and therefore cannot be serviced and must be renewed if defective.

Brushgear:

Dismantle the motor sufficiently to expose the brushgear. If the armature has to be removed, or the brush holder pulled off the commutator, take great care not to damage the brushes on the end of the commutator or to allow them to pick up grease from the shaft bearings.

Check the brushes for wear and renew them if they are excessively worn. Check that the brushes move freely in their holders and that the springs press them firmly down against the commutator. Renew the springs if they are weak or damaged. If the brushes stick, remove each one in turn from its holder and polish its sides on a fine file. Clean the brush holder with a piece of cloth moistened with fuel and install the brush so that it is facing in its original direction (so as to preserve the bedding-in).

Brushes are usually supplied with their ends ground to shape but if further bedding-in is required, make up a wooden mandrel of the same diameter as the commutator. Wrap the mandrel with glasspaper (never use emerycloth as this will leave particles imbedded in the brush) and rotate the mandrel with the brushes lightly pressed against the glasspaper.

When new brushes have to be attached by soldering, grip the connector with a pair of pliers to prevent solder from creeping up the connector and making it stiff.

Commutator:

Normally the commutator should have a darkened polished surface which is free from scores or burn marks. Cleaning can be carried out by wiping the commutator with a piece of lint-free cloth dipped in denatured alcohol or gasoline. Light burn marks or scores can be polished out using a strip of fine glasspaper. Never use emerycloth on the commutator as it leaves hard particles imbedded in the copper.

Deeper damage may be machined off in a lathe. Use the highest speed possible and a very sharp tool to remove enough stock to clean off the damage. On some motors the insulation between the segments is undercut. Either use a special file or else grind a hacksaw blade to

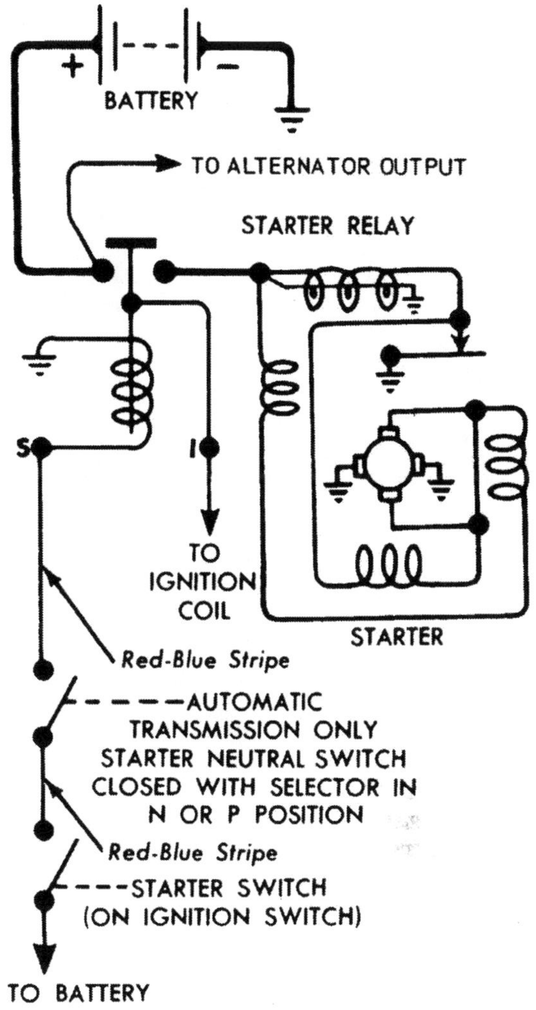

FIG 12:1 Starter circuit

the exact width of the insulation and use it to undercut the insulation squarely to a depth of $\frac{1}{32}$ inch. Take a light final cut using a diamond or carbide-tipped tool. If such a final cut cannot be taken, because of the lack of the correct tool, polish the commutator as it revolves with a fine-grade of glasspaper.

Armature:

Apart from reconditioning the commutator very little can be done to the armature. A service garage can check the windings for shortcircuits, though these can be suspected if individual commutator segments are badly burnt.

Check the laminations and segments for loosening, often caused by overspeeding of the motor. Check the laminations for scoring. If the scoring is all round the laminations it is likely to have been caused by loose field coils or polepieces or by defective bearings. If the scoring is on one side only then it is likely to be caused by a bent armature shaft. Do not attempt to machine or straighten the armature and install a new one if it is defective.

Field coils:

On some smaller motors a permanent magnet is fitted instead of field coils. Care must be taken when dismantling such motors as the armature will be drawn out with the cover, with possible damage to the brushes if they are not

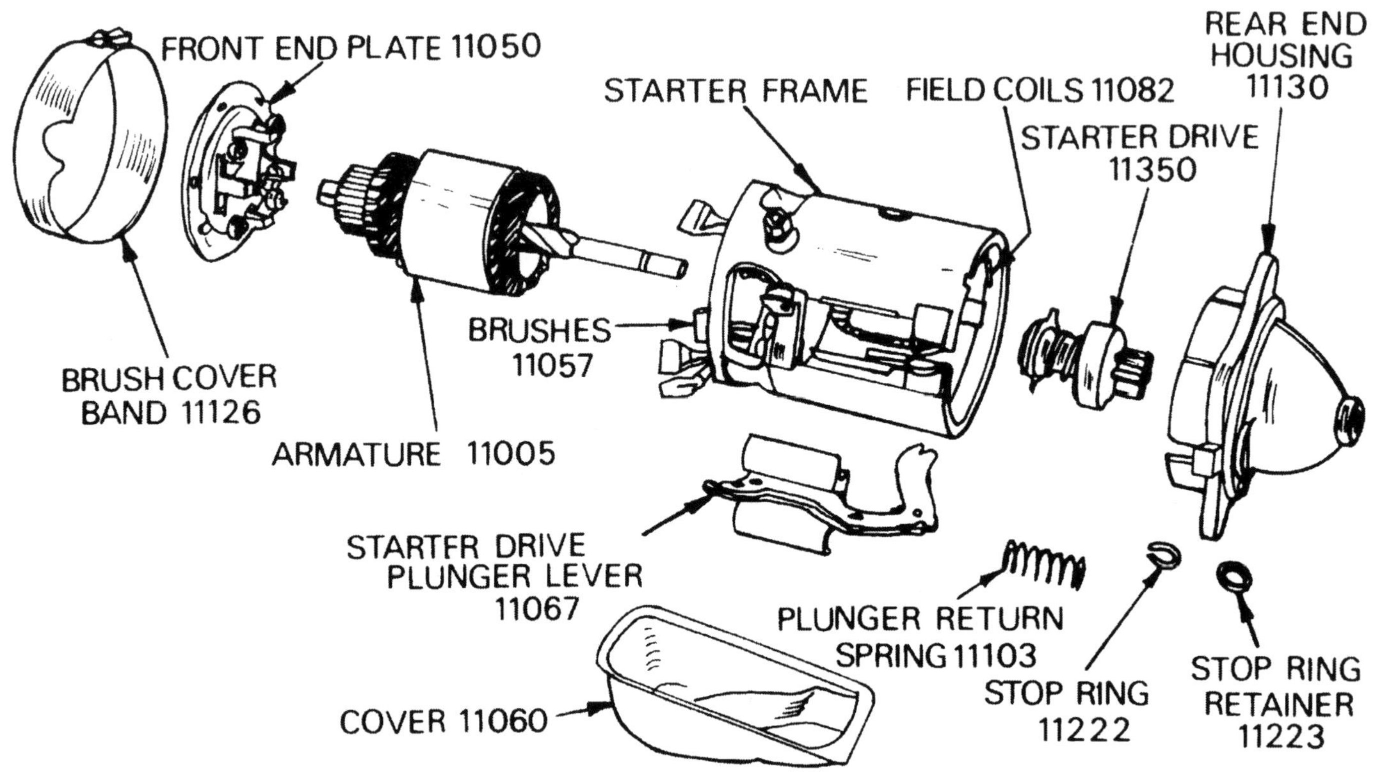

FIG 12:2 Starter motor components

eased off the commutator. Care must also be taken to install the cover correctly as the motor will run in reverse if the cover is 180 deg. out.

The field coils may be tested for continuity by connecting a test lamp and battery across them. The resistance can be measured using a resistance meter or it can be calculated using an ammeter and 12-volt supply. Note the current that flows through the coils and use Ohm's law to calculate the resistance (Resistance=Voltage divided by Current).

The field coils on the starter motor are held in place by the polepieces which are in turn held in place by special screws. These screws can only be removed and installed using a machine screwdriver and therefore if the field coils are defective they should be renewed by a service garage.

Insulation:

Wipe away all carbon and metal dust from the parts. If necessary use a cloth moistened with denatured alcohol or gasoline. The insulation may be checked by connecting a test lamp and 12-volt battery across it, though it is more effective to use a 110 AC supply and neon bulb. The lamp should not light, and if it does light then the insulation is defective.

12:4 The starter motor (positive engagement)

A typical starting circuit is shown in **FIG 12:1** and the components of the starter motor in **FIG 12:2**. The starter relay is only energized when the ignition switch is turned to the start position and, on models with automission, when the selector is in the N or P position. When the relay is energized, the coils draw in the plunger and close the heavy-duty contacts, allowing current to flow to the starter motor. One of the field coils is connected directly to ground and therefore a heavy current will flow through it. The field coil pulls across the movable pole shoe and this presses the drive pinion into engagement with the teeth on the flywheel ring gear. When the movable polepiece is fully seated, the linkage opens the contacts which ground the field coil. A holding coil keeps the movable pole in position while current flows through the motor.

Starter fails to operate or operates sluggishly:

1 Check the condition and charge of the battery, paying special attention to the connectors and battery posts. Check the ground points for corrosion or dirt that could cause a high resistance. If need be, take the car to a service garage so that the voltage drop in the starter circuit can be checked.

2 The starter may fail to operate because the pinion is jammed in mesh. On vehicles with manual-shift transmission, select a gear and rock the vehicle backwards and forwards. If this fails, or on models with automatic transmission, remove the cover from the movable pole and try to tap the pole back. If all fails, remove the starter motor for further checks and repair. Worn pinion or ring gear teeth can cause the pinion to jam.

3 If the motor fails to operate at all, switch on some lights that can be seen from the driver's seat and again try to operate the motor. If the lights go dim, the motor is taking current and will have to be removed for examination.

4 If the lights do not go dim the motor is not taking any current. Trace through the wiring using a voltmeter or test lamp to find the defective component or wire. If power is available at the starter relay, short across the heavy-duty terminals with a thick piece of wire, and if the motor now operates satisfactorily the relay is at fault and must be renewed.

Removal:

Raise the front of the vehicle on a hoist or jacks. Disconnect the battery and then disconnect the cable from the starter motor terminal. Remove the starter motor attachment bolts and then take the starter motor out of the car.

Install the motor in the reverse order of removal. Hold the motor firmly in place while snugging up the bolts, to ensure that the motor is mounted squarely against its mounting surface.

Servicing:

The components of the starter motor are shown in **FIG 12:2**. Slacken off the brush cover retaining screw and remove brush cover and starter drive cover. Note the positions of the brushes and remove all four of them from their brush holders. Remove the through-bolts (not shown in the figure) and take off the drive housing and plunger return spring. If needle roller bearings are fitted, install a dummy shaft into the bearing to keep the rollers in place. Drive out the pivot pin for the starter drive plunger lever and remove the lever. Slide the armature assembly out of the frame. Remove the brush end plate from the frame.

If the starter drive requires removal, remove the stop ring retainer so that the stop ring can be taken off and discarded.

Renew the brushes if they are worn down to less than $\frac{1}{4}$ inch length. New contacts are supplied in the service kit. The brush springs should be renewed if they apply less than 40 ozs pressure to the brushes. Check all insulation for charging or signs of overheating. The brush positions are shown in **FIG 12:3**. If new brushes have to be installed, cut the old connectors as short as possible and solder the new ones to the tags, using a 300-watt iron.

The motor is reassembled in the reverse order of disassembly. Lubricate the armature shaft splines with a thin coating of Lubriplate 777 and apply a little grease to the needle roller bearings. A fiber thrust washer is fitted to the armature shaft between the commutator and end plate. Take care not to pinch the brush leads when the end plate is installed and be sure that the springs act centrally on the brushes.

Starter drive:

Check that the drive moves freely along the splines of the armature shaft. Dirt will cause the drive to jam. The teeth must mesh to a depth of at least $\frac{1}{2}$ inch to ensure trouble-free operation. Light damage to the teeth may be smoothed down with an oilstone or fine file but if the teeth are milled on the end, worn, or broken, a new drive must be installed. Examine the teeth on the starter ring gear through the aperture for the starter mounting while slowly cranking the engine over.

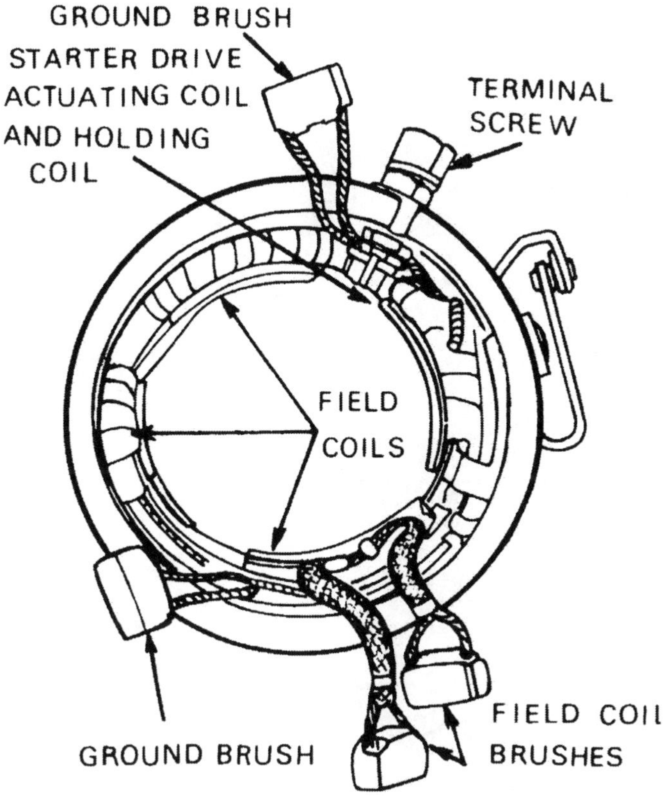

FIG 12:3 Starter motor field coils and brushes

To check the drive with the motor installed, flood the engine by pumping the accelerator 8 to 10 times. Set the ignition switch to start and hold it there. The engine should fire but not start while the starter motor continues to crank the engine. If possible, remove the drive cover so that the action of the drive plunger can be observed. If the engine stops turning and the starter motor rotates at high speed then there is a fault in the drive.

Starter motor tests:

The circuit for testing the starter motor on the bench is shown in **FIG 12:4**. The starter motor should be held down firmly as it will kick vigorously on start-up. Heavy-duty leads must be used to connect the battery to the starter as even on a no-load condition the starter may take up to 70-amps. With the battery connected as shown the motor should spin freely and at high speed.

12:5 The alternator

Typical alternator charging circuits are shown in **FIGS 12:5, 12:6** and **12:7**. Typical alternator components are shown in **FIG 12:8**.

The alternator produces an AC current. Heavy stator windings carry the main current, and as these are stationary no allowance need be made for centrifugal force and the coils can be solidly constructed. The rotor carries only a light excitation current. The excitation current does not need to change direction so it is led to the rotor by simple brushes and slip rings, which give far longer service than a commutator because of the low current. The rotor can be designed for high rotational speeds as only light windings are required.

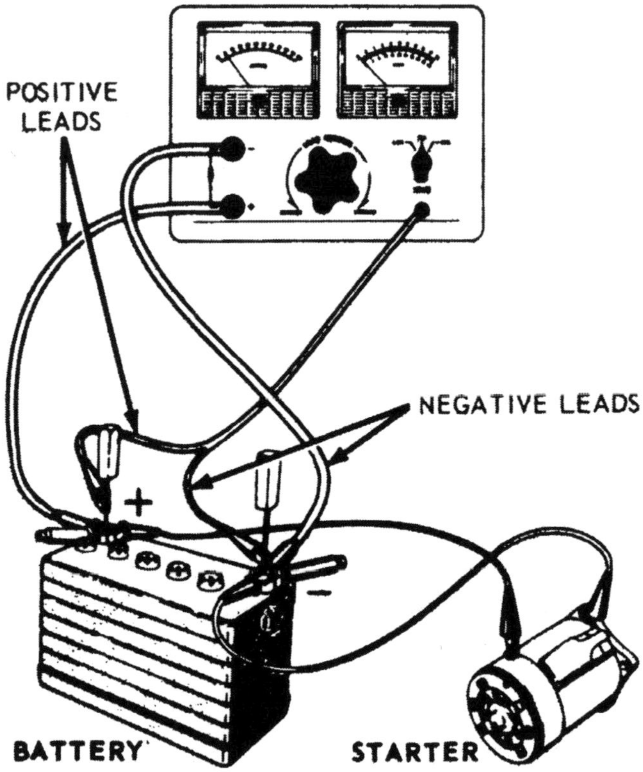

FIG 12:4 Circuit for testing the starter motor under no load

Diode rectifiers change the AC current produced to a DC current acceptable to the battery. The diodes also prevent a reverse current flow so no cutout is required and a regulator controls the output. It should be noted that a defective diode will lower the maximum output current by approximately 10 amps and it can also cause the alternator to operate with a whining noise.

Alternator fails to charge:

1 Check the drive belt tension (see **Chapter 4, Section 4:4**) and be sure that the belt is not slipping or excessively glazed.

2 Be sure that all the alternator connections are clean and tight. The charging circuit must not be made or broken while the engine is running. The connections vary slightly from year to year so if they are disconnected, careful note must be made of the cable colors to ensure that they will be correctly reconnected.

3 Typical output tests on the alternator are shown in **FIG 12:9** and **FIG 12:10**. If an ammeter and voltmeter are used in place of the proper test equipment, switch on the headlamps and other lights to apply load to the alternator. The engine should be run for at least 20 minutes before carrying out the tests, to ensure that the alternator is at its normal operating temperature. Adjust the load until the voltmeter reads 15 volts with the engine running at 2500 rev/min, and if the alternator is satisfactory the ammeter will be reading the same as the alternator rated output.

Disassembly and checking:

After removing the alternator, lightly mark both end housings and stator so that they will be reassembled in the same relative positions. Take out the long throughbolts and remove the front end housing and rotor from the stator and rear end housing. Take off the nuts and washers so that the rear end housing can be removed. Take out the brush holder mounting screws and remove the brush holder assembly.

If it is necessary to remove the stator or diodes, carefully note the positions of the wires. Unsolder them rapidly with a hot iron. When unsoldering wires from the diodes, grip the diode pins lightly with needle-nosed pliers so that the pliers act as a heat sink and prevent heat from reaching the diodes themselves.

Brushgear and slip rings:

Check the brushes for wear and renew them if they are excessively worn. Check the springs and renew the assembly if the springs are weak.

The slip rings should only require cleaning with a piece of cloth dipped in gasoline but light burn marks or scores can be polished off with a fine-grade of glasspaper. If

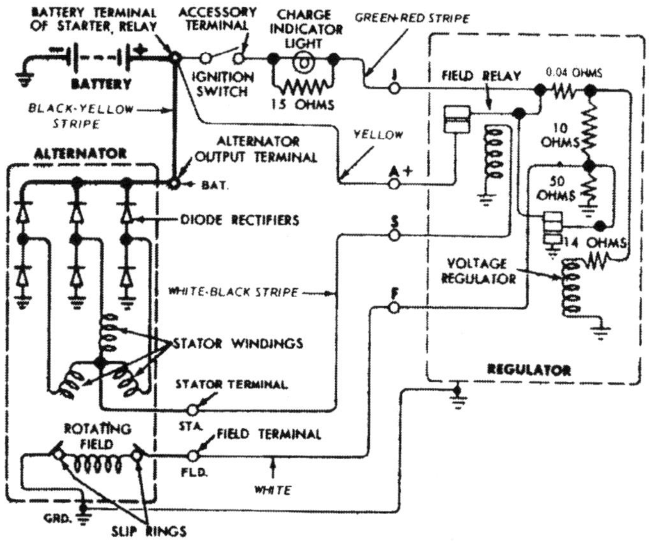

FIG 12:5 Charging circuit with indicator light

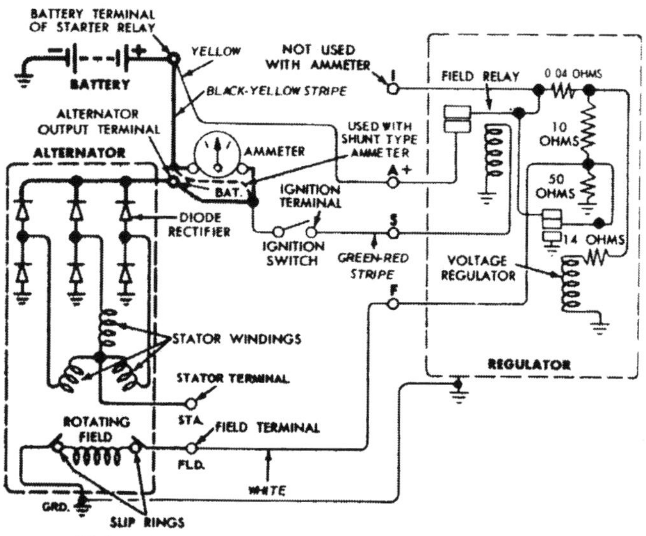

FIG 12:6 Charging circuit with ammeter

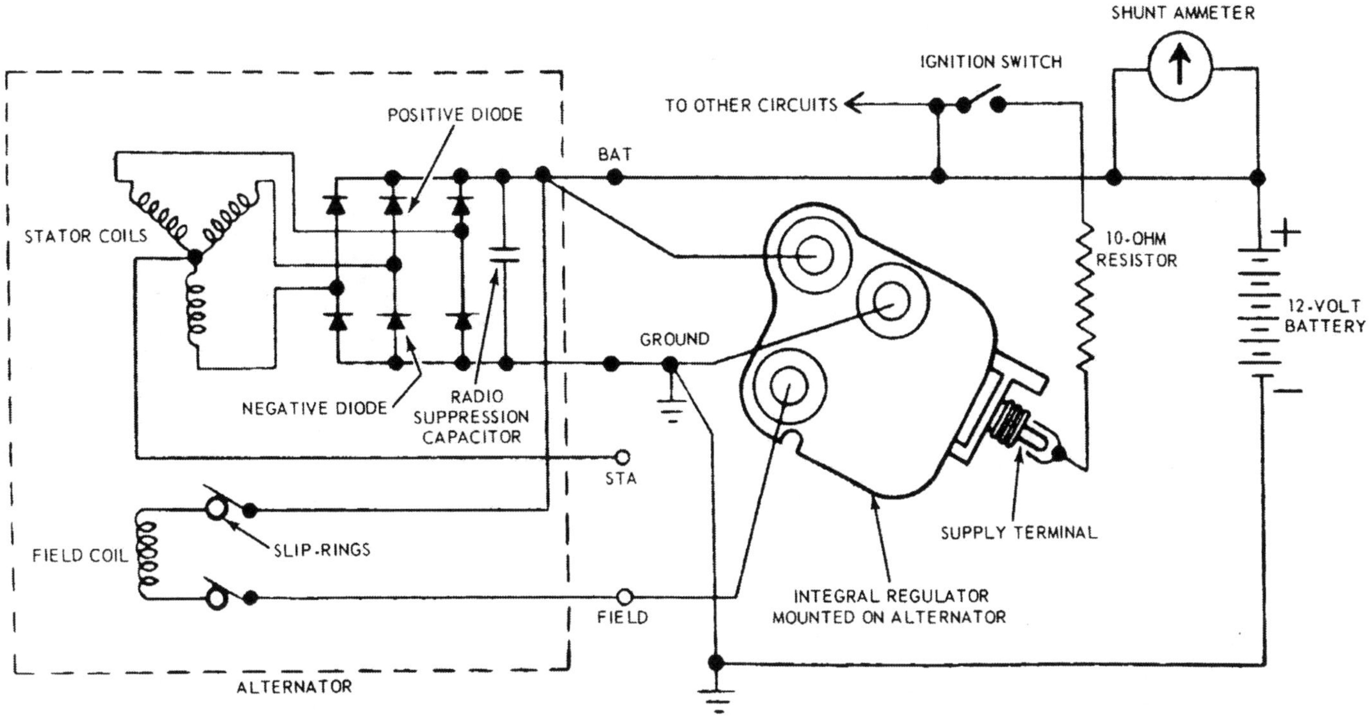

FIG 12:7 Charging circuit with regulator integral in the alternator

badly burnt or scored they may be machined down, provided that the diameter is not machined down to less than 1.22 inch.

Rotor and stator:

Check the continuity of the stator, using a test lamp and battery between any pair of stator wires, after unsoldering them from the diodes. Repeat the test between one of the wires tested and the third wire. In both cases the lamp should light.

Measure the resistance of the field coils between the slip rings on the rotor.

Blow away loose dust and dirt with an airline and give both the rotor and stator a light wipe with a piece of cloth moistened with denatured alcohol. Check the insulation using a 110 AC supply and neon test lamp.

Diode tests:

Connect a test lamp and battery, or resistance meter, to two probes. Press one probe firmly through the varnish

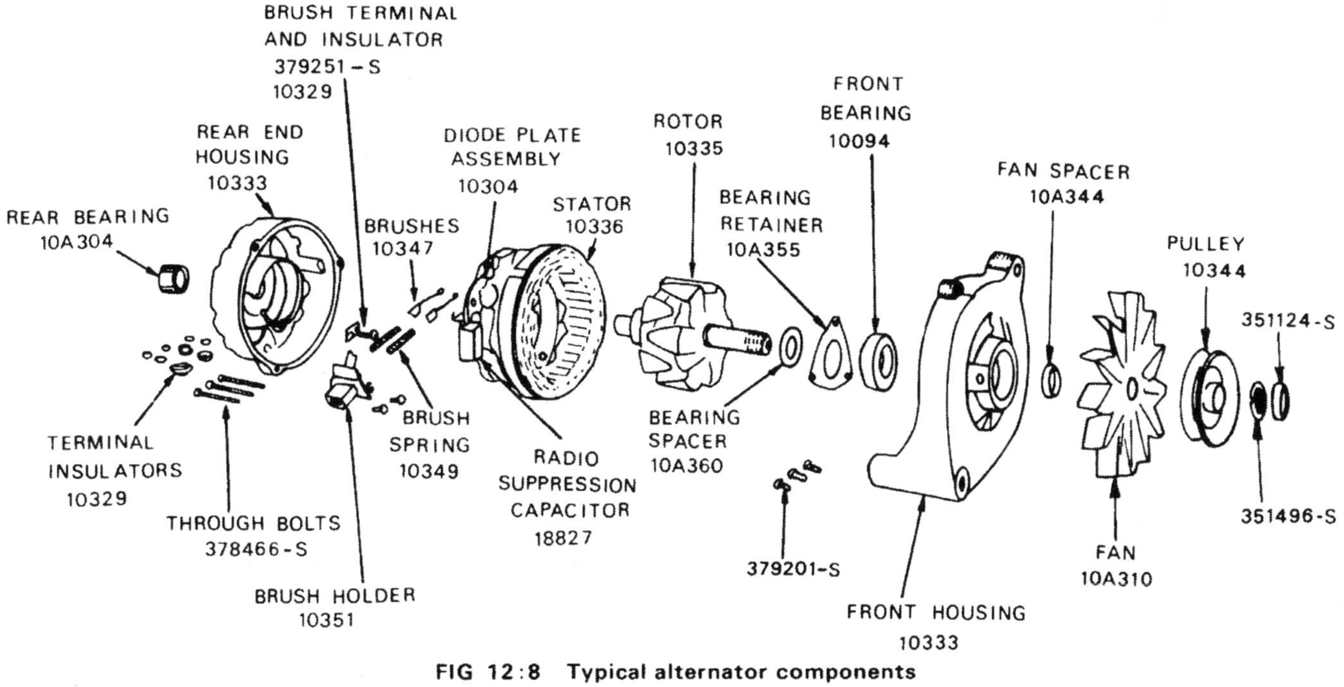

FIG 12:8 Typical alternator components

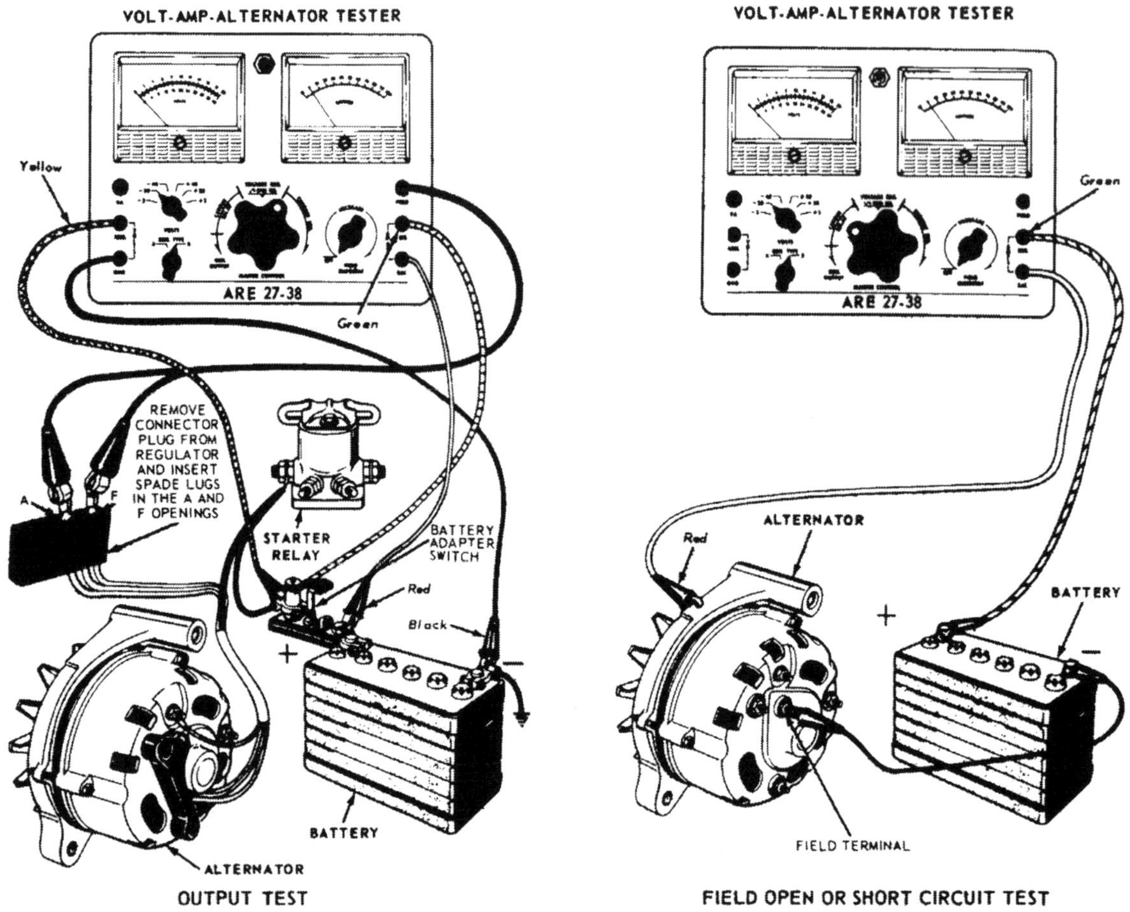

FIG 12:9 Alternator test circuits

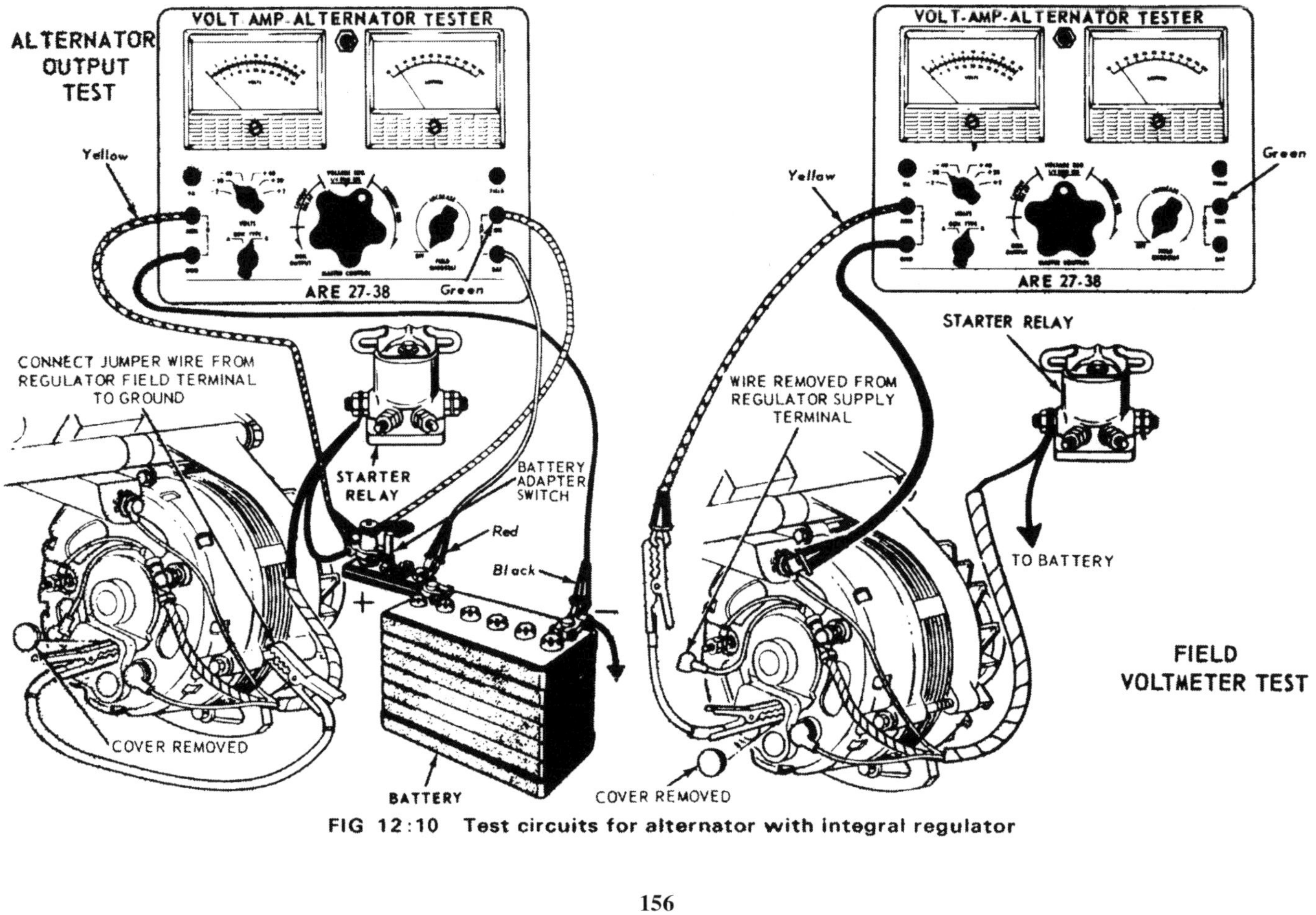

FIG 12:10 Test circuits for alternator with integral regulator

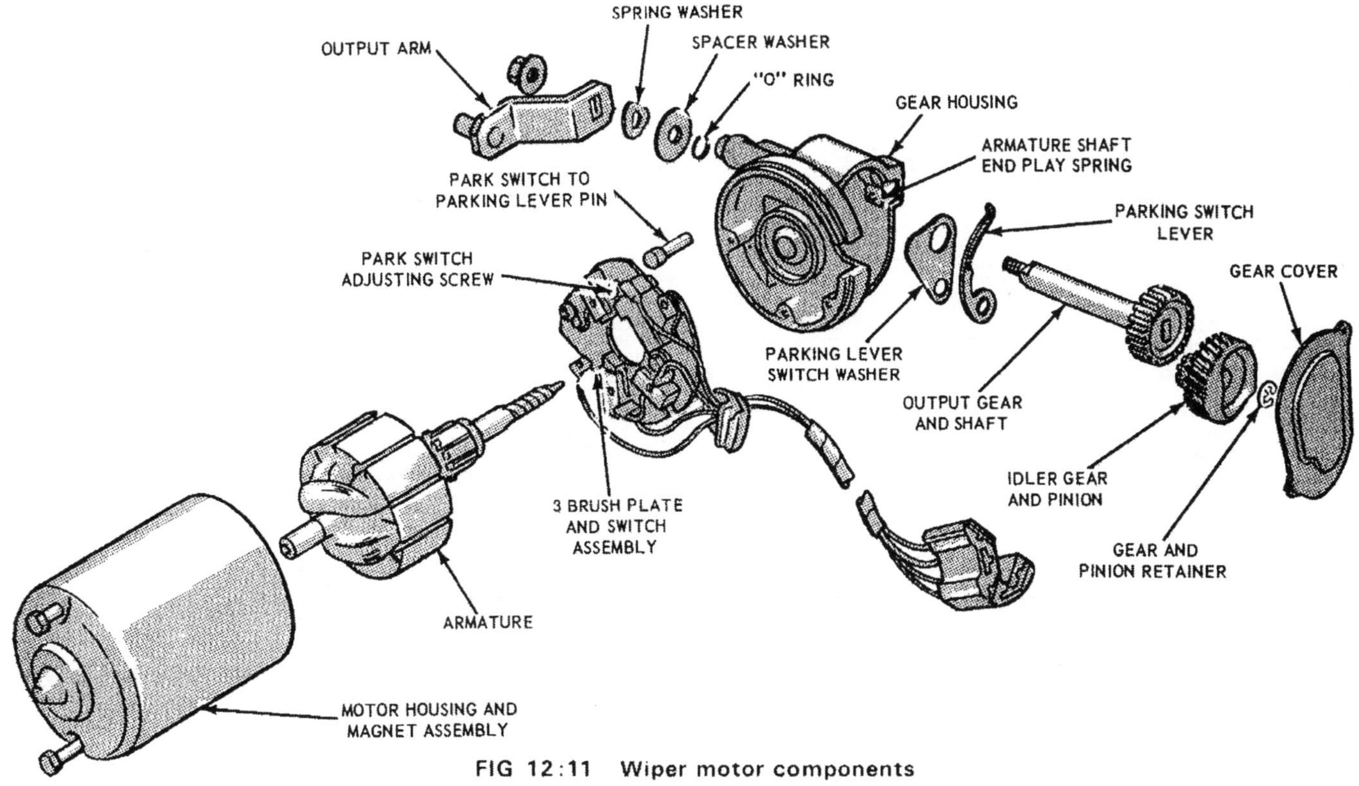

FIG 12:11 Wiper motor components

to make good contact with the baseplate and press the other probe to the diode pin. Repeat the test but this time reverse the polarity of the circuit. The diode is satisfactory if it passes current in one direction only. If the diode passes current in both directions or fails to pass it in either then the diode is defective.

Regulator contacts:

Dirt on these can cause erratic or poor charging. Clean them with a folded strip of worn 400 grade silicone carbide paper. Partially wear the strip by folding and rubbing the abrasive together and then fold it so that the abrasive is outwards. Clean the contacts by pulling the strip through them. Blow away all loose dust and dirt.

12:6 The windshield wiper

The components of a typical two-speed motor are shown in **FIG 12:11** and the motor is situated under the cowl top grille panel.

Parking:

The motor is self-parking and when it is switched off the wiper arms should lie flat along the bottom edge of the windshield. If the arms are not flush when parked, free them from the wheel box spindles and operate the motor briefly to ensure that it returns to the parked position. Install the wiper arms back into the correct position.

Typical connections for testing the park of the motor are shown in **FIG 12:12** and the test circuit for the high-speed circuit and low-speed circuit is shown in **FIG 12:13**. If the motor operates satisfactorily on the test connections then there is a fault in the wiring.

If the motor fails to park correctly on the test connections, remove the through-bolts and carefully take off the motor cup and armature. The cover should be partially withdrawn and the brushes then lifted off the commutator so that they are not damaged as the armature is withdrawn. Rotate the output shaft until the contacts, shown in **FIG 12:14**, are firmly closed. Rotate the adjusting screw clockwise until the contacts just open and then rotate the screw anticlockwise by one complete turn. Check

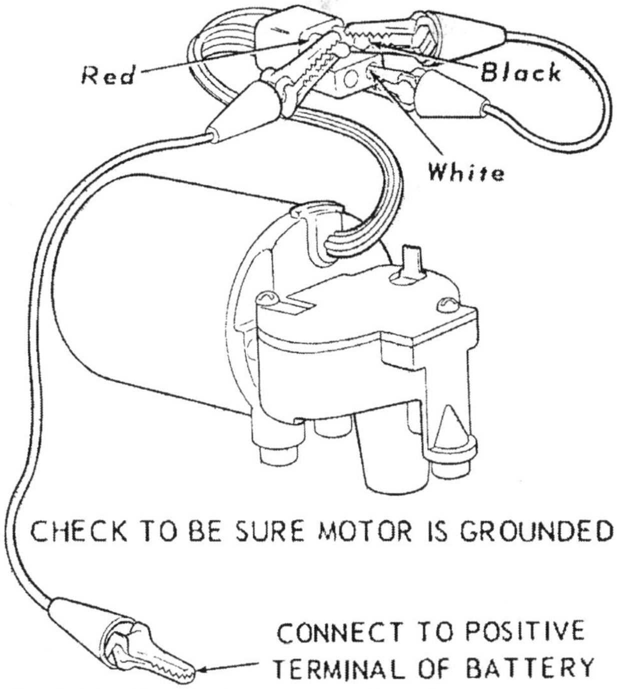

FIG 12:12 Circuit for testing the wiper motor park function

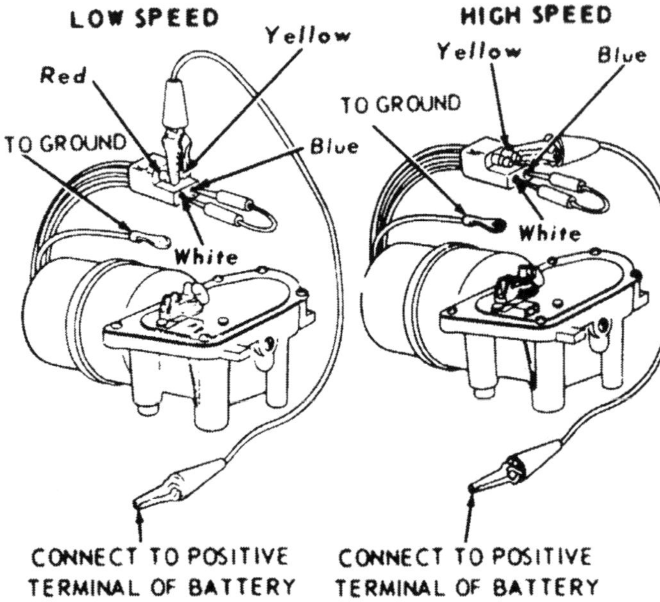

FIG 12:13 Circuit for testing the wiper motor operation

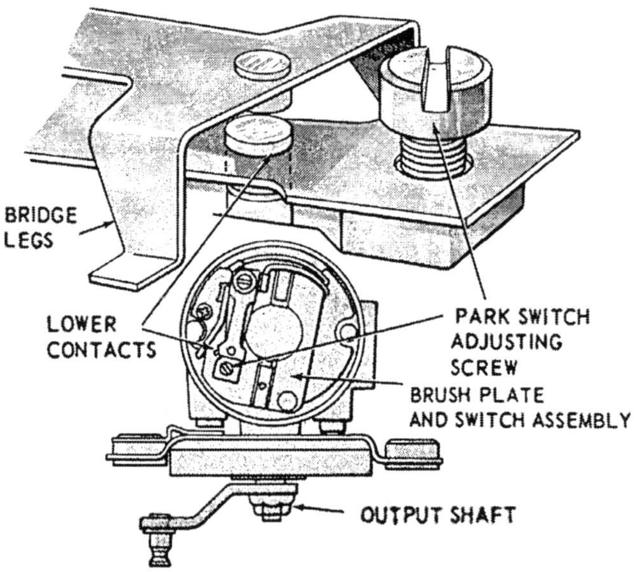

FIG 12:14 Adjusting the wiper motor park contacts

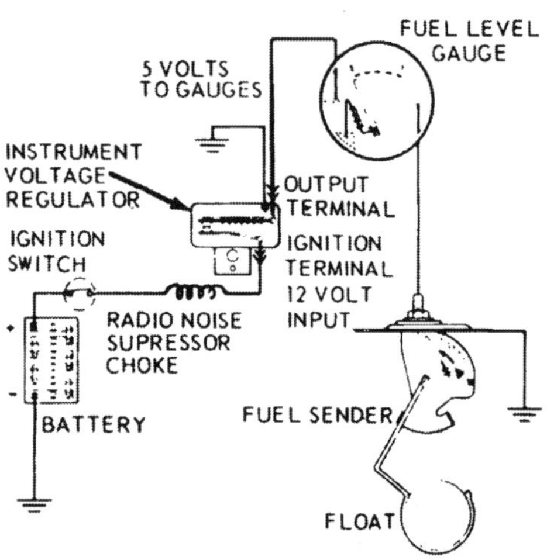

FIG 12:15 Fuel gauge circuit

the bridge to ensure that legs are contacting the brush plate when the contacts are closed. Install the armature and cup in the reverse order of removal and check that the motor now operates satisfactorily.

Poor wiper operation:

Note that the blades will drag if the windshield is dry, so wet it before carrying out tests.

Disconnect the linkage from the motor and run the motor separately so that it can be checked separately. Move the linkage by hand and check for binding or defects. If the motor is operated with the gearbox cover removed, take great care as the gears have enough torque to cause serious injuries to the fingers. Likely causes of poor motor operation are defective brushgear, insufficient armature end float (should be approximately.003 inch), insufficient lubrication, and worn or damaged gears.

12:7 The temperature and fuel gauges

The gauges operate by the bending of a bi-metallic strip as it is heated by a coil around it. The higher the current flowing through the coil the greater will be the heating effect and therefore the greater will be the deflection on the needle. To ensure that fluctuations of voltage in the system do not cause the gauges to read incorrectly, and also to ensure that they give a steady reading which does not fluctuate as the fuel sloshes in the tank, the supply is taken from a voltage stabilizer unit (mounted on the back of the instrument panel). The unit has a bi-metallic strip which is opened and closed by the heating and cooling of a coil. In use the unit gives the current flow as if a constant voltage of 5 volts were applied. Special instruments are required to test the components and if an ordinary voltmeter is used to measure the output it will fluctuate between 0 and 10 volts if the unit is operating correctly.

Intermittent or incorrect readings can be caused by defective wiring so trace through the wires and check for breaks, faulty insulation, or poor connections. No reading indicates a defective voltage stabilizer unit. The circuit for the fuel gauge is shown in **FIG 12:15**.

If the fuel gauge operates incorrectly, borrow a new sender unit. Disconnect the wires from the unit in the tank and connect them to the new unit. Connect a jumper lead between a good ground and the body of the unit. Switch on the ignition and leave the new sender unit float arm in the lowest position. The new needle in the gauge should settle on the Empty mark. Raise the float arm gently to its highest position and wait a minute for the needle to climb and settle, after which it should read Full. If the reading is still incorrect, and the wiring is satisfactory, the gauge is defective and must be renewed. If the readings are satisfactory, renew the tank unit.

It should be noted that on some models a low-fuel level warning is fitted. A thermistor is fitted to the tank unit, and connected to the system by a third connector. When the fuel is so low that it uncovers the thermistor, the thermistor warms up slightly and operates the warning light, using a relay.

If a temperature gauge is fitted, the system can be checked using two resistors. Disconnect the wire from the sender unit, on the engine, and connect it to the

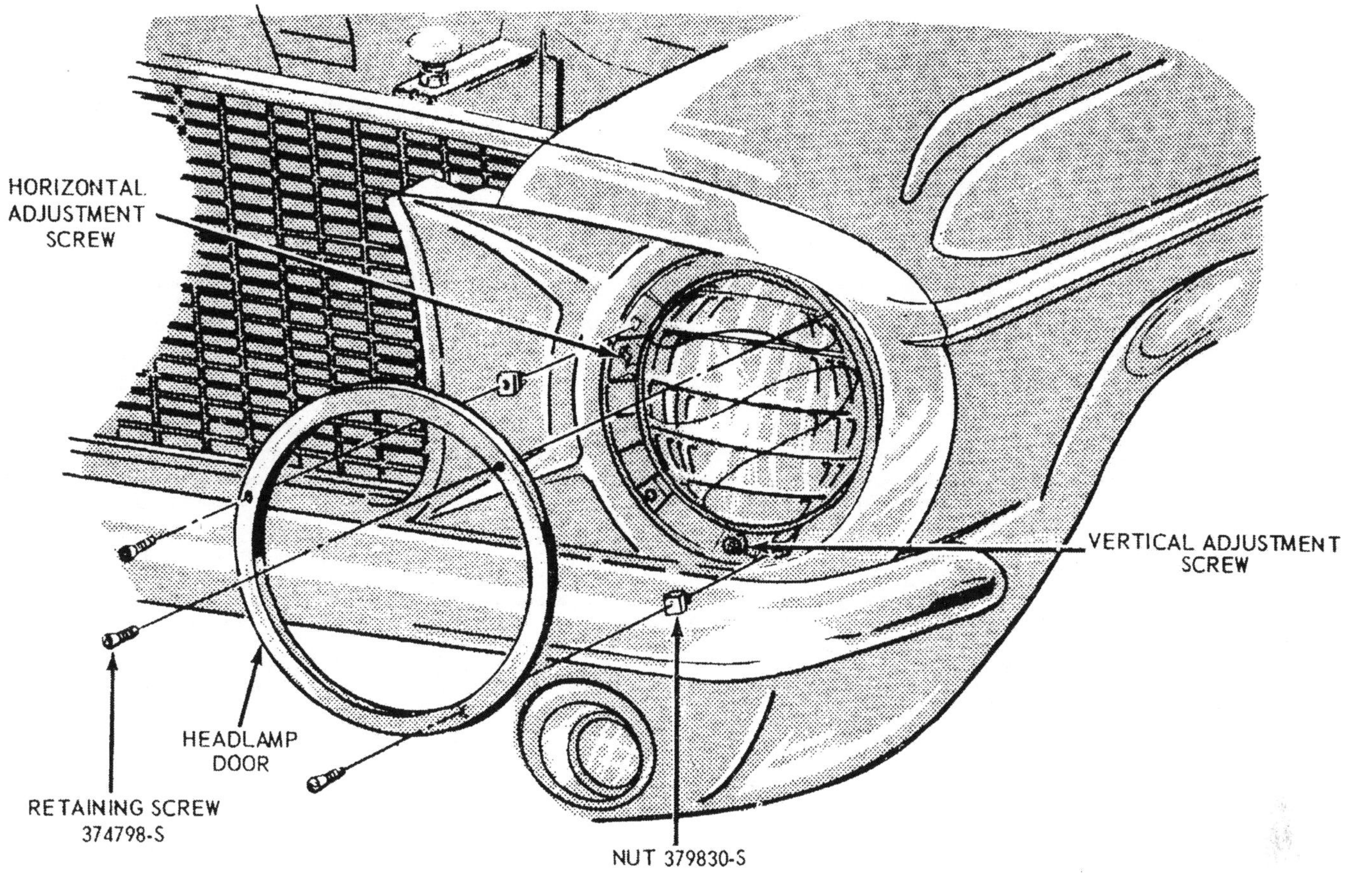

FIG 12:16 Headlamp attachments

ground through a 10-ohm resistor and the gauge should read High. Substitute a 73 ohm resistor and the gauge should give a low reading.

If temperature warning lights are fitted, they operate from full battery voltage. The sender unit contains a bi-metallic strip which closes one contact when the temperature is low and closes another contact when the temperature is high.

The temperature sender unit must be sealed in place using a waterproof sealer which is also electrically conducting.

Full battery voltage must never be applied across units operated from the voltage stabilizer as they will be damaged by this type of overload.

12:8 The turn indicators

The turn indicators use the dual-filament parking lights to indicate the direction of turn. A flasher unit located in the back of the instrument panel makes-and-breaks the circuit, under the action of a bi-metallic strip heating and cooling, to provide the flashing. A separate emergency flasher unit comes into operation when the emergency button is pressed and this makes all four lights flash together. The flasher units are delicate and liable to damage if dropped or incorrectly connected.

If the indicators do not operate correctly, check the bulbs as a blown or damaged filament will alter the flash rate and may produce all sorts of odd effects. Use the emergency flasher as a test.

Reach under the instrument panel, and push in and then twist the unit to free it from its mounting. Disconnect the wires and use a voltmeter or bulb to check that full battery voltage is reaching the unit with the ignition on. Switch off the ignition and connect the wires together. Switch on the ignition and operate the turn indicator switch. If no lamps light, suspect the switch but still check all wiring. If only one lamp on a side operates, check the wiring and ground point of the lamp that fails to operate. If all the lamps operate correctly, the flasher unit is defective and a new unit should be installed.

12:9 The headlights

A typical headlight attachment is shown in **FIG 12:16**. Sealed beam units are used where the filaments are integral in the reflector and lens, forming one large bulb. The unit can be removed after taking off the headlight door. When removing or installing the unit, do not alter the beam adjustment screws.

Hold the headlamp assembly by hand, after taking off the door, and unhook the spring from the retaining ring. Remove the retaining ring by unhooking it from the opposite side to the spring. Pull out the bulb and disconnect it from the wiring assembly plug.

Install the headlight in the reverse order of removal.

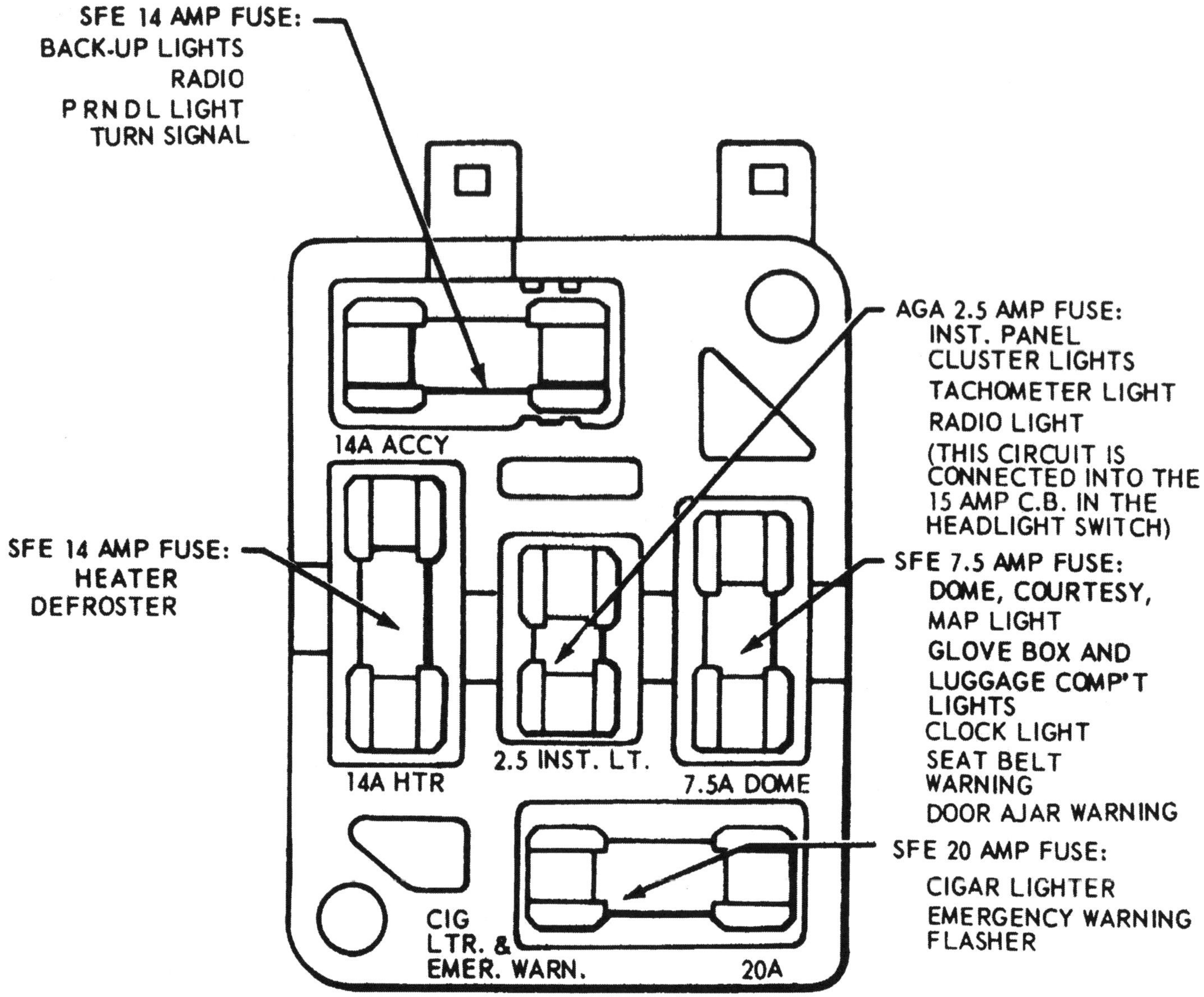

FIG 12:17 Fuse panel (early type)

① (20 AMP.) HORNS.

② (14 AMP.) (COURTESY LAMPS) INSTR. PANEL, DOOR, "C" PILLAR, GLOVE BOX, MAP LAMP, LUGGAGE COMPARTMENT LAMP, DOME, DOME LAMP, CIGAR LIGHTER FEED, IGN. KEY BUZZER AND AUTO. SEAT BACK LATCH AND CLOCK FEED

③ (4 AMP.) (INSTR. PANEL ILLUM.) HEATER SWITCH, HEADLIGHT SWITCH, W/S WIPER–WASHER SWITCH, ASH TRAY, RADIO, CLOCK AND GAGE ILLUM.

④ (14 AMP.) (INDICATOR LAMPS) SEAT BELT AND EMISSION CONTROL, DOOR AJAR, PARKING BRAKE, DUAL BRAKE

⑤ (14 AMP.) HEATER-DEFROSTER NOTE: (30 AMP.) REQUIRED FOR AIR CONDITIONER

⑥ (15 AMP.) BACK-UP LAMPS, RADIO, W/S WASHER, LOW FUEL AND TURN SIGNAL INDICATOR

⑦ (20 AMP.) (ACCESSORY FEED)

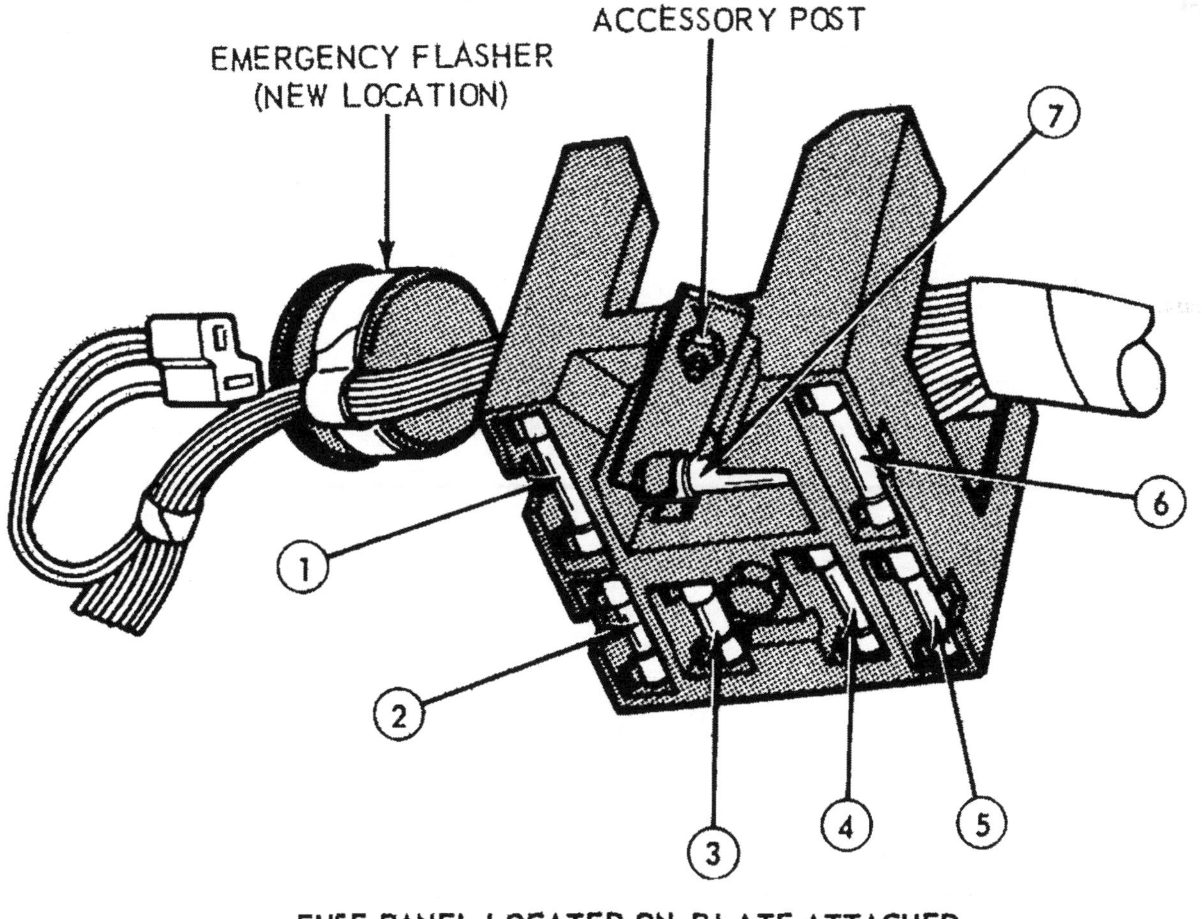

FUSE PANEL LOCATED ON PLATE ATTACHED TO LOWER RIGHT HAND FLANGE OF BRAKE PEDAL SUPPORT

FIG 12:18 Fuse panel (later type)

Beam setting:

Two adjusting screws are fitted to each headlight and they are used to set the beams correctly. Rough adjustments can be made by standing the car squarely to and some distance from a plain wall. To ensure that the lights meet all legal requirements the car should be taken to a service garage where the lights will be set with accuracy using special beam-setter equipment.

12:10 Fuses

Circuits are protected by fuses in a fuse panel and in case of a shortcircuit or overload the fuse element melts and cuts the circuit. The early fuse panel is shown in **FIG 12:17** and the later type in **FIG 12:18**.

A fuse blowing is not always an indication of a fault as fuses do weaken with age. If a fuse does blow, briefly examine the circuits it protects and install a new fuse, of the correct rating. Switch on each circuit in turn. If the new fuse blows when a circuit is switched on, check that circuit thoroughly. Shake the wires in case insulation is defective and there is an intermittent ground. Do not install a fuse of higher rating than called for (or wrap the fuse with foil) as the fuse is designed to be the weak link in the circuit. If a fuse of too high a rating is installed the fuse may not blow and the wiring may catch fire instead.

Circuit breakers:

These are fitted to some circuits (such as headlights and windshield wipers) in place of fuses to ensure that the essential systems will operate at least intermittently if there is a fault in the system. The circuit breakers can be tested using a 12-volt supply, ammeter and variable resistance. The current is adjusted to the normal value and the circuit breaker should not operate. When the current is adjusted to a high value, the circuit breaker should open within a minimum of 30 seconds and then close again as it cools (only to operate again). Renew defective circuit breakers as they cannot be repaired.

12:11 Trouble shooting guide

(a) Battery discharged

1 Loose or dirty terminal connectors
2 Insufficient charging current
3 Shortcircuit in part of circuit not protected by fuse
4 Defective alternator, slipping belt or defective regulator
5 Battery internally defective
6 Accessories left on overnight

(b) Battery will not hold charge

1 Low electrolyte level
2 Electrolyte lost through spillage or leakage
3 Battery plates sulphated or separators ineffective

(c) Alternator output low or nil

1 Broken or slipping drive belt
2 Defective or incorrectly set regulator
3 Loose connectors or defective wiring
4 Defective diode
5 No battery supply
6 Defective rotor or stator windings

(d) Starter motor lacks power or will not operate

1 Battery discharged or loose and dirty connectors
2 Starter pinion jammed in mesh
3 Defective ignition switch
4 Defective starter relay
5 High resistance in heavy-duty leads
6 Defective brushgear or commutator
7 Defective armature or field coil windings
8 Engine abnormally stiff

(e) Starter runs but does not turn engine

1 Badly worn or broken teeth in drive

CHAPTER 13

THE BODYWORK

13:1 Bodywork repairs

Large scale repairs to bodywork should be left to the experts. Even small dents can be tricky as injudicious hammering will cause the metal to stretch and will make the dent worse instead of better.

Acrylic finishes are used as standard on the exterior and nitrocellulose lacquers must therefore not be used on top of standard finishes.

Spotting or minor blemishes on the paintwork can be removed by polishing. This method can also be used to restore the finish if it has suffered by weathering and lost its gloss. Machine polishing will make the task much easier. Remove the defects using 600 grit paper with white spirits or water as a lubricant. Take care to remove only sufficient paint to take out the defect. Paint on a white or light colored medium-grit machine polishing compound (such as Sno-Flake No. 16) with a brush. Polish the entire surface, at 1850 rev/min, using a carpet pad ($\frac{5}{8}$ inch nap) or lambswool pad. Remove the compound and buff the surface with a clean lambswool pad. Wax polish is not essential on acrylic surfaces but there are no reasons against its use.

Self-spraying cans and touch-up kits in matching colors are available. The touch-up kit should be used regularly on stone chips and minor scratches. Polishing down after hardening can be carried out with a mild metal polish.

For deeper dents, filling and then spraying is the best method of repair available to the owner. It is better to spray a complete panel rather than just a patch as any mismatch in the paint will not be so readily obvious. Remember that the finish fades and alters color slightly with age and the new paint will be an exact match for the car when new.

Remove all trim and handles from the area if possible. The final result will be better if the trim is removed rather than masking over it. Wash off all wax polish with white spirits. If silicone-based polishes have been used it is essential to use a special cleansing fluid to remove all traces of them. Traces of polish will leave the surface with pits, craters and pop-ups. Lightly scuff the area to give a good key for the new finish. If there are any traces of corrosion, rub down to bright metal and apply a coat of primer.

Apply paste stopper or filler as required to bring the damaged area just slightly proud of the original finish. Leave the filler to harden and then rub it down using 400 grit paper and water as a lubricant. Add further paste or

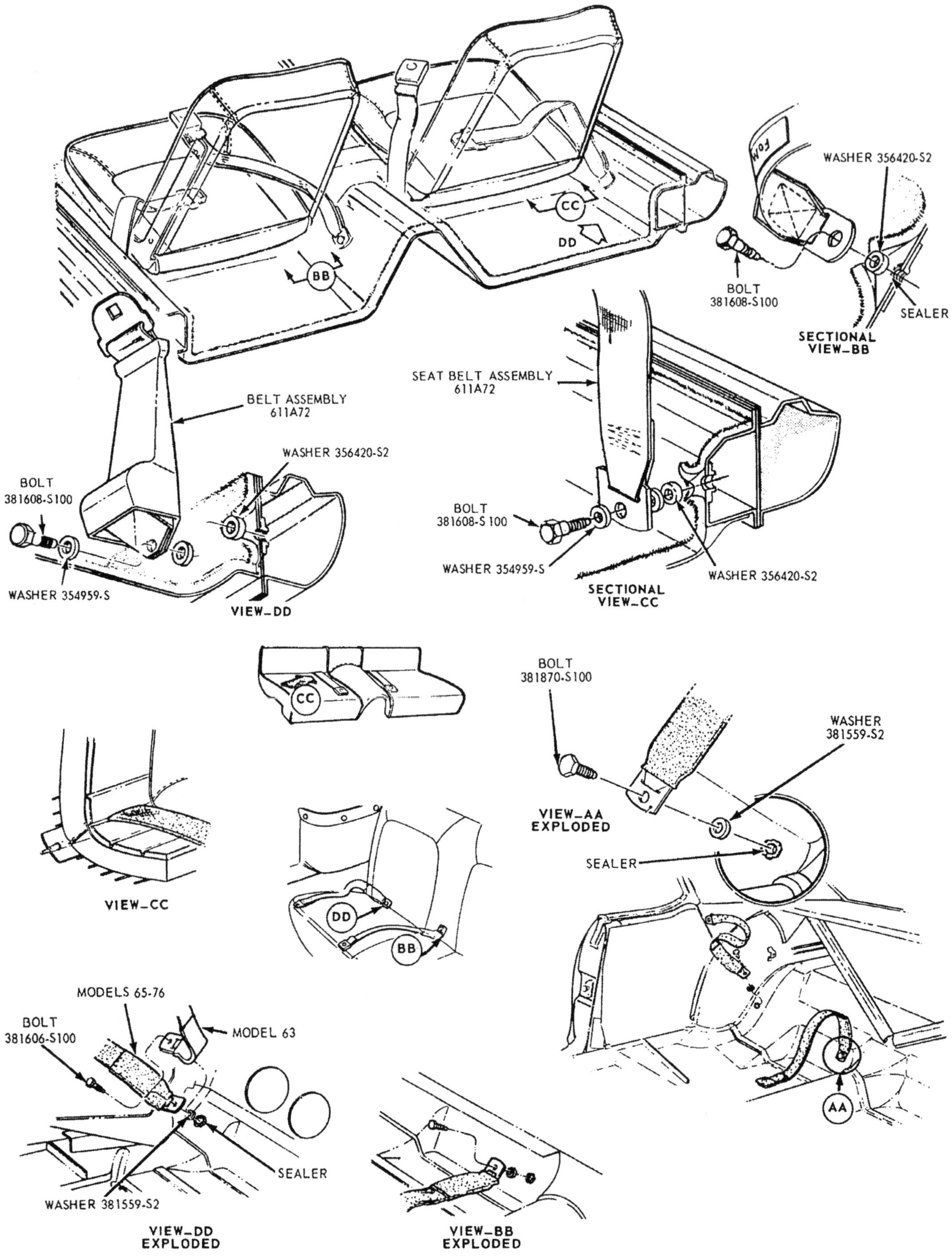

FIG 13:1 Seat belt attachments

filler as required to fill in any hollows or blemishes. Spend plenty of time and patience on this filling and rubbing-down stage as this finish will govern the final finish, when small blemishes will stand out glaringly in the gloss. Wash down the surface with plenty of clean water and leave it to dry. Some slurry will probably have been missed in the first wash so give a second wash to ensure that the surface is perfectly clean. When it has fully dried again, wipe it over with a proprietary 'tacky' cloth to remove any dust or specks.

Mask off the surrounding areas with paper and masking tape. Small, difficult-to-mark parts can be smeared over lightly with a thin film of grease instead of masking them, but take great care to avoid touching the area to be sprayed with greasy fingers.

Spray evenly over the surface, taking care not to let the paint build up on the edges of the panel. Apply the paint in two or more thin coats, gently rubbing down between coats, rather than one thick coat which will easily sag or run. Remove the masking and leave to dry for at least 24 hours. Polish the surface as described earlier to give a high gloss. Cutting compound may be used in place of the machine polishing, followed by careful polishing with a mild metal polish. Leave the paint to harden for several weeks before applying metal polish.

13:2 Seat belts

Typical seat belt installations are shown in **FIG 13:1**. Only Ford approved belts should be used and they must be correctly installed. **If the owner has any doubts whatsoever about his ability to make a safe installation the work should be carried out by the dealer.**

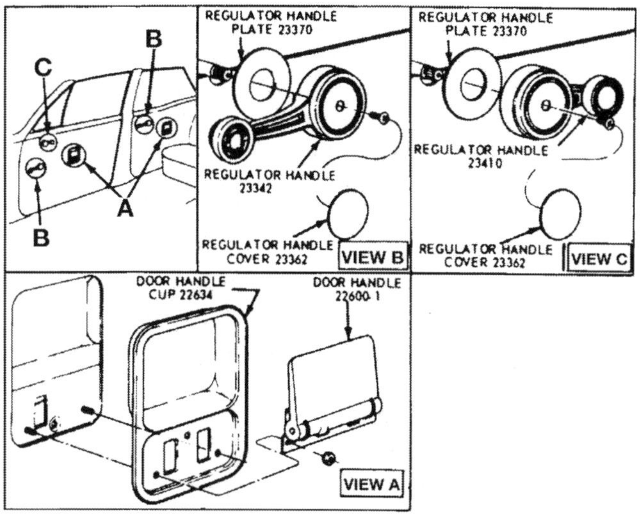

FIG 13:2 Typical door and window regulator attachments

Cleaning:

Belts that are soiled should only be cleaned with a solution of commercial soap or detergent in warm water. **Do not use carbon tetrachloride, naptha, trike or any such solvent.** Leave the belts laid out to air-dry away from direct sunlight. Bleaching or re-dyeing the belts is not recommended as it may weaken the belts.

13:3 Door components

The door glass is of the one-piece ventless type controlled by a regulator.

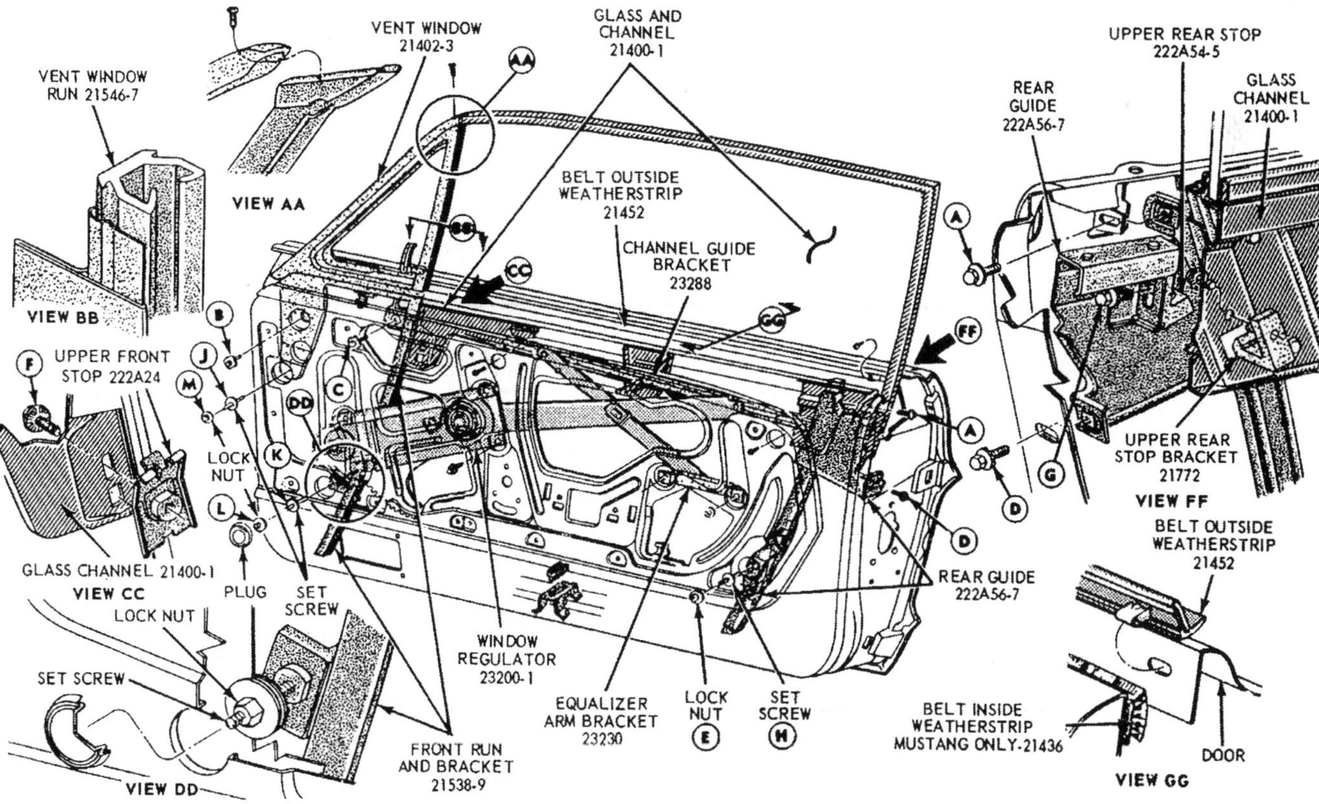

FIG 13:3 Typical door glass and regulator attachments

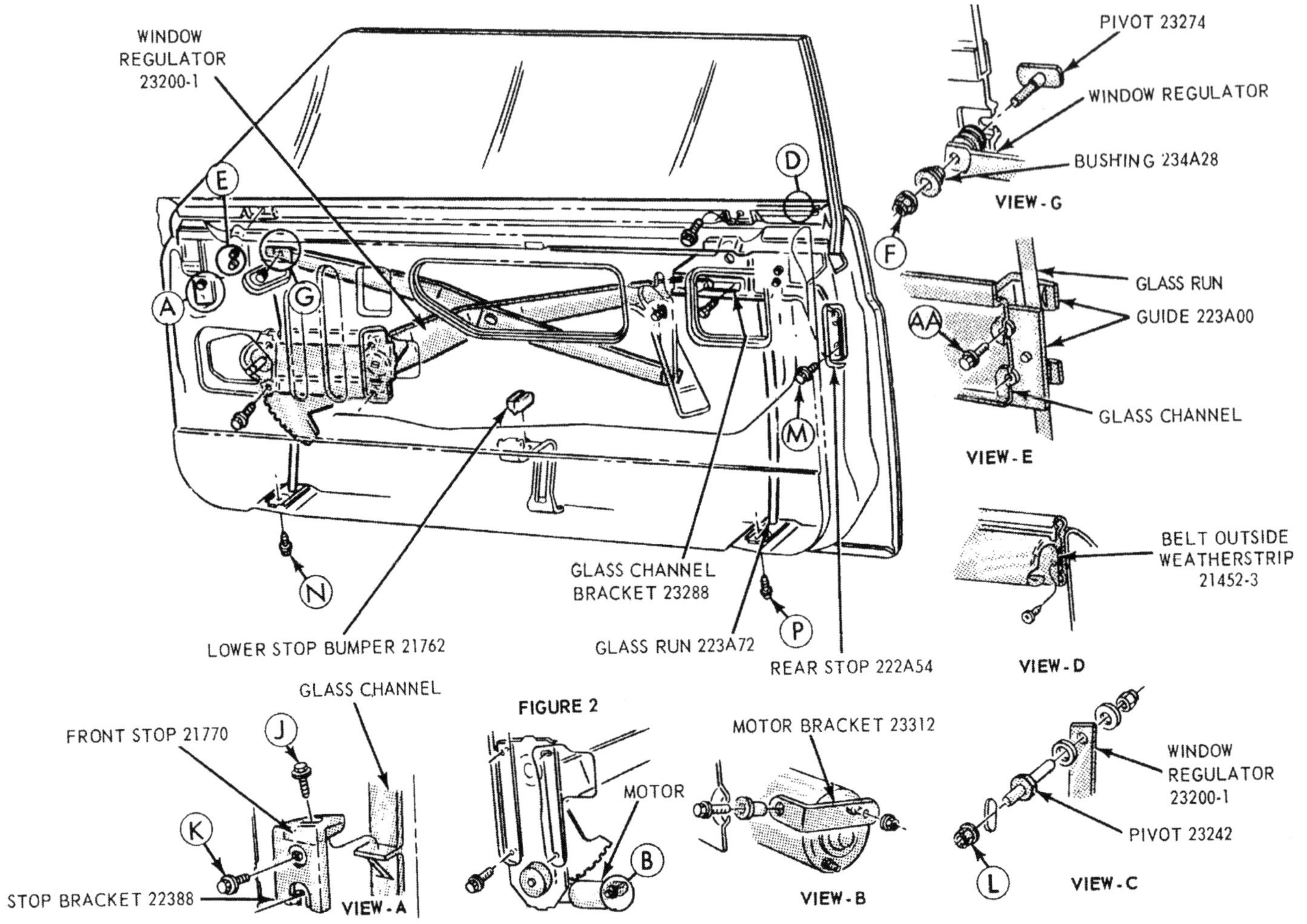

FIG 13:4 Door mechanism without vent panes

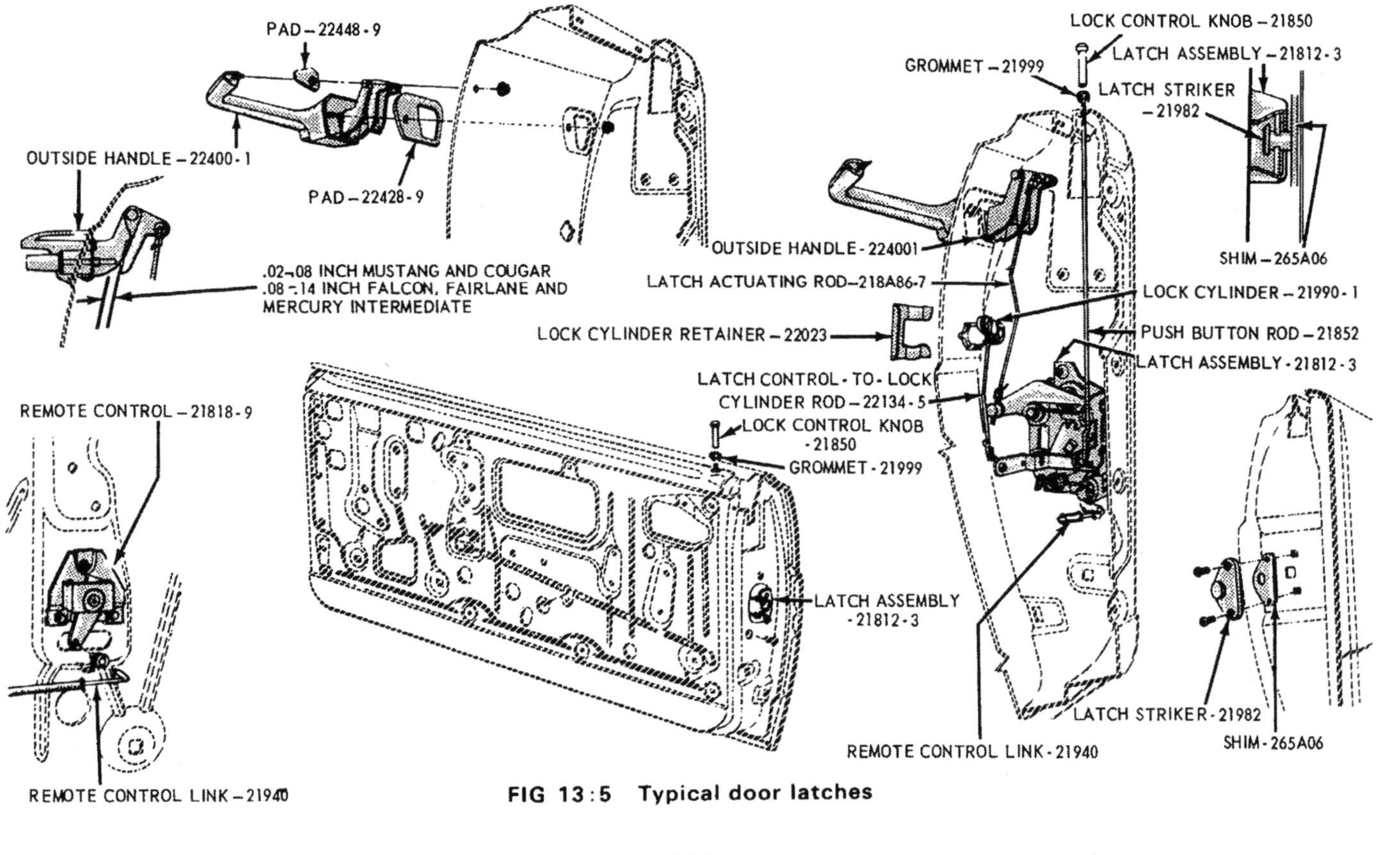

FIG 13:5 Typical door latches

Trim panel:

Pry the cover from the regulator handle and then take out the securing screw so that the handle can be removed, as well as the plate under it. Unscrew and remove the door latch control. Take out the armrest retaining screws and remove the armrest with its applique. If a rear view mirror is fitted to the door, unscrew the nut for the control and push the control toward the inner door panel.

Slide a putty knife or steel rule between the trim panel and door, taking care not to scratch the paint, and ease out the clips that secure the panel. Keep the knife as near to the clip as possible to prevent the clip being torn from the panel. When all clips are free lift the panel off from the door. Carefully peel off the plastic watershield from the inner door panel.

The parts are installed in the reverse order of removal. The watershield should be attached to the inner panel with sealer. Daub a little AB.19560.A sealer over each retaining clip hole and also around door handles, regulator shaft and other apertures. The regulator handle should be installed so that it is straight downwards when the window is closed.

Window regulator:

The attachments are shown in **FIG 13:2**. Remove the panel. Take out the five bolts securing the regulator to the inner door panel, while holding the glass in position with a rubber wedge. Disengage the regulator arms from the door channel and remove the regulator through the aperture in the door panel.

The regulator is installed in the reverse order of removal. Before installation, lubricate the gears with grease and rollers with polyethylene grease, as shown in **FIG 13:3**. The polyethylene grease is available in spray cans with extension nozzles.

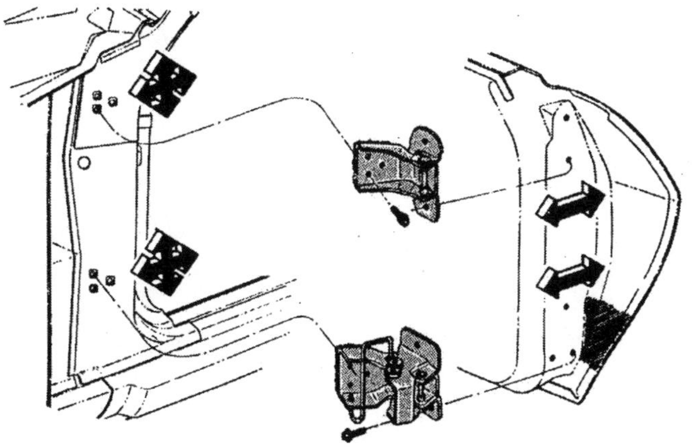

FIG 13:6 Door hinges

Power operated regulators (see FIG 13:4):

Electrically operated windows are available on some later models. The window ventilation is controlled by a switch secured to the door trim: the switch energizes a small electric motor which operates the regulator arm.

To remove the switch control, first remove the trim panel. Disconnect the wires from the switch, remove the attaching screws and detach the switch.

To remove the regulator and motor, remove the trim panel and watershield. Remove the regulator motor bracket and disconnect the wires. Disconnect the window regulator arms from the glass channel and the door inner panel. Disconnect the remote control rod from the door latch and remove the rear weatherstrip cap. Take off the upper rear stop and remove the belt weatherstrip and moulding. Back out the window regulator attaching screws and detach the regulator from the door.

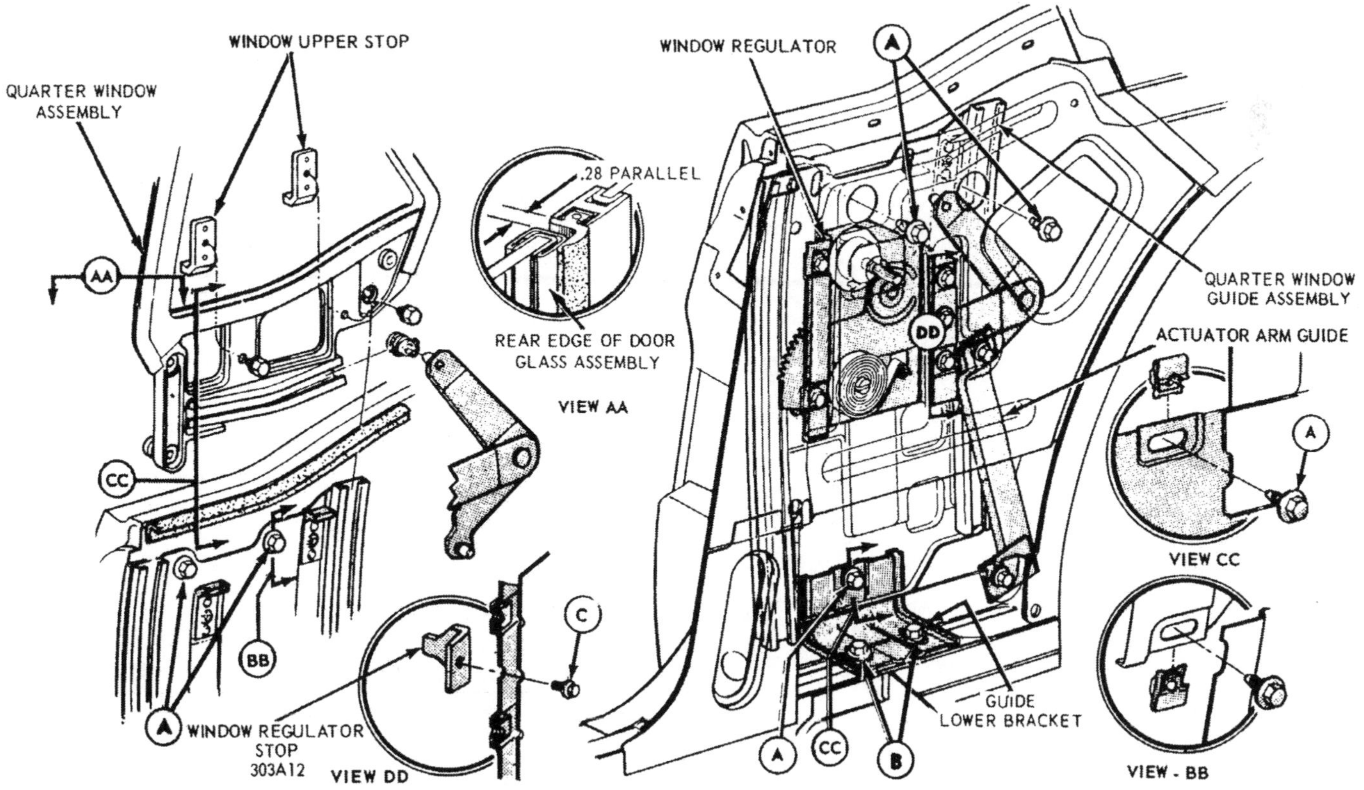

FIG 13:7 Typical quarterlight mechanism

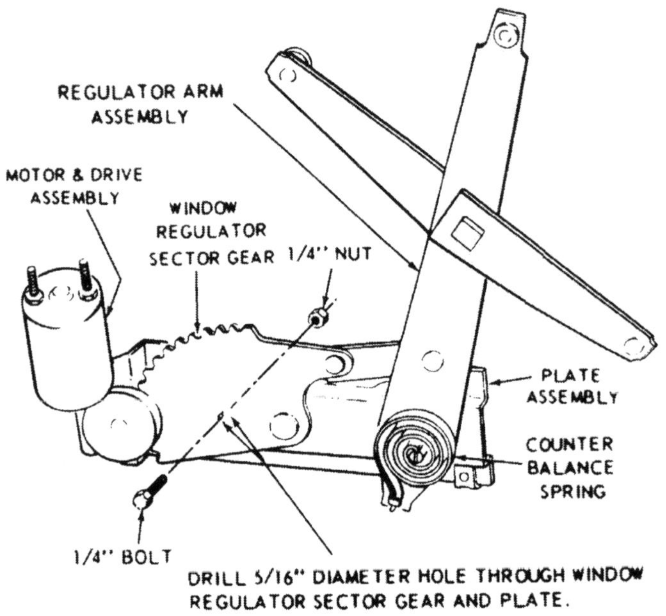

FIG 13:8 Power window motor and drive assembly

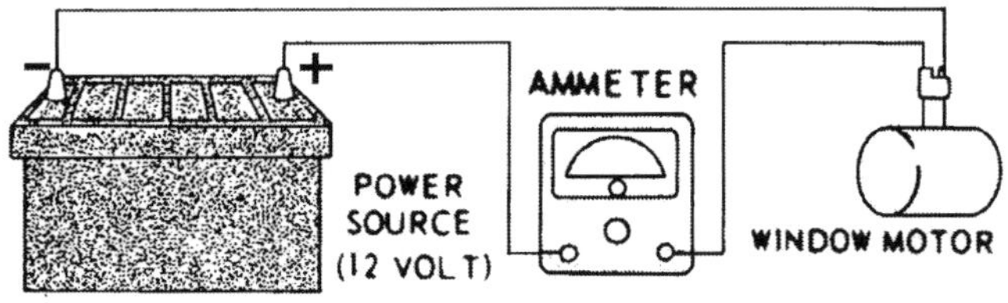

FIG 13:9 Current draw test for power window motor

Refer to **FIG 13:8** and mount the regulator assembly in a vise. To prevent the sector gear from moving when the motor and drive assembly is removed, the sector gear and regulator plate should be locked together. Drill a hole as shown and fit a $\frac{1}{4}$ inch bolt and nut. Remove the motor assembly from the regulator.

Reassemble in the reverse order of dismantling and remove the nut and bolt when the motor assembly has been fitted.

To test a window motor refer to the system illustrated in **FIG 13:9**.

Window glass:

Remove the regulator as just described. Lower the glass to the bottom of the door. Depress the tangs on both the inner and outer belt weatherstrips so that they can be removed. Lift the glass assembly up and out of the door.

The glass channel is bolted to the door. Install the glass in the reverse order of removal.

Door locks:

A typical door lock installation is shown in **FIG 13:4**. The striker on the door pillar is adjustable for position, though it must not be used to correct a sagging door. Shims should be inserted between the striker and pillar to ensure that the clearance between striker and door is correct. **Do not slam the door while the striker is out of adjustment**. Slacken the attachment bolts until the striker can just be moved under firm pressure. Adjust its height so that the door closes smoothly and gently shut the door. Pull or push on the door until it is in alignment

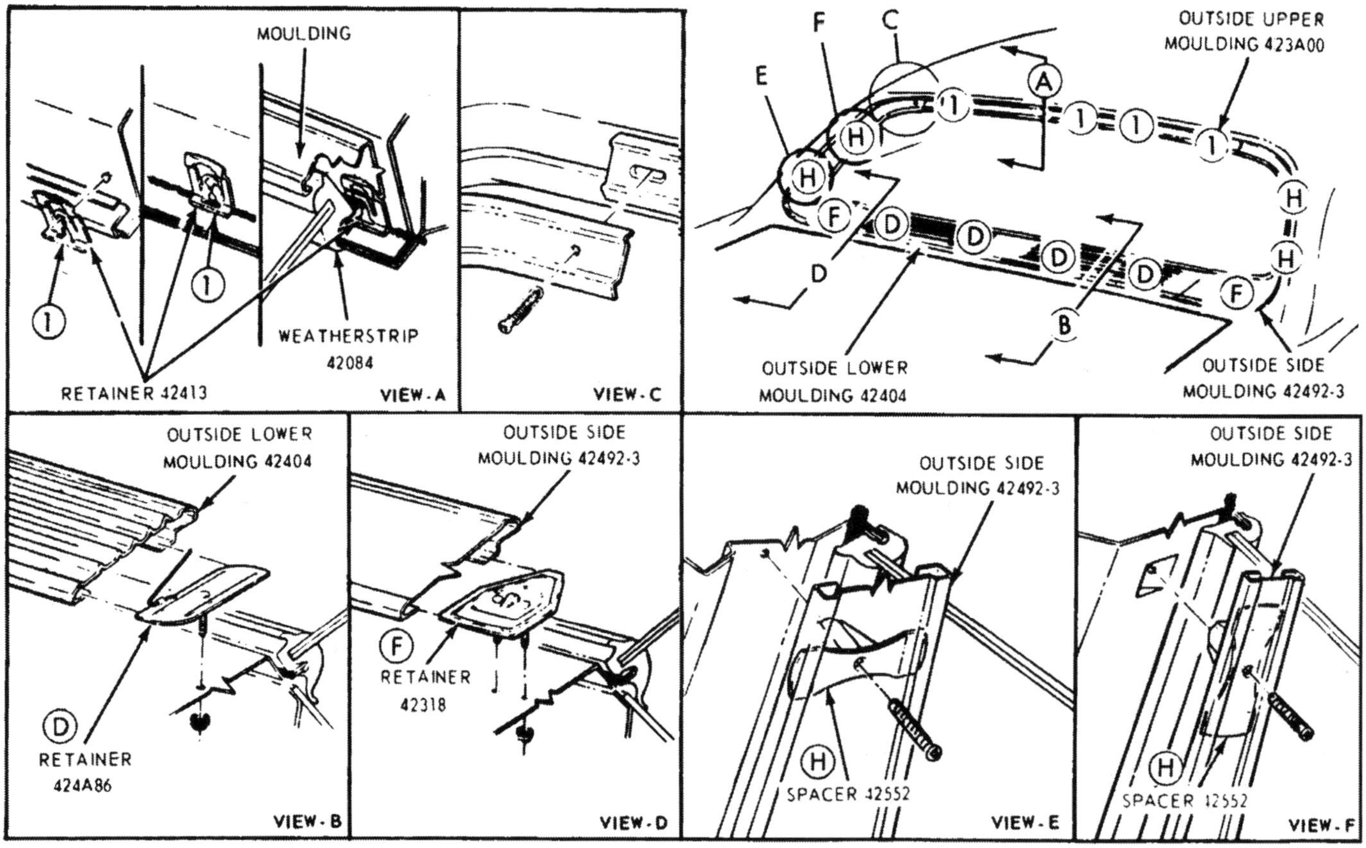

FIG 13:10 Typical back window installation using Butyl tape

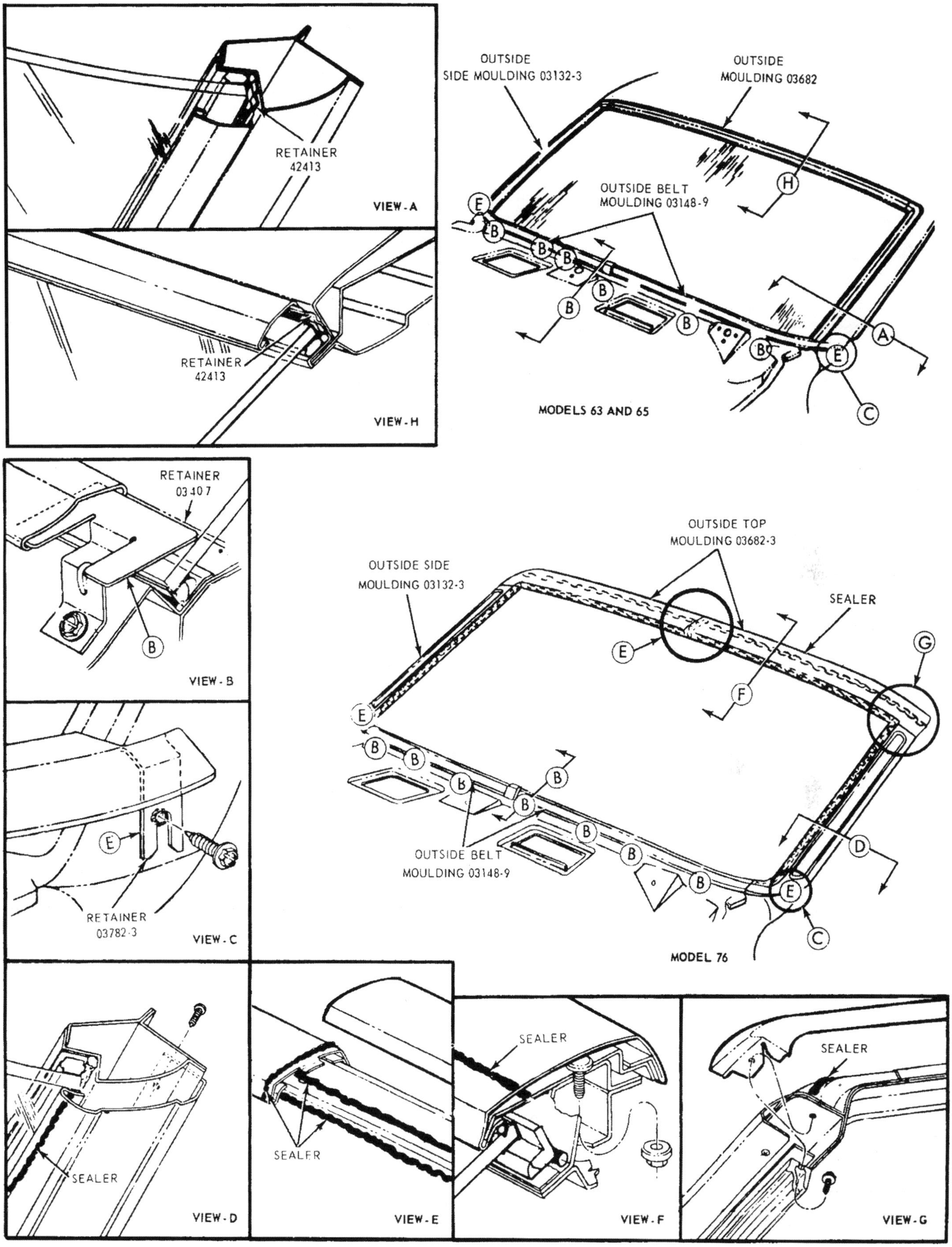

FIG 13:11 Typical windshield installation using Butyl tape

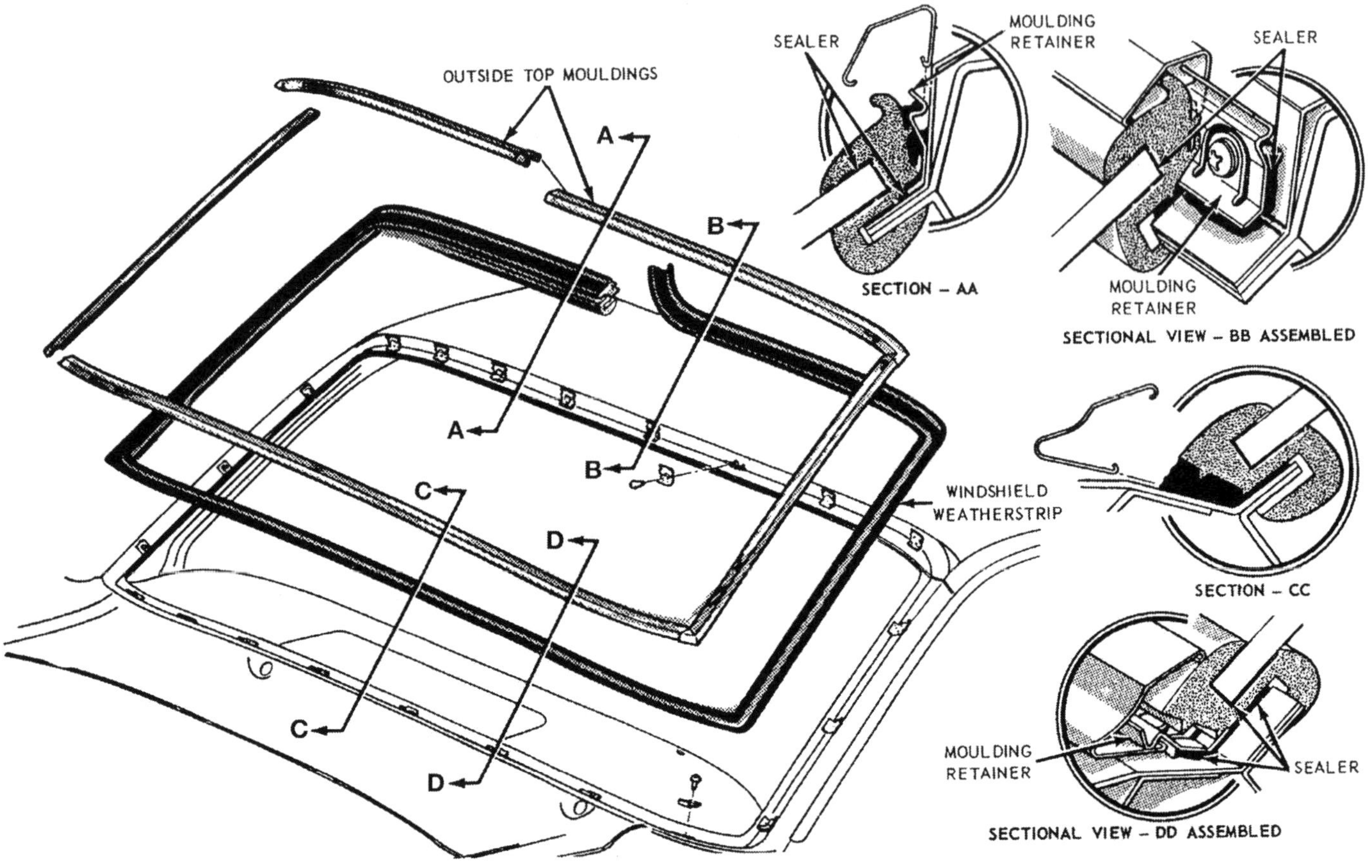

FIG 13:12 Typical windshield installation using a weatherstrip

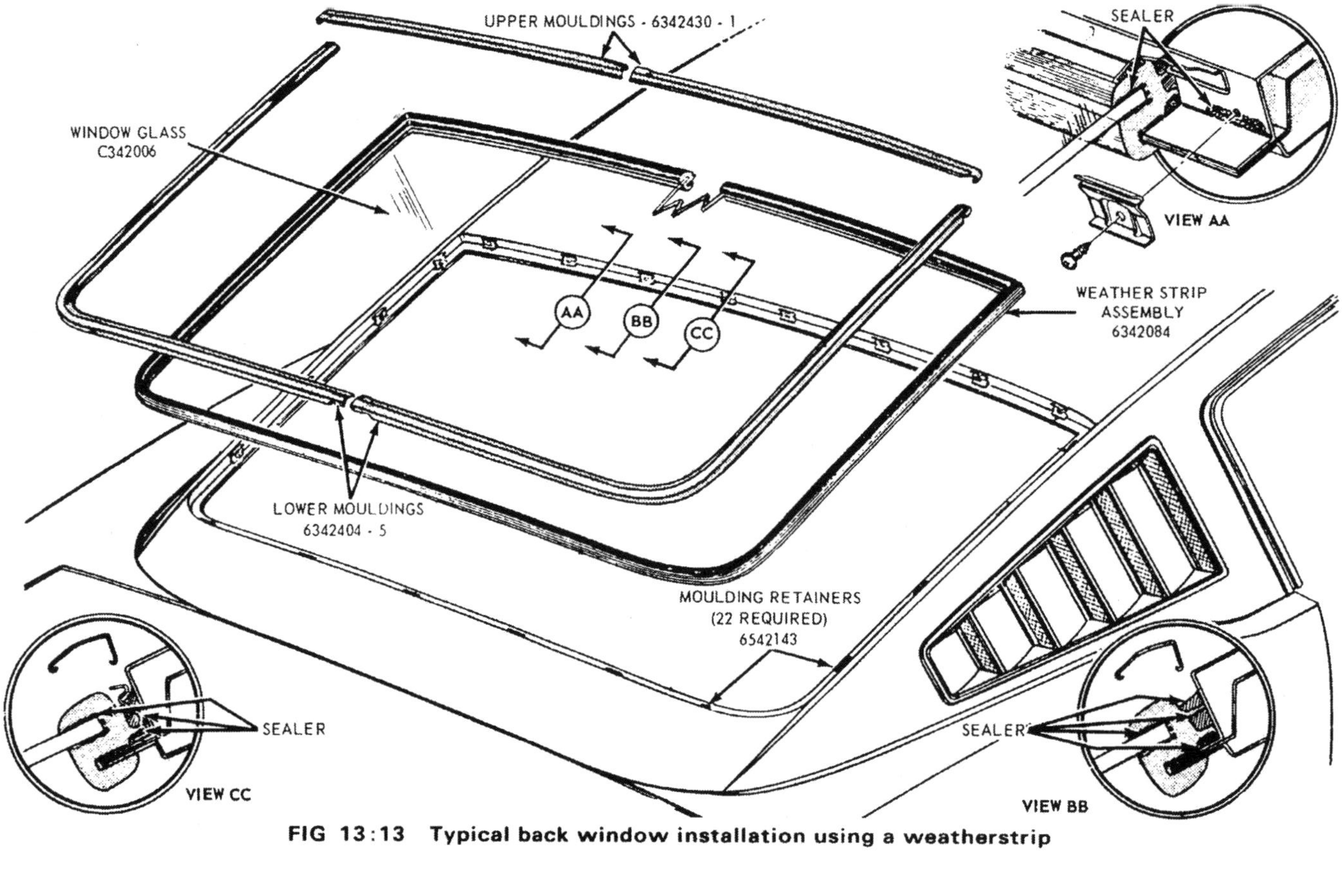

FIG 13:13 Typical back window installation using a weatherstrip

with the body and then carefully open the door. Fully tighten the striker attachments. The door inside handle and remote control assembly installation is shown in **FIG 13:5**.

Door hinges:

The installation of the door hinges is shown in **FIG 13:6**. The attachment holes are slotted to allow the door to be accurately aligned with the bodywork.

Quarterlight:

The quarterlight attachments are shown in **FIG 13:7**. To remove the glass, pry off the latch cover bracket and remove the screws retaining the bracket to the body. Peel back the weatherstrip and take out the screws securing the hinge to the body. Once the assembly has been removed, the three-piece moulding can be removed from around the glass.

Install the parts in the reverse order of removal.

13:4 Windscreen and back window

If the glass has broken but is still in place, it will be found easier to remove it if paper is stuck to both sides of the glass. It is advisable to lay padding over the hood or luggage locker lid to prevent damage to the paintwork. If the glass has actually broken, be sure to remove all particles, preferably with a powerful vacuum cleaner. Particular attention must be paid to the windshield defogger and defroster system as if particles remain in the system they may be forcibly blown out by the action of the blower motor, causing injury to front seat occupants.

Once all the parts have been removed, check the aperture flange on the body. Dents should be dressed down with a block of metal on one side while gently tapping them down, and protrusions should be filed smooth. If the flange is left in a damaged state it will allow leaks around the glass and may even set up stress points which will cause the glass to fail in service.

Rear view mirror:

On some models the mounting for the rear view mirror is bonded to the windshield. Heat from a hot air gun is used to soften the vinyl mounting pad for the installation or removal of the mounting. Provided that the pad is clean and free from foreign bodies it may be used again.

Insulate any chips or scratches with wet rags or putty. Blow hot air over the mounting area for approximately 8 to 10 minutes and remove the mounting by gently levering on the rear view mirror. Gradually reduce the heat and then allow the glass to cool naturally.

Preheat the glass in the area and then heat the button. Press the button firmly into place and keep heating until vinyl can just be seen extruding from around the button. Keep applying pressure and maintain it for at least two minutes while reducing the heat and allowing the glass to cool naturally. Leave plenty of time for the glass to cool and do not remove the insulation while the glass is hot.

Glass removal:

Typical mouldings for the back window are shown in **FIGS 13:10** and **13:13** and those for the windshield in **FIGS 13:11** and **13:12**.

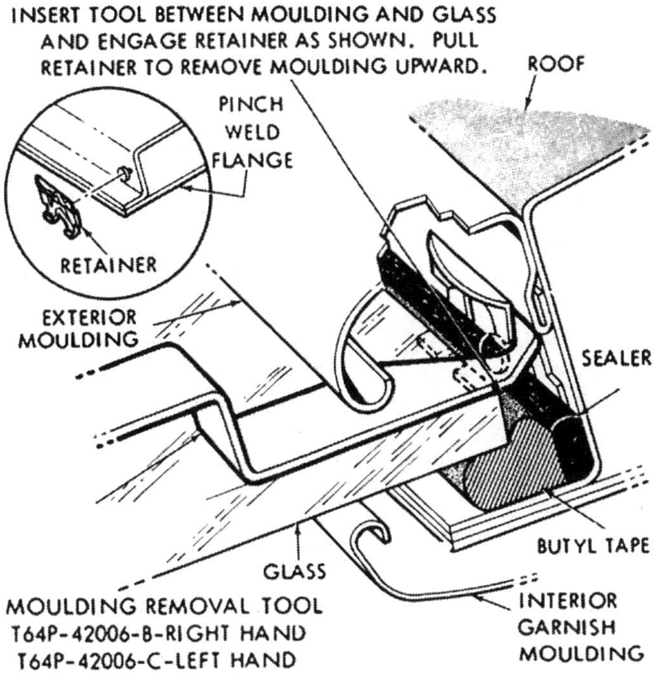

FIG 13:14 Method for removing windshield or back window moulding

FIG 13:15 Draw cord installed in the weatherstrip

Remove the mouldings and on the windshield also remove the wiper arms. Windshield glass fitted with weatherstrip type seals should be removed using the special tool T64P.42006 as shown in **FIG 13:14**. To remove shields sealed with butyl tape insert a two foot length of piano wire between the glass and the body and then bend the ends around two pieces of wood dowelling to act as handles. With the aid of an assistant cut the seal all the way around the glass using a sawing motion. Remove the glass and pry away the remains of the Butyl strip from a corner of the aperture. Grasp the strip near the flange and pull it steadily away from the flange. Collect the two spacers from the lower edge of the aperture.

Installation:

Clean the glass thoroughly by washing with soap and water or a suitable solvent. Check all the mouldings and renew any that are damaged. Install the spacers in the bottom of the aperture. If a cloth covered roof is used, protect the material using masking tape to prevent primer from getting onto it.

Temporarily position the glass in the aperture and move it to the best position for even overlap. Rubber suckers will make the handling of the glass easier. When the glass is in the best position, lightly mark it with wax crayon and then remove it again.

Apply the primer, supplied with the kit, all around the flange in the aperture where the tape will lie, and also apply it to the inside and edge of the glass where it will contact the tape. Allow the primer to dry for 10 minutes.

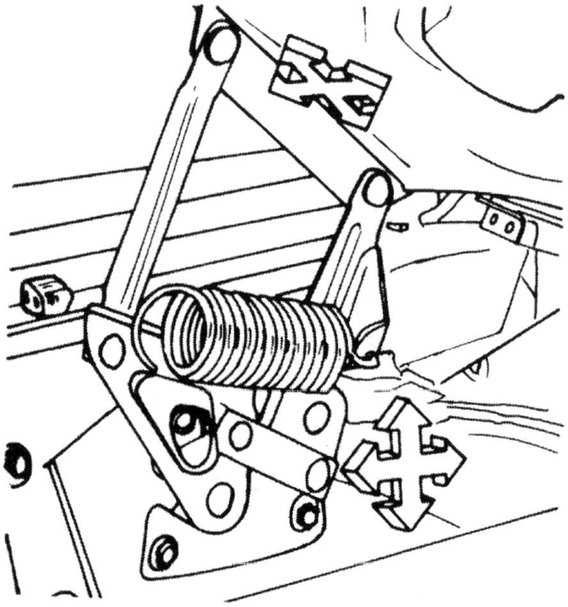

FIG 13:16 Typical hood hinge installation and adjustments

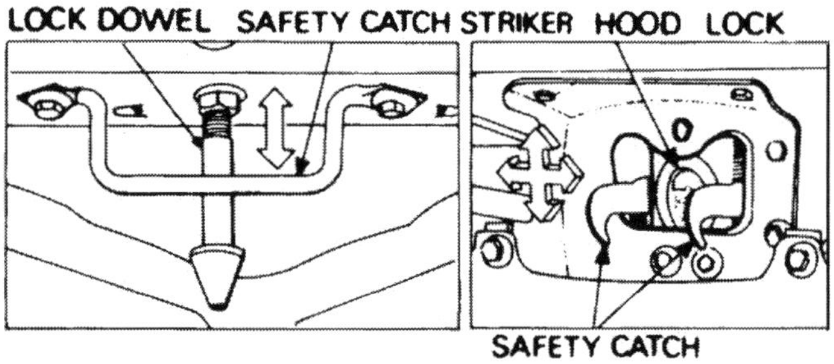

FIG 13:17 Hood lock mechanism

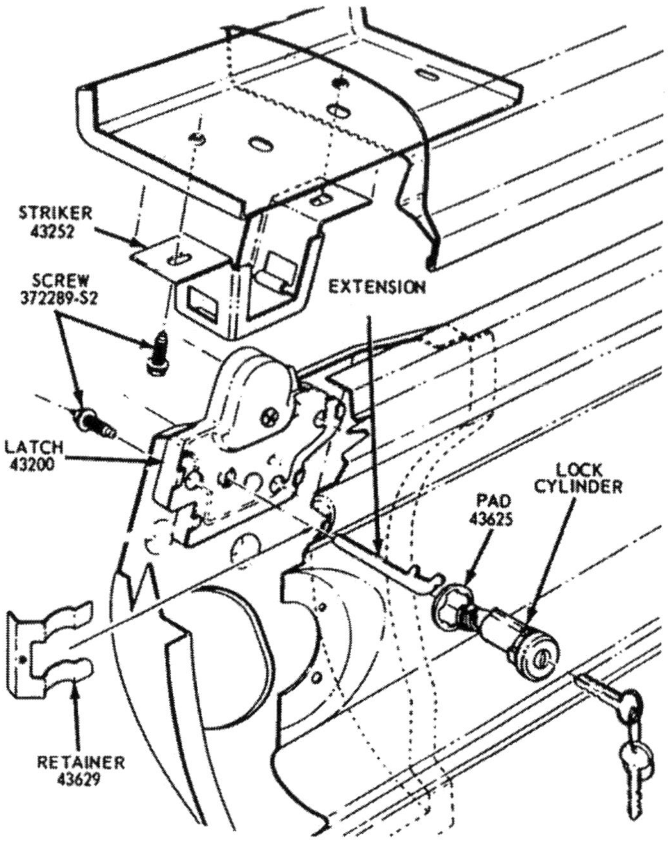

FIG 13:19 Trunk lid latch and lock

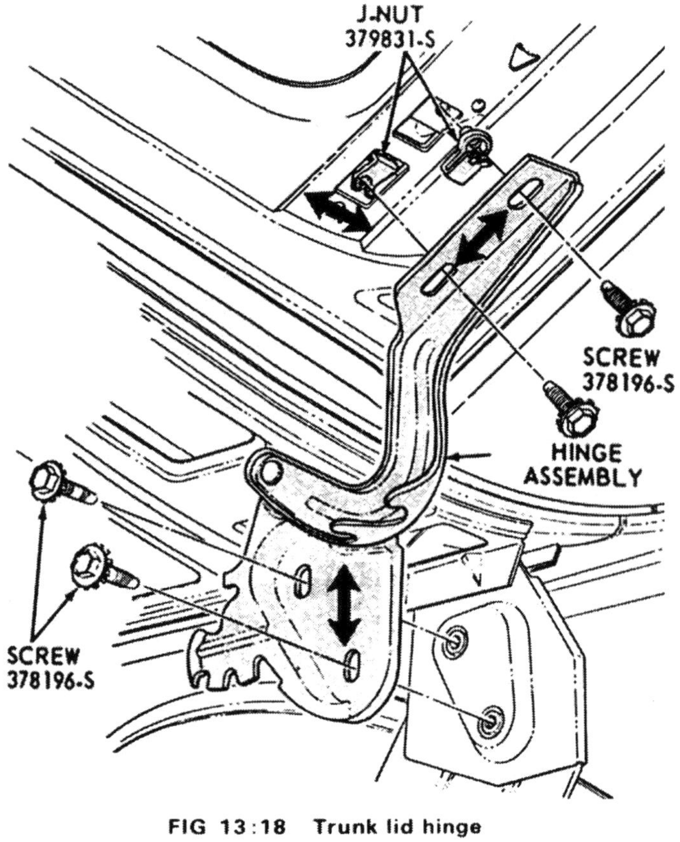

FIG 13:18 Trunk lid hinge

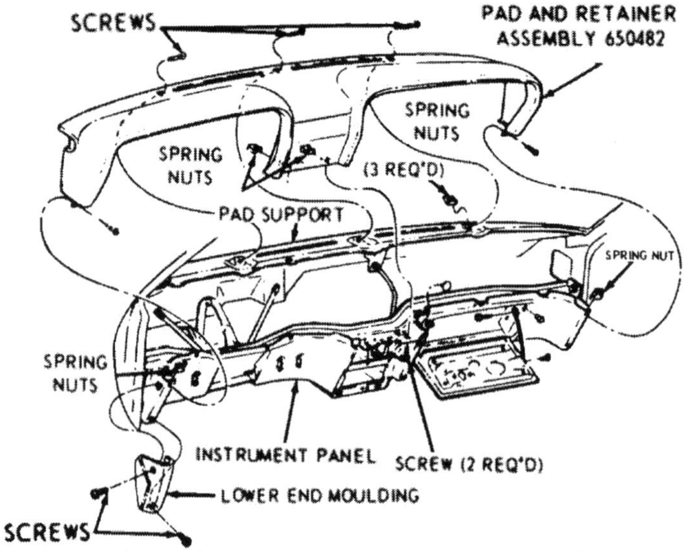

FIG 13:20 Instrument panel pad installation

Position the Butyl tape around the flange. The edge of the tape must be within $\frac{1}{16}$ inch of the edge of the flange and it must not be allowed to overhang. **Take care not to stretch the tape at the corners but guide it round instead.** Carefully splice the two ends of the tape where they meet.

Place the glass against the aperture, using the previously made marks to align it accurately. When the glass is in position, press it firmly into place using hand pressure. If the tape appears dull through the glass at any point it is not in contact with the glass, so apply hand pressure until the tape appears shiny and has fully seated.

When fitting the glass with weatherstrip type seal, insert a draw cord in the pinch weld opening of the weatherstrip and overlap the cord about 18 inch at the lower center of the glass as shown in **FIG 13:15**. Tape the ends of the cord to the glass to prevent them trailing and becoming trapped during the windshield positioning. Position the glass to the body, aligning it to the previously made marks. With a second operator pressing the glass firmly from the outside pull on the ends of the cord from the inside to draw the weatherseal over the body flange.

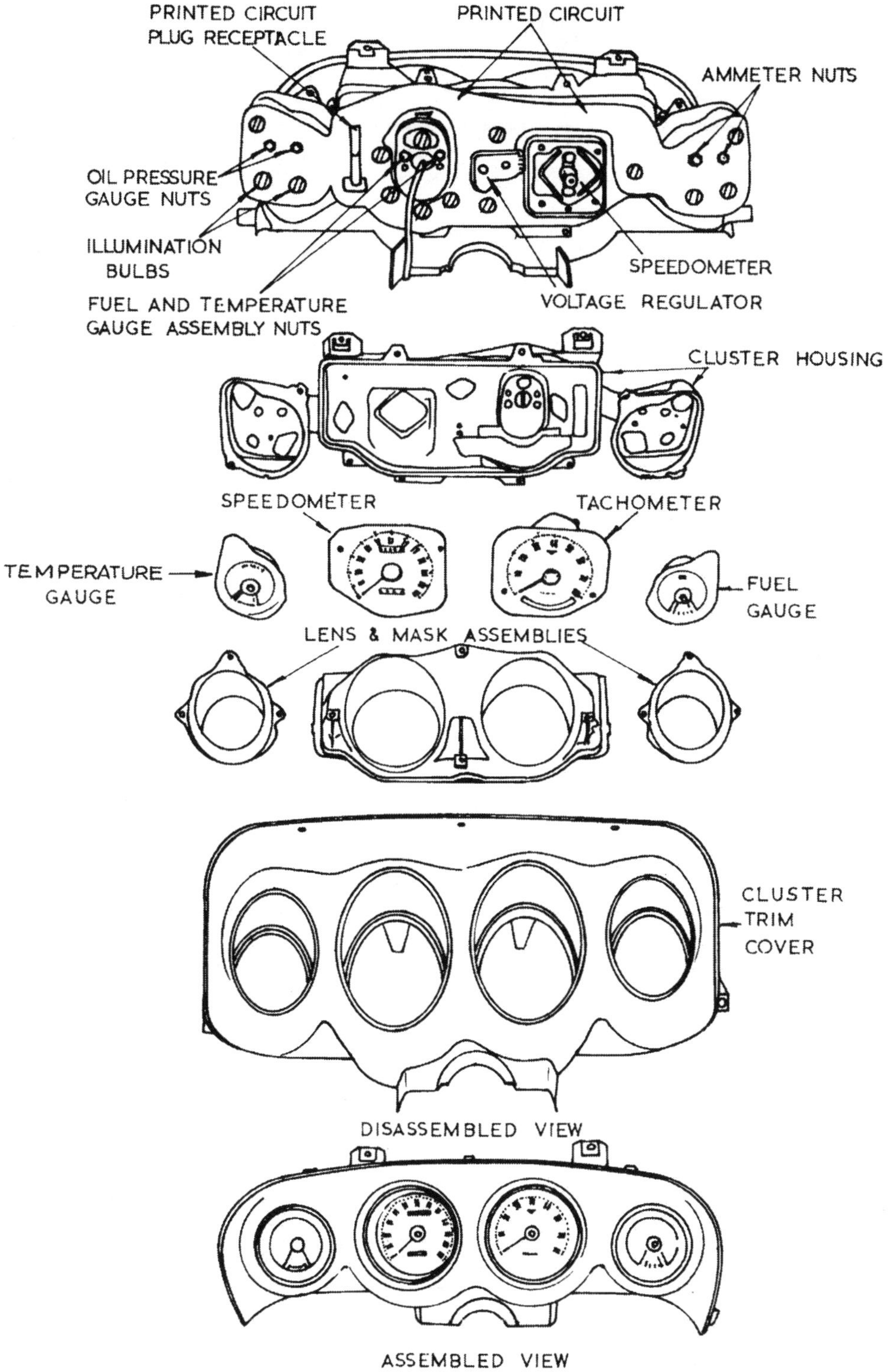

FIG 13:21 Typical instrument cluster components

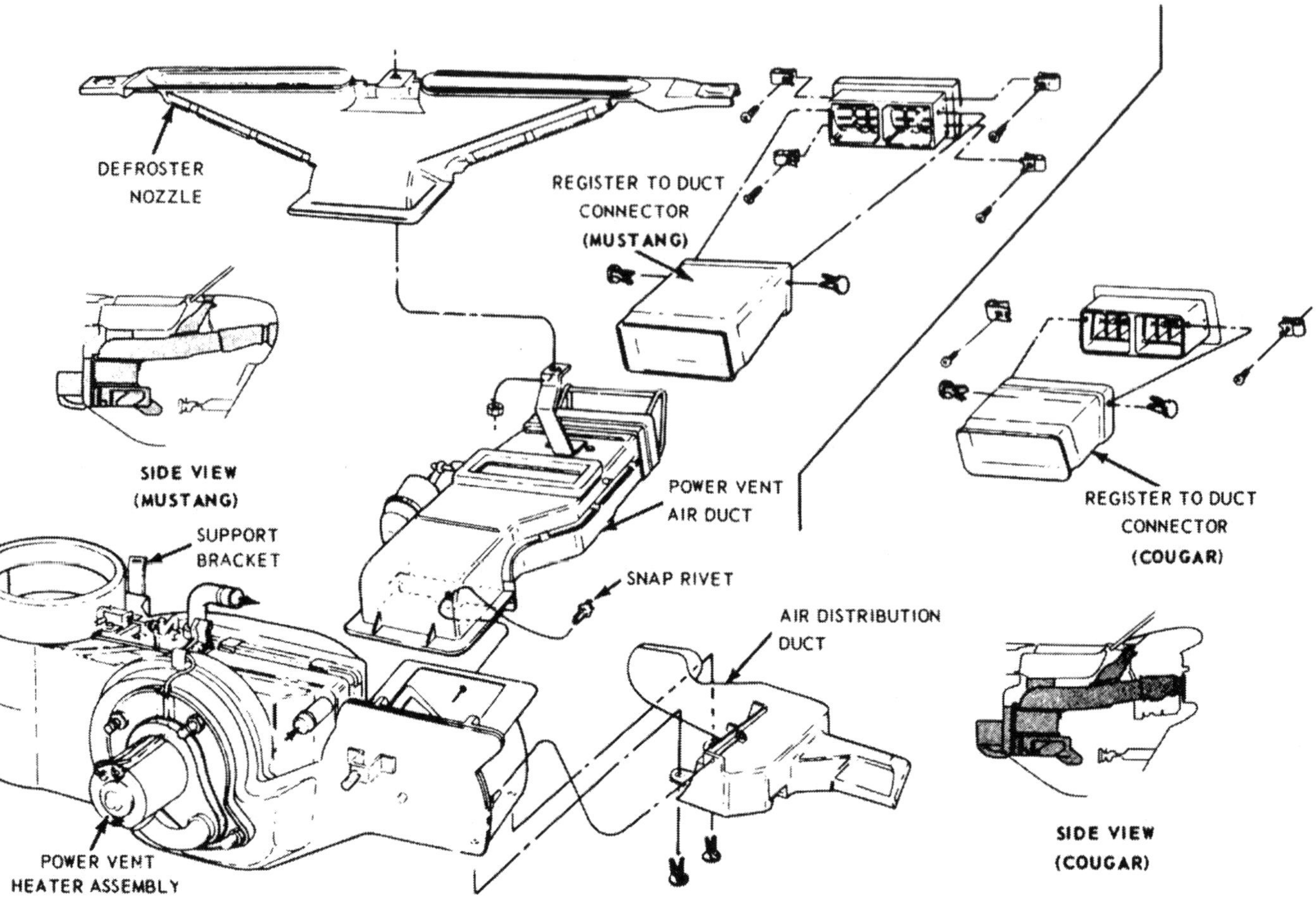

FIG 13:22 Power ventilation and heating system

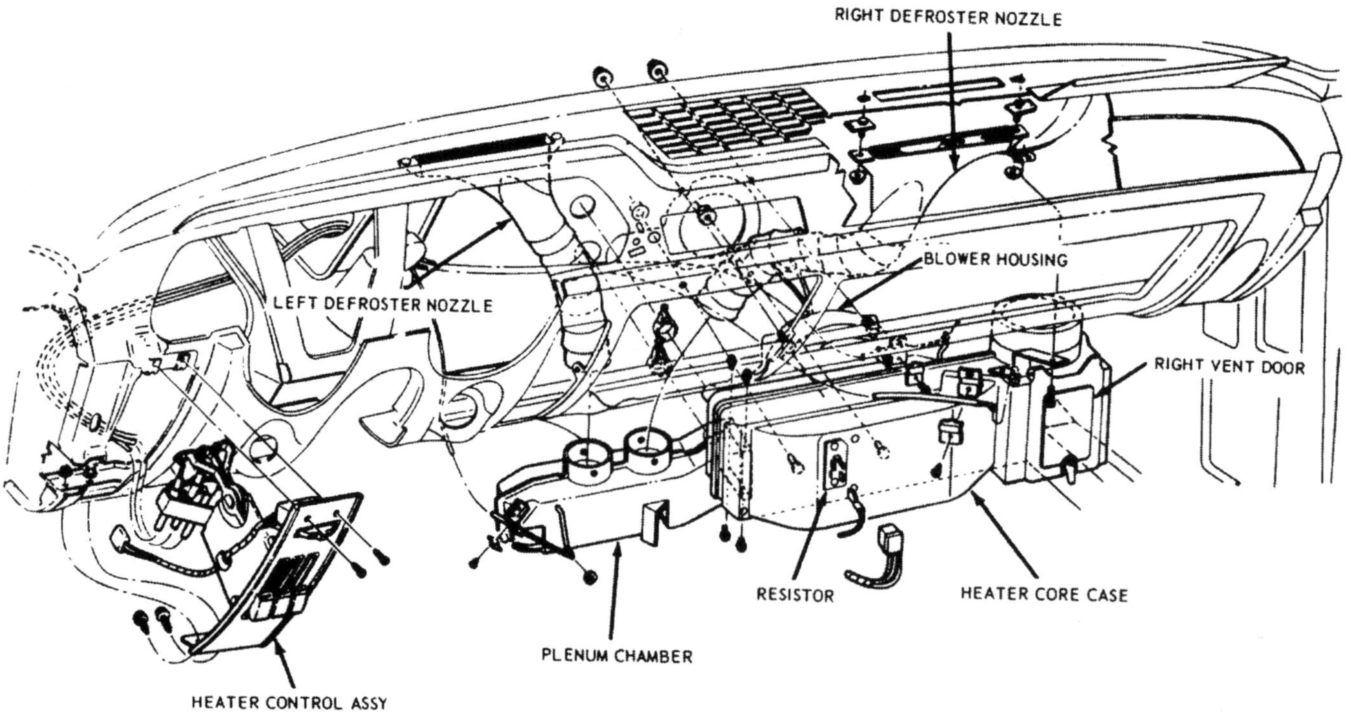

FIG 13:23 Heater installation

Ensure that no more than .20 inch clearance exists between the surface of the glass and the pinch weld flange.

Remove the surplus primer from the glass with a razor blade and wipe clean with a cloth moistened with naptha. **Do not wet the rag as this may allow naptha to creep in between glass and tape.**

Pour water over the outside or preferably use a hose-pipe to check for leaks. Leaks can be repaired by applying liquid Butyl around the edge of the glass.

Install the mouldings and any other parts removed.

13:5 The hood

The hood hinge and its adjustments are shown in **FIG 13:16**. When removing or installing the hood have an assistant to support it and pad the corners with rags so that they do not damage the paintwork. Pencil around the hinges to make the task of aligning the hood after installation that much easier. With the hood open, remove the bolts that secure it to the hinge and lift it off.

Install the hood in the reverse order of removal. If the hood is not in alignment, slacken off the bolts that secure it or the hinge, so that the hood can be moved with firm pressure but does not move under its own weight. Move the hood into alignment and fully tighten the hinge bolts. Adjustable rubber buffers are fitted. The bumpers should be raised, after slackening their locknuts, to eliminate any looseness of the hood when it is closed.

A typical hood latch is shown in **FIG 13:17**. The bolts securing the latch should be slackened and the latch moved sideways until it aligns accurately with the latch on the hood.

Slacken the locknut on the hood latch dowel and adjust the dowel up or down until the top of the hood is flush with the fenders when closed fully. Tighten the locknut on the dowel and adjust the bumper stops.

13:6 The luggage locker lid

Torsion bars are used to counterbalance the weight of the door, as shown in **FIG 13:18**, and the ends of the torsion bars can be fitted into alternate slots to alter their effect.

The hinges, latch and latch portion on the door are all adjustable to ensure that the door lies flush with the body, giving a tight seal, and that the latch parts are in alignment (see **FIG 13:19**).

13:7 Instruments and instrument panel

The attachments of the instrument panel are shown in **FIG 13:20**. Disconnect the battery and then reach up underneath the instrument panel so that the speedometer drive cable can be disconnected from the instrument head, by pressing on the flat surface of the plastic connector and pulling the cable away from the instrument. Remove the two retaining screws at the top of the cluster and swing it down. New instrument panel bulbs can be fitted at this stage. If the cluster is being removed, disconnect the multi-pin connector and disengage the brackets on the lower edge from the slots in the panel.

Install the cluster in the reverse order of removal, making sure that the brackets are correctly in the slots.

The instrument cluster broken down into its main components is shown in **FIG 13:21**.

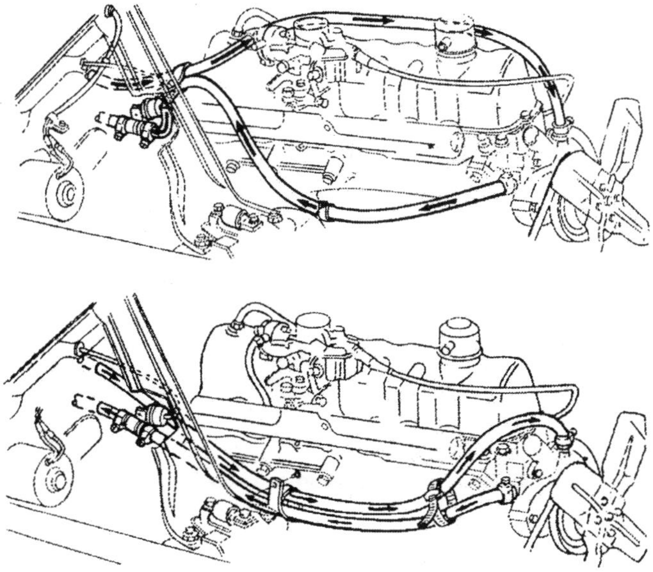

FIG 13:24 Heater hose routing

The instrument panel pad can be removed after the instrument cluster has been removed and controls disconnected.

13:8 The heater

The fresh air heater is shown in **FIG 13:22** and the controls are shown in **FIG 13:23**. Removal is from inside the car but the cooling system must be drained and the heater hoses disconnected from the heater in the engine compartment. Heater pipe runs are shown in **FIG 13:24**.

Once the heater has been removed from the car, the blower motor can be removed as a separate unit. The two halves of the cover are held together by clips and these can be removed by expanding them using a suitable pair of thin-nosed pliers between the lugs. Once the cover parts have been separated the heater core can be removed.

13:9 Air-conditioning

A typical air-conditioning system is shown in **FIG 13:25** and the actual parts of the heater assembly are shown in **FIG 13:26**. The system is charged with a non-toxic refrigerant (R.12). The compressor is driven by a belt from the engine, through a magnetic clutch. The magnetic clutch only comes into operation when the system is selected and the internal temperature rises above the set value. With the system operating, the compressor draws gas from the evaporator and pressurizes it. The operating pressure from the compressor varies between 80 to 300 lb/sq inch, depending on the ambient air temperature and compressor operating speed. The high pressure gas, hot because of compression, passes from the compressor to the condenser, where it is cooled by the passage of air over the fins. The refrigerant changes from gas to liquid in the condenser and passes from there through the receiver. From the receiver the liquid refrigerant, still at high pressure, passes through a sight glass and hoses to the expansion valve. The receiver contains water-absorbing material to ensure that the refrigerant is perfectly free from water. The expansion valve controls the amount of liquid passing through and at the same time reduces its pressure to between 12 to 50 lb/sq inch, depending on conditions. The low-pressure

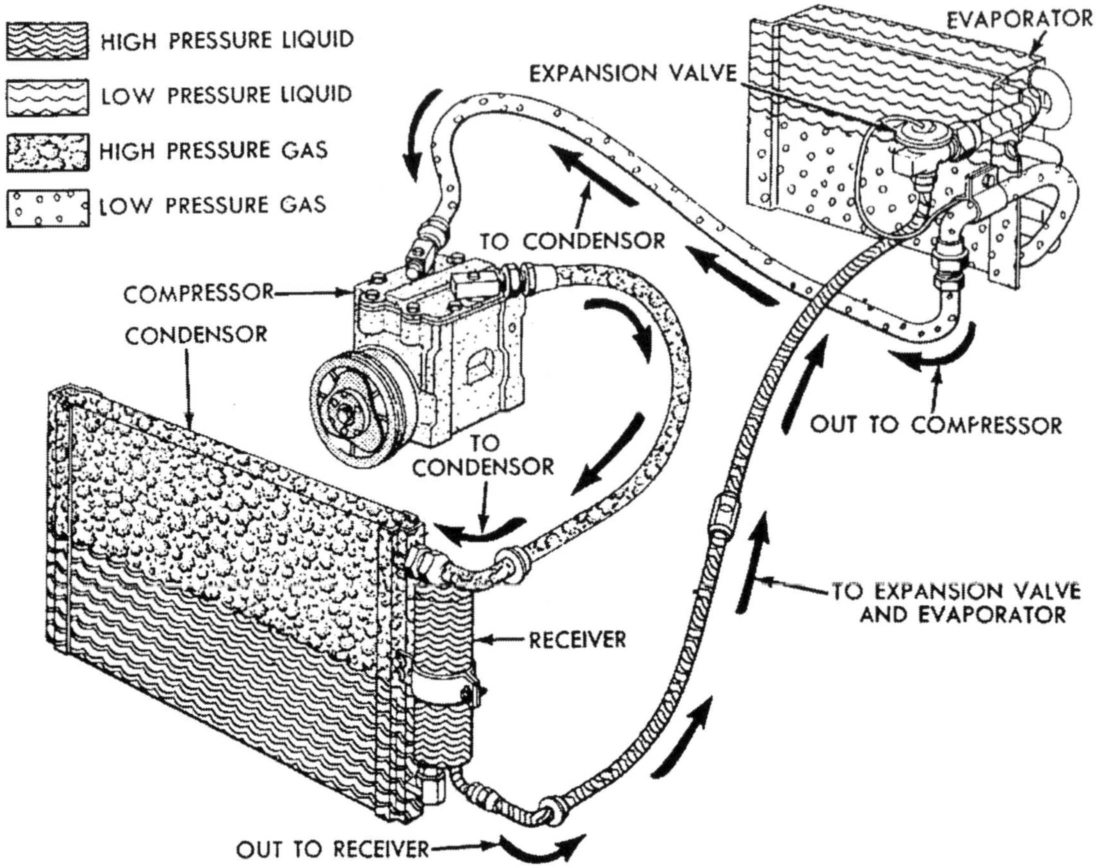

FIG 13:25 Schematic layout of air-conditioning system

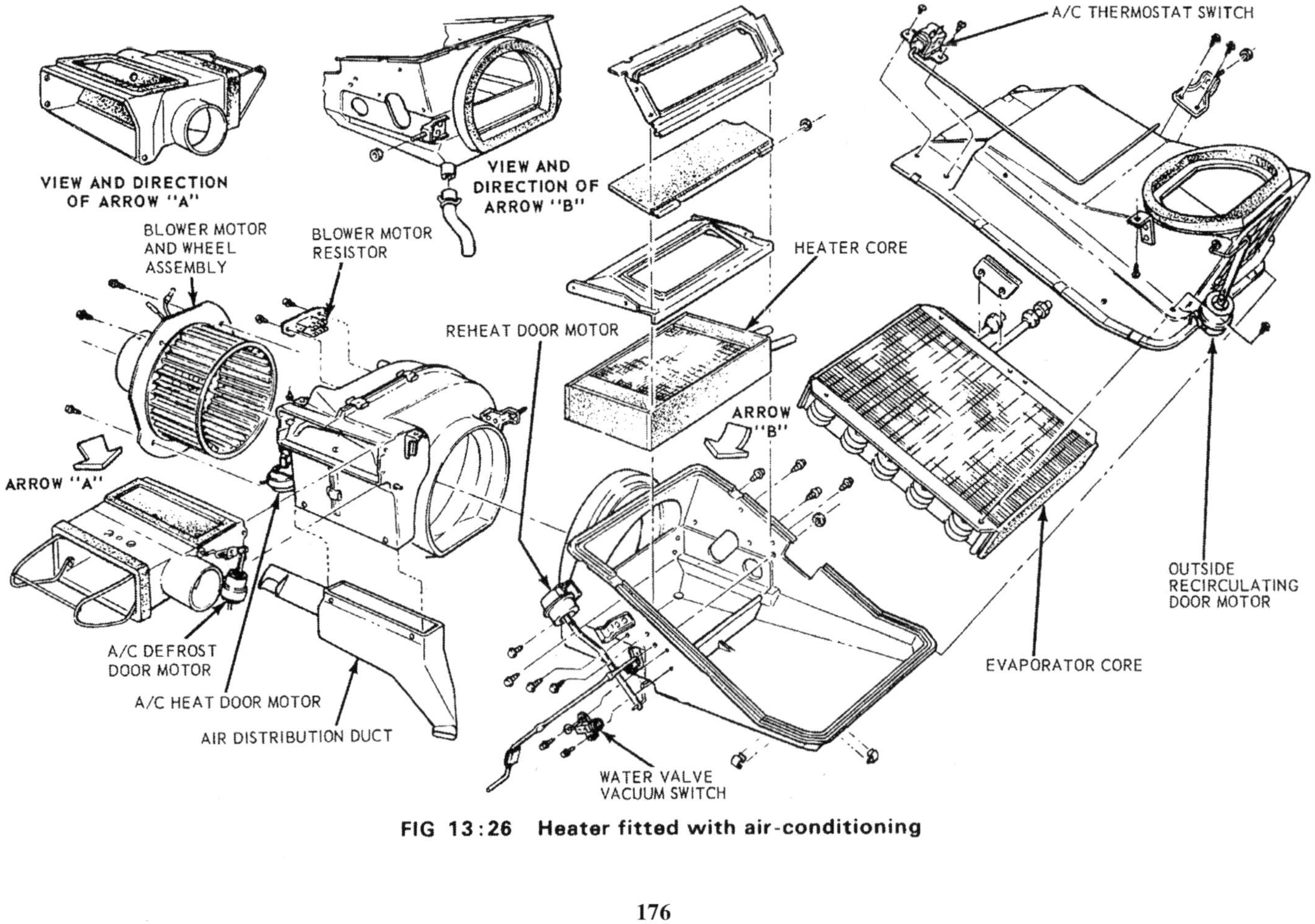

FIG 13:26 Heater fitted with air-conditioning

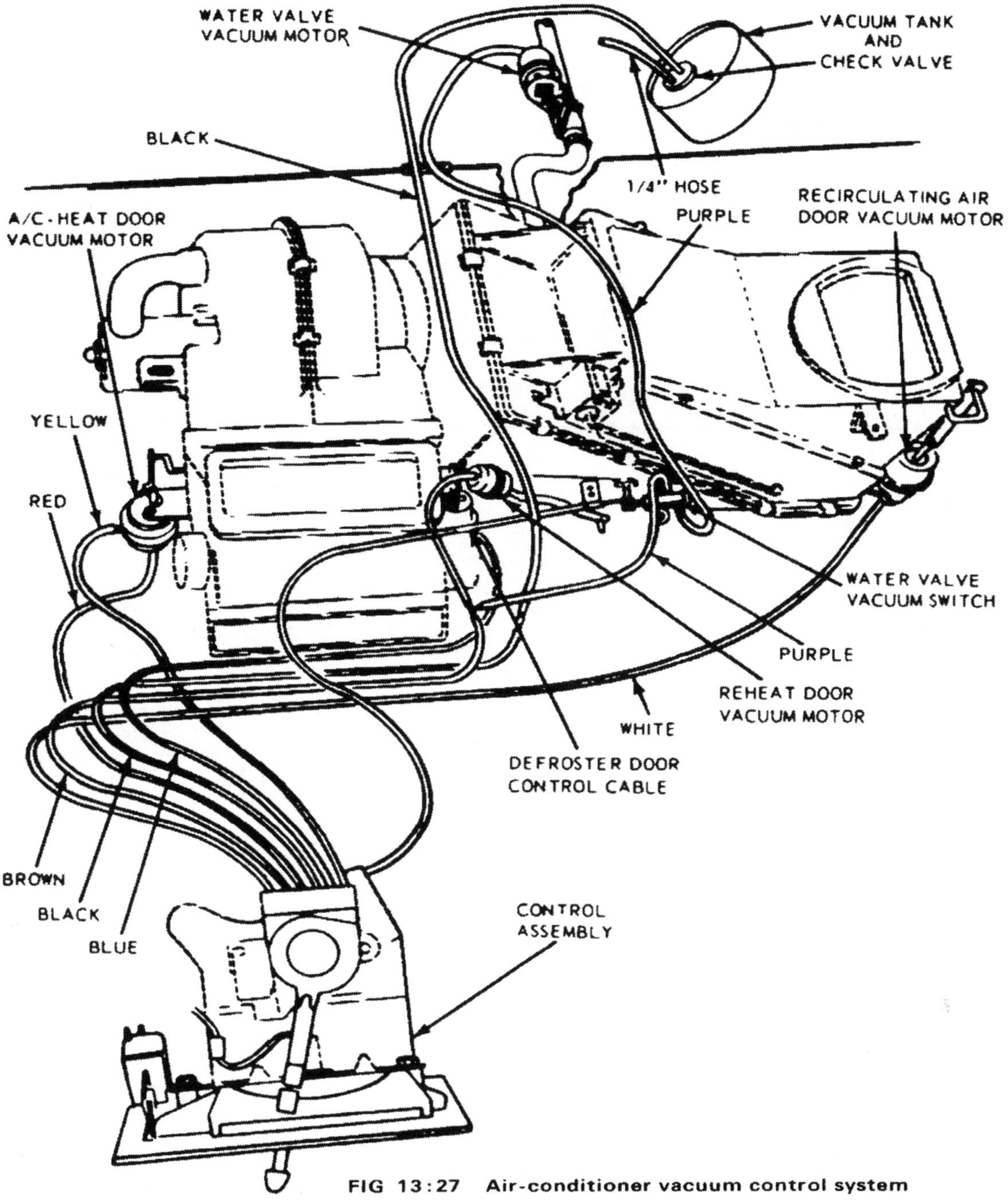

FIG 13:27 Air-conditioner vacuum control system

liquid passes into the evaporator where it changes back into gas, absorbing heat from the air passing through the evaporator as it does so. The warm low-pressure gas is then cycled back to the compressor.

Servicing of the system should only be carried out by trained personnel as, though the refrigerant is non-toxic, serious injury or even death can result from incorrect methods or failing to observe safety precautions.

Controls:

Vacuum is used for the operation of the system and the connections are shown in **FIG 13:27**.

Safety precautions:

At normal temperatures and pressures the refrigerant evaporates practically instantaneously, chilling whatever it is in contact with. If liquid refrigerant is spilled onto the same manner as chilblains. Safety glasses should always be worn to prevent splashing into the eyes, as this is serious. If liquid refrigerant splashes into an eye, use a few drops of sterile mineral oil to wash out the eye (as the refrigerant is readily absorbed by this) and then clean the eye with a weak solution of boric acid. **Seek medical aid even if the irritation has ceased.**

If the system is discharged, do so slowly to avoid flushing the oil out of the compressor, and discharge directly into the open. If the refrigerant is discharged into a confined space, the rapid evaporation will displace the air and though the vapor is non-toxic asphyxiation can result from lack of oxygen.

Do not heat the system, such as when steam-cleaning or welding, as the rise in pressure could cause a component to burst.

Do not disconnect any hoses without first emptying the system. The owner should take the car to a service agent for such work.

The refrigerant itself produces very poisonous fumes if passed through a naked flame. When using the leak tester, which has a small copper disc heated by a flame, avoid breathing the fumes from it.

Sight glass:

Normally the glass should appear perfectly clean and free from bubbles. Check by running the engine at 1500 rev/min and cycling the magnetic clutch. Bubbles should appear when the clutch is off. If no bubbles appear at all, the system is empty and this will be confirmed by the lack of cooling. If the system is empty switch off the air-conditioning and do not use it until it has been recharged and the leak found.

Continuous bubbles indicate a low charge in the system but in conditions of high ambient temperatures bubbles or foam may appear when the system is operating.

Compressor removal:

This is the only disconnecting that should be carried out by the owner.

Turn both high and low pressure valves fully clockwise, as shown in **FIGS 13:28** and **13:29**. Loosen the cap on the high pressure valve, by a small amount only, and allow the refrigerant in the compressor to escape. Disconnect the lines from the compressor and remove it from the engine.

The compressor is installed in the reverse order of removal. Before opening the valves again, evacuate the compressor using a vacuum pump. After installation the owner will probably find it most convenient to drive the vehicle to a service agent, without using the system, so that the compressor can be evacuated, and the oil level checked.

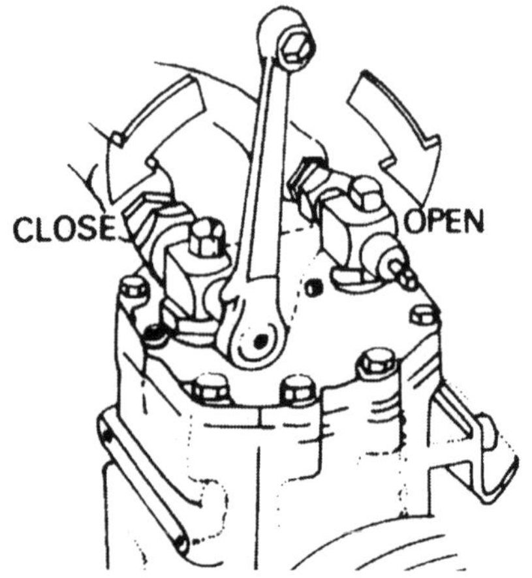

FIG 13:28 Compressor low-pressure port

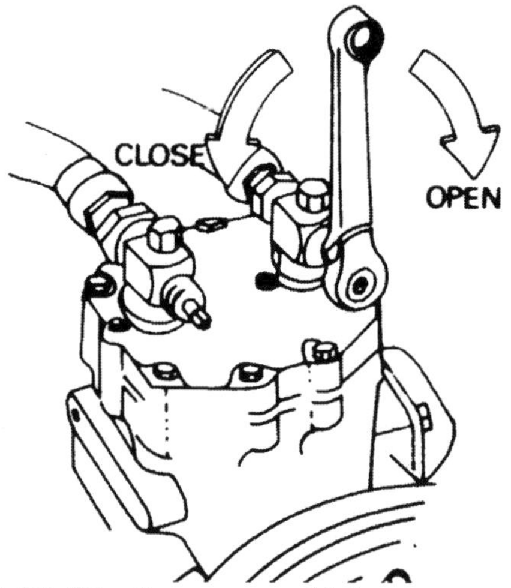

FIG 13:29 Compressor high-pressure port

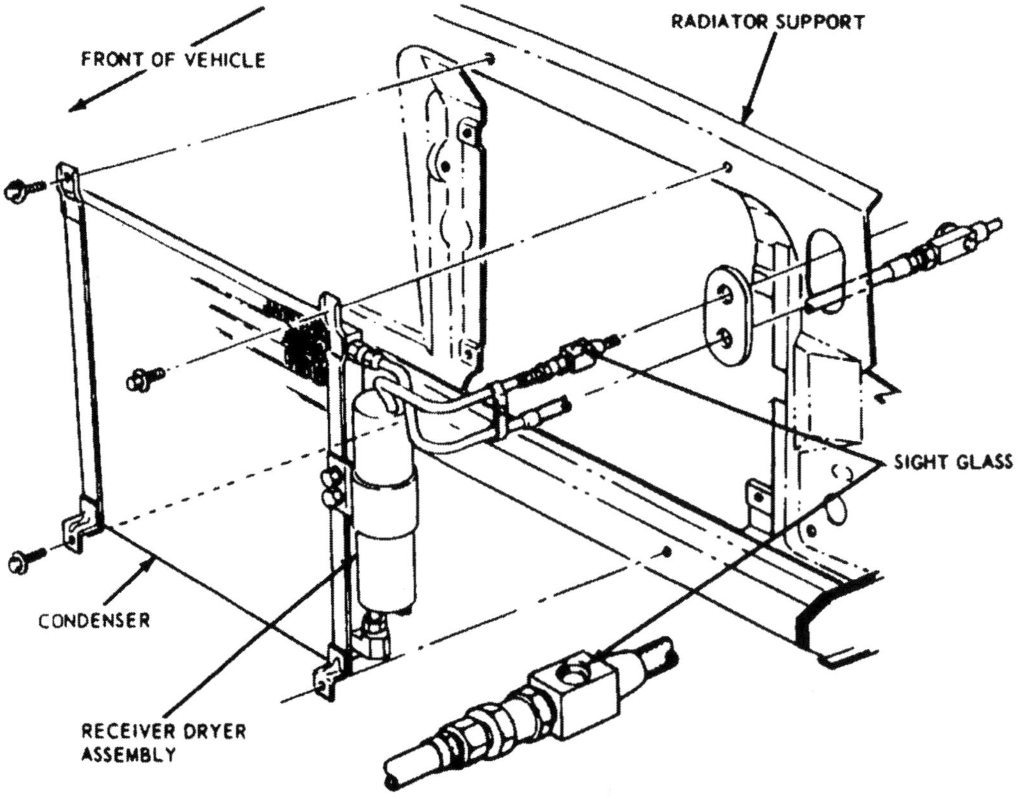

FIG 13:30 Condenser installation and sight glass

APPENDIX

TECHNICAL DATA

Unless otherwise stated, all dimensions are given in inches

ENGINE

Type	Straight-six, watercooled. OHV usingrockerarmsandhydraulically operated valve lifters
Capacity	200 and 250 cu in
Compression ratios	
200	8.7:1
250	9.0:1
Bore	
200	3.68
250	3.68
Stroke	
200	3.126
250	3.91
Firing order	1-5-3-6-2-4
Vacuum in manifold	17 in HG minimum at sea level and at idling speed (16 for dual diaphragm distributor engines)
Oil pressure	35 to 60 lb/sq in at 2000 rev/min with engine hot
Crankshaft	
Main bearings	Renewable steel-backed
200 and 250	7 main bearings
End play	.004 to .008 (wear limit .012)
Journal diameter:	
200	2.2482 to 2.2490
250	2.3982 to 2.3990
Crankpin diameter, all types	2.1232 to 2.1240
Journal to bearing clearance:	
Desirable	.0005 to .0015
Allowable	.0005 to .0022
Crankpin to bearing clearance:	
Desirable	.0010 to .0015
Allowable	.0008 to .0024
Crankpin maximum taper	.003 per inch
Journal and crankpin maximum out of round	.0004
Flywheel OD runout:	
Standard	.018
Automatic transmission	.020
Connecting rod	
Length centre to centre:	
200	4.7135 to 4.7165
250	5.8785 to 5.8815
Maximum bend*	.004
Maximum twist*	.008

**Using 8 inch Mandrel and Measuring 4 inches on either side of rod*

Side clearance when fitted to crankshaft:	
Standard	.035 to .0105
Wear limit	.014
Small-end diameter	.9107 to .9112
Big-end diameter	2.2390 to 2.2398

Piston pin

Length	3.010 to 3.040
Clearance to piston	.003 to .005 (wear limit .0008)
Diameter:	
Standard	.9119 to .9124
.001 oversize	.9130 to .9133
.002 oversize	.9140 to .9143

Pistons

Clearance to bore	.0014 to .0020		
Piston pin bore	.9122 to .9125		
Diameter	Measured at piston pin bore centerline at 90 deg. to piston pin		
	Coded Red	*Coded Blue*	*.003 oversize*
200 and 250	3.6790 to 3.6796	3.6802 3.6808	3.6808 to 3.6814
Ring groove widths:			
Compression rings	.080 to .081		
Oil ring	.1880 to .1890		

Piston rings

Width	.077 to .078 compression rings
Side clearance	.002 to .006
Fitted gap:	
Compression rings	.010 to .020
Oil ring	.015 to .055 (steel rail snug)

Camshaft

Lobe lift:	
Intake, 200	.2320
250	.2450
Exhaust, 200	.2320
250	.2450
Maximum loss of lift	.005
Theoretical valve lift:	
200	.3480
250	.3680
End play	.001 to .007 (wear limit .012)
Journal to bearing clearance	.001 to .003 (wear limit .006)
Camshaft sprocket maximum runout	.006
Crankshaft sprocket maximum runout	.006
Maximum timing chain deflection	.5
Journal diameter	1.8095 to 1.8105

Cylinder block

Cylinder bore diameter:		
	Std.	*.003 O.S.*
200 and 250	3.68000 to 3.6824	3.6824 to 3.6836
Maximum bore out of round	.001	
Bore wear limit	.005	
Tappet bore diameter	.8752 to .8767	

Cylinder head

Valve guide bore diameter	.3115 to .3125 standard
Valve seat width:	
Intake	.060 to .080
Exhaust	.070 to .090
Valve seat angle	45 deg.
Valve arrangement	E-I-I-E-I-E-E-I-E-I-I-E
Surface flatness	.003 in any 6 inches, .007 overall

Valve rocker assembly

Rocker shaft O.D.	.780 to .781
Arm to shaft clearance	.002 to .0045

Rocker arm bore	.783 to .784
Valve lift ratio	1.5:1
Valve lifter:	
Diameter	.8740 to .8745
Clearance in bore	.0007 to .0027

Valve springs

Approximate free length	1.79
Assembled height	$1\frac{9}{16}$ to $1\frac{39}{64}$
Test length	1.59 inch at 51 to 57 lb 1.222 inch at 142 to 158 lb
Wear limit	1.59 inch at 46 lb 1.222 inch at 128 lb

Valves

Stem diameter:	
Intake	.3100 to .3107 standard
Exhaust	.3098 to .3105 standard
Oversizes	.003, +.015, +.030
Clearance to valve guide:	
Intake	.0008 to .0085
Exhaust	.0010 to .0027 (wear limit .0047)
Head diameter:	
Intake	1.642 to 1.657
Exhaust	1.381 to 1.396
Seat angle	44 deg.

Oil pump

Relief valve spring:	
200	1.078 inch at 9 to 10.1 lb
250	2.490 inch at 20.6 to 22.6 lb
Drive shaft to housing clearance	.0015 to .0029
Relief valve clearance	.0015 to .0029
Rotor assembly end clearance	.0011 to .0041
Outer rotor to housing radial clearance	.006 to .013

FUEL SYSTEM

Fuel pump

Static pressure	4.0 to 6.0 lb/sq in
Delivery	1 pint to 30 to 36 seconds

Carburetor application

200 engine	Carter YF 1-V or Autolite 1100
250	Carter RBS or Autolite 1101

Curb idle speed

Manual transmission	750 rev/min
Automatic transmission	550 rev/min

Fast idle speed

Carburetor	*Transmission*	
Early Carter YF	Manual	.030 in throttle clearance
	Automatic	.035 in throttle clearance
Later Carter YF	Manual	1750 rev/min
	Automatic	2000 rev/min
Carter RBS	All types	1600 rev/min
Autolite 1100	Manual	1400 rev/min
	Automatic	1500 rev/min
Autolite 1101	Manual	1400 rev/min
	Automatic	1600 rev/min

Anti-stall dashpot setting

Carter YF		.10 in
Autolite 1101		.08 in
Autolite 1100	Manual	2 turns in $\pm \frac{1}{4}$ turn
	Automatic	3 turns in $\pm \frac{1}{4}$ turn

Float level setting		
Carter YF to 1967	$\frac{7}{32}$ in	
Later Carter YF	$\frac{3}{8}$ in	
Carter RBS	$\frac{9}{16}$ in	
Autolite 1100 and 1101	$1\frac{3}{32}$ in ($\pm \frac{1}{32}$)	
Choke setting		
Autolite 1100	Manual	.15 in clearance
	Automatic	.13 in clearance
Autolite 1101		.036 in gauge wire
		.20 in clearance
Carter RBS		.026 in gauge wire
	Manual	.270 in clearance
	Automatic	.190 in clearance

IGNITION SYSTEM

Firing order	1-5-3-6-2-4
Spark plug gap	.032 to .036
Spark plug type	Autolite BF-82 Champion UF-14Y AC 85TS
Ignition coil	
Primary resistance	1.40 to 1.54 ohms at 75°F
Secondary resistance	7600 to 8800 ohms at 75°F
Current draw:	
Engine stopped	4.5 amps
Engine at idle	2.5 amps
Primary circuit resistance	1.30 to 1.40 ohms at 75°F
Distributor point gap	.027
Distributor points dwell angle	35 to 40 deg. at idle
Breaker arm tension	17 to 21 oz
Capacitor	.21 to .25 microfarads
Initial timing settings	
Pre-1968 without emission control:	
Manual transmission	6 deg. BTDC
Automatic transmission	12 deg. BTDC
Pre-1968 with emission control:	
Manual and automatic	5 deg. BTDC
1968 and later, all models	6 deg. BTDC

COOLING SYSTEM

Antifreeze	Ethylene Glycol, should always be used—even in tropics—to give protection down to 0°F in order to raise boiling point and inhibit corrosion	
Radiator pressure cap	12 to 15 lb/sq in	
Belt tension	New 140 lbs, used (after 10 minutes running) 110 lbs. Minimum 90 lbs	
Thermostat	*Starts opening*	*Fully open*
Low temperature	155°F to 162°F	182°F
High temperature	188°F to 195°F	210°F to 212°F
Air inlet duct	Test by soaking for 5 minutes in water	
Valve in full hot position	100°F and below	
Valve in heat off position	135°F	

CLUTCH

Operation	Mechanical
Adjustment	Minimum $\frac{1}{2}$ inch free travel of pedal at 3000 rev/min. Normal 1 inch
Release lever adjusting nut to swivel sleeve	.136
Clutch diameter	9 inch
Number of springs in cover assembly	6
Number of springs in disc	6
Color identification of cover:	
Pressure plate	Orange
Cover	Purple
Springs	Aluminium stripe
Color identification of disc:	
Linings	Pink
Springs	Blue

MANUAL-SHIFT TRANSMISSION

Ratios:	*First*	*Second*	*Third*	*Fourth*	*Reverse*
3 speed (1965 only)	2.79:1	1.70:1	1:1	—	2.87:1
3 speed	2.99:1	1.75:1	1:1	—	3.17:1

AUTOMATIC TRANSMISSION

Type	C4
Stall speed	
200	1400 to 1700 rev/min
250	1600 to 1800 rev/min

REAR AXLE

Type and ratio	
WDV-J	2.83:1
WDV-K	3.08:1
WDV-L	3.20:1
WDW-M	2.79:1
WDW-L	3.00:1
WDW-T	2.79:1
WDW-U	3.00:1
WFS-C	3.00:1

FRONT SUSPENSION

	Minimum	*Maximum*	*Optimum*
Toe-in (inches)	$\frac{1}{16}$	$\frac{5}{16}$	$\frac{3}{16}$
Caster (deg.)	$-\frac{3}{4}$	$+1$	$+\frac{1}{4}$
Camber (deg.)	0	$+1\frac{1}{2}$	$+\frac{3}{4}$

STEERING

Manual:	
Model	SMA.R. Recirculating ball
Ratio	22:1
Gear turn, lock to lock	$5\frac{3}{4}$
Worm bearing preload	4 to 5 lb/inch
Center mesh load	9 to 10 lb/inch
Power steering	
Ford non-integral	SMB.K. Recirculating ball
Ratio	16:1
Gear turns, lock to lock	$3\frac{3}{4}$
Worm bearing preload	3 to 4 lb/inch
Center mesh load	8 to 9 lb/inch

Saginaw	Recirculating ball and torsion bar
Ratio, constant	17.5:1
Ratio, variable ratio	16:1 on center, 13:1 at lock
Wheel turns, lock to lock, constant	4
variable	$3\frac{1}{8}$
Sector shaft end play	.002 maximum
Preload torque limits	
A—Wormshaft drag	1 to 4 lb/inch
B—Thrust bearing	A +1 to 3 lb/inch*
C—Center mesh load	B +4 to 8 lb/inch**

* Total not to exceed 8 lb/inch
** Total not to exceed 18 lb/inch

Worm ball sizes	
Code 6 size	.28112 to .28122
Code 7 size	.28120 to .28130
Code 8 size	.28128 to .28138
Code 9 size	.28136 to .28146
Code 10 size	.28144 to .28154
Code 11 size	.28152 to .28162

BRAKES

Type:	Hydraulically operated drum brakes, disc brakes optional on front only
Adjustments:	All service brakes self-adjusting. Park brake linkage adjusted at equalizer
Shoe wear limit:	
Drum brakes	$\frac{1}{32}$ from rivet heads
Disc brakes	.030
Drum brake diameter:	
New	10.000
Maximum	10.060
Brake fluid	Ford Heavy Duty

ELECTRICAL SYSTEM

Polarity:	Negative earth
Voltage:	12 volts
Battery:	Lead/acid
Charging	Alternator
Starter motor:	
No load current draw	70 amps
Brush length	
New	.5 inch
Minimum	.25 inch
Brush tension	40 ozs
Test data	
Normal current (Amps)	Stall current and torque (battery down to 5 volts)
1965 to 1967:	
4 inch diameter, 100 to 150	460 amps at 9 lb ft
$4\frac{1}{2}$ inch diameter, 250	670 amps at 15.5 lb ft
1968 onwards:	
4 inch diameter, 150 to 200	460 amps at 9 lb ft
$4\frac{1}{2}$ inch diameter, 150 to 200	670 amps at 15.5 lb ft

LUBRICANT SPECIFICATIONS

Item	*Part Name*	*Ford Part No.*	*Ford Specifications*
Motor oil	Ford 6000-mile motor oil (MS sequence tested)	C5AZ-19579-D, E, K, L, M (Canadian CC7AZ-19579-A) (Canadian CC5AZ-19579-G)	ESE-M2C101-B
Engine oil filter	Autolite oil filter (6000 mile type)	C1AZ-6731-A (Canadian CC1AZ-6731-A)	ES-C8AF-6714-4, C
Manual-shift transmission	Ford standard transmission fluid	C3RZ-19C547-B	ESW-M2C83-B
Automatic transmission shift linkage	Ford chassis lube	C1AZ-19590-B	ESA-M1C75-B
Automatic transmission Power steering (pump reservoir)	Ford automatic transmission fluid	C1AZ-19582-A, C, D	ESW-M2C33-F
Steering arm stops	Ford steering arm stop pad lube	C7AZ-19590-A	ESA-M1C25-A
Steering gear housing (manual)	Ford steering gear lubricant	C3AZ-19578-A	ESW-M1C87-A
Rear axle	Ford hypoid gear lube	C6AZ-19580-B	ESW-M2C105-A
Brake master cylinder	Ford brake fluid extra heavy duty	C6AZ-19542-A	ESA-M6C25-A
Front wheel bearings Power steering valve ball stud Front suspension, ball joints	Ford ball joint and multi-purpose lube	C1AZ-19590-B	ESA-M1C75-B
Body hinges hood latch and auxiliary catch	Ford polyethylene grease	C4AZ-19584-A, B	ESB-M1C93-A, ESB-M1C106-B
Lock cylinders	Ford lock lubricant	B4A-19587-A	ESB-M2C20-A
Speedometer cable	Ford speedometer cable lube	B5AZ-19581-A	ESF-M1C135-A

CAPACITIES

Automatic transmission

200 .. .. 8 quarts ($6\frac{1}{2}$ Imp quarts)
250 .. .. 9 quarts ($7\frac{1}{4}$ Imp quarts)

Cooling system

200 standard .. .. 8.7 quarts
200 with air conditioning or emission control .. 8.9 quarts
250 with air conditioning only .. .. .. 11.2 quarts

Engine .. .. $4\frac{1}{2}$ quarts ($3\frac{1}{2}$ Imp quarts) which includes 1 quart for filter replacement

Rear axle .. .. 4 pints ($3\frac{3}{8}$ Imp pints), $\frac{1}{4}$ inch below filler plug hole

Steering

1965 and 1966 .. .. .5 lb
1967 onwards .. .. .55 $\pm$.05 lb

Transmission

Threespeed .. .. $3\frac{1}{2}$ pints
Fourspeed .. .. 4 pints

TORQUE WRENCH SETTINGS

All given in lb ft unless otherwise stated

Engine

Cylinder head bolts	55 then 65 and finally 70 to 75
Exhaust manifold to head	13 to 18
Flywheel bolts	75 to 85
Oil pan attachments	7 to 9
Connecting rod nuts	19 to 24
Main bearings bolts	60 to 70
Camshaft sprocket	35 to 45
Camshaft thrust plate	12 to 15
Valve rocker shaft to head	30 to 35
Puller to damper:	
UBS bolts	25 to 35
Place bolts	35 to 45
Front bracket to insulator	18 to 25
Front insulator to support bracket	24 to 34
Support bracket to crossmember	20 to 30
Rear insulator to transmission	30 to 60
Spark plugs	15 to 25

Clutch

Transmission assembly to flywheel housing bolts	37 to 42
Pressure plate and cover assembly to flywheel attaching bolts	12 to 20
Clutch release equalizer frame bracket bolts	15 to 24
Clutch pedal free travel adjustment nut	15 to 20
Flywheel housing to cylinder block bolts 6-cylinder	23 to 33
Clutch pedal assist spring bracket to brake pedal support bracket	12 to 17

Transmission

Extension housing bolts	42 to 50
Input shaft bearing retainer-to-transmission case bolts	30 to 36
Speedometer cable retainer-to-transmission extension	3.0 to 4.5
Transmission-to-flywheel housing bolts	37 to 42
Gearshift control levers to cam and shaft assembly locknuts	18 to 23
Transmission cover bolts	14 to 19
Filler plug-to-case	10 to 20
Drain plug-to-case	20 to 30
Shift fork-to shift rail	10 to 18
Gear shift control assembly to transmission extension	8 to 12
Shift selector levers to shift rods	10 to 20

Automatic transmission

Converter to flywheel	23 to 28
Converter housing to transmission case	28 to 40
Front pump to transmission case	28 to 40
Overrunning clutch race to case	13 to 20
Oil pan to case	12 to 16
Rear servo cover to case	12 to 20
Stator support to pump	12 to 20
Converter cover to converter housing	12 to 16
Intermediate servo cover to case	16 to 22
Diaphragm assembly to case	15 to 23
Distributor sleeve to case	12 to 20

Extension assembly to transmission case	28 to 40
Pressure gauge tap	9 to 15
Band adjustment screw locknut to case	35 to 45
Yoke to output shaft	60 to 80
Reverse servo piston to rod	12 to 20
Converter drain plug	20 to 30
Manual valve inner lever to shaft	8 to 12
Downshift lever to shaft	12 to 16
Filler tube to engine	20 to 25
Transmission to engine 6-cylinder	23 to 33
End plates to body	20 to 35
Inner downshift lever stop	40 to 55
Reinforcement plate to body	40 to 55
Screen and lower to upper valve body	40 to 55
Neutral switch to column	20
Accumulator plate to body	50 to 90
Control assembly to case	80 to 120
Governor body to collector body	80 to 120
Cooler line fittings	80 to 120

Rear suspension

Spring shackle bars to body and spring	15 to 22
Shock absorber to upper mounting bracket	15 to 25
Shock absorber to spring clip plate	15 to 25
Spring to axle U-bolt nut	30 to 45
Wheel lug nut, 4-lug wheel	55 to 85
Wheel lug nut, 5-lug wheel	70 to 115
Spring to spring front hanger—nut	70 to 100
Universal joint U-bolt nut	10 to 15

Front suspension

Shock absorber upper attachment	10 to 15
Shock absorber upper bracket to body pre-1968	8 to 16
1968 and later	20 to 30
Front suspension compression bumper to body	12 to 17
Brake backplate to spindle (drums)	28 to 35
Brake backplate to spindle (discs)	9 to 14
Upper arm and inner shaft to body	75 to 100
Spring seat to upper arm pre-1968	17 to 25
1968 and later	30 to 40
Shock absorber to spring seat pre-1968	12 to 17
1968 and later	8 to 12
Ball joint to spindle (upper and lower)	60 to 90*
Wheel bearing adjusting nut	17 to 25**
Strut to lower arm pre-1968	55 to 70
1968 and later	70 to 105
Lower arm to underbody	75 to 100
Stabilizer mounting bracket to underbody pre-1971	17 to 25
1971 and later	6 to 12
Strut to underbody	60 to 80
Stabilizer to lower arm	6 to 11
Early disc brakes:	
Caliper adaptor to spindle	28 to 42
Caliper adaptor to caliper	45 to 60
Later disc brakes:	
Caliper to spindle (upper)	100 to 120
Caliper to spindle (lower)	55 to 75

**Torque to minimum specification: then tighten nut to next cotterpin slot and insert cotterpin*

***See Chapter 9, Section 9:3.*

Steering linkage

Cylinder mounting bracket to underbody or frame, side hole	28 to 35
bottom hole	35 to 43
Power cylinder to bracket	18 to 24
Power cylinder to bracket locknut	3 to 5
Steering tie rod end to spindle arm	30 to 40*
Idler arm mounting bracket to underbody or frame	28 to 35
Pitman arm to control valve (power steering)	35 to 47*
Steering gear to side rail or frame	50 to 65
Pitman arm to sector shaft	150 to 225
Idler arm to Pitman arm to idler arm rod	60 to 80*
Spindle connecting rod clamp to adjusting sleeve	10 to 14
Pitman arm to steering arm-to -idler arm rod	35 to 47*
Steering spindle arm connecting rod to arm to idler arm rod	30 to 40*

**Torque to low limit of specifications; then, tighten the nut to the nearest cotterpin slot and insert the cotterpin*

Steering column

Column lower bracket to column upper bracket	8 to 15
Flex coupling to steering gear	24 to 37
Flex coupling to steering shaft flange	10 to 18
T-plate to instrument panel	5 to 15
Steering wheel attaching nut	30 to 40
Brake pedal support bracket to brake pedal support	13 to 25

Steering gear
Manual and power assisted

Sector shaft coverbolts	15 to 22
Mesh load adjusting screw locknut	32 to 40
Ball return guide clamp screw	18 to 42 lb in
Preload adjuster locknut	45 to 60
Lubricant fill plug and vent	3 to 9[1]

[1]*Minimum of one thread must remain exposed when installed.*

Brakes

Parking brake control assembly mounting nuts and bolts	8 to 12
Master cylinder to dash panel nut	13 to 20
Wheel cylinder to backing plate screws (9 inch brake)	5 to 7
Pressure differential valve to fender apron	7 to 11
Rear brake backing plate to axle housing	20 to 40
Front brake backing plate to spindle	28 to 35
Wheel cylinder bleeder screw	32 to 65 lb in
Brake hose connection to front wheel cylinder	12 to 20
Brake line connection to rear axle housing	12 to 19
Hydraulic tube connections*	
$\frac{3}{8}$ x 24	10 to 15
$\frac{7}{16}$ x 24	10 to 15
$\frac{1}{2}$ x 20	10 to 17
$\frac{9}{16}$ x 18	10 to 17
Wheel to hub and drum	55 to 85

**All hydraulic lines must be tightened to the specified torque value and be free of fluid leakage*

WIRING DIAGRAMS

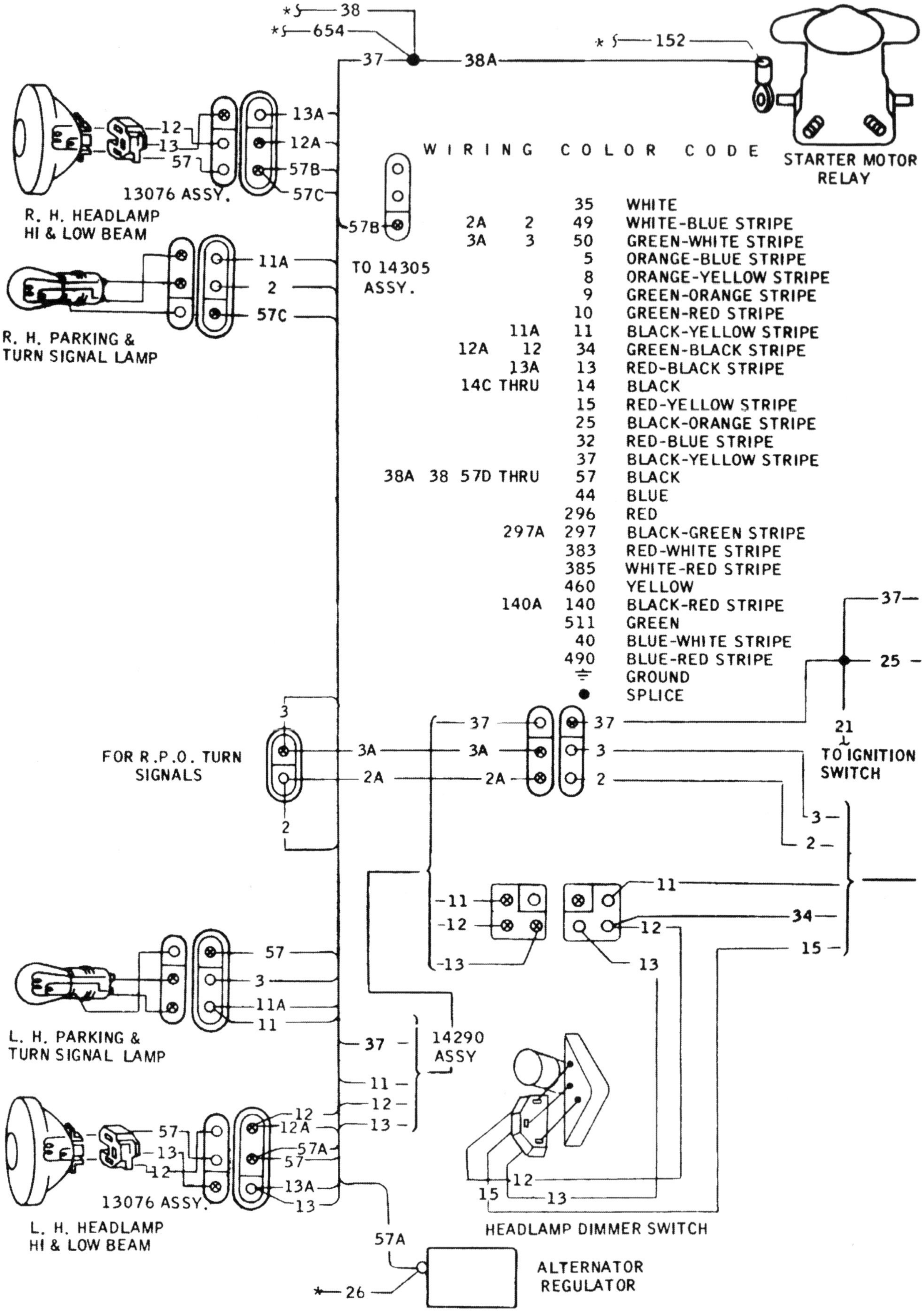

FIG 14:1 Exterior lighting and turn signals

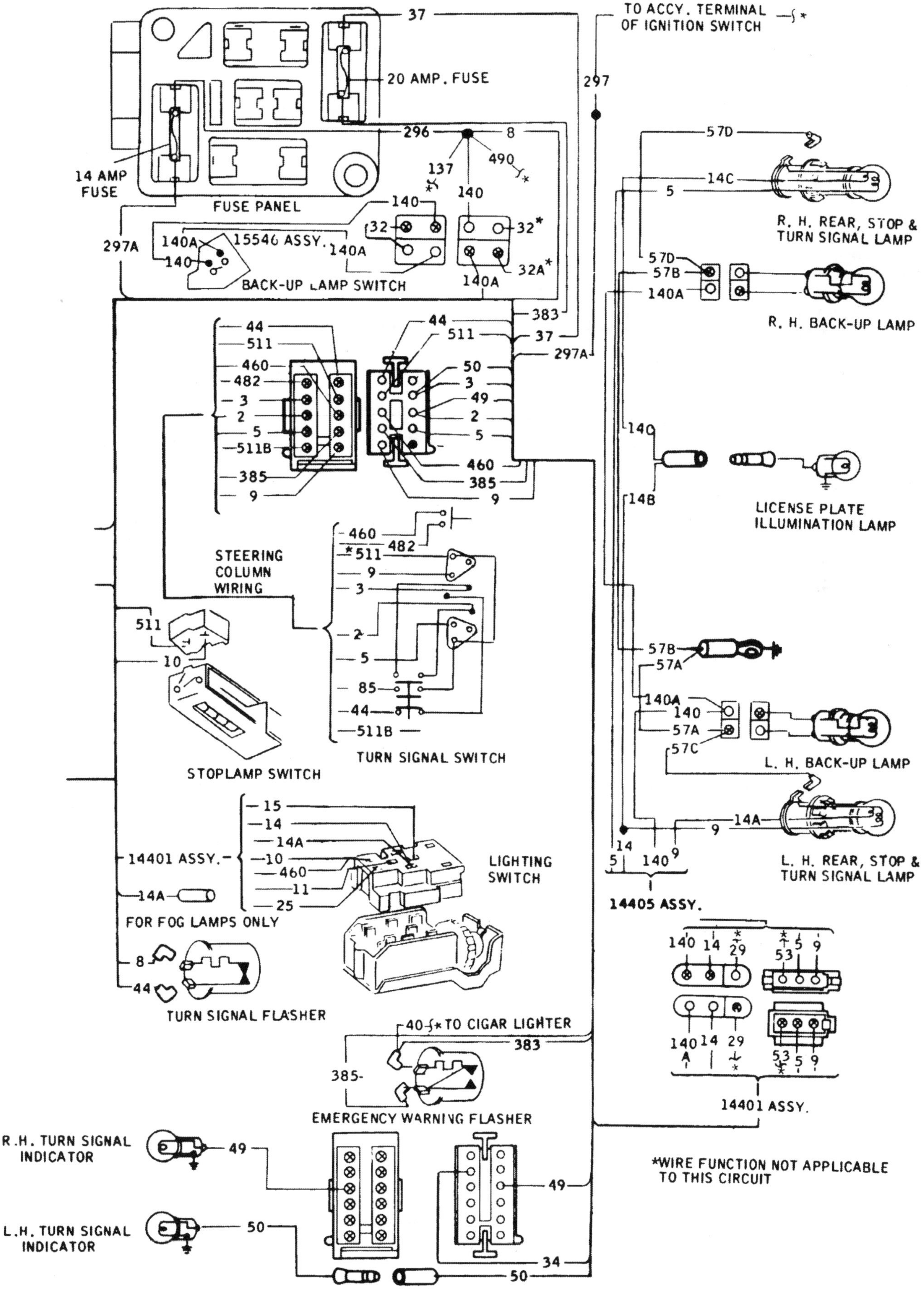

FIG 14:2 Exterior lighting and turn signals continued

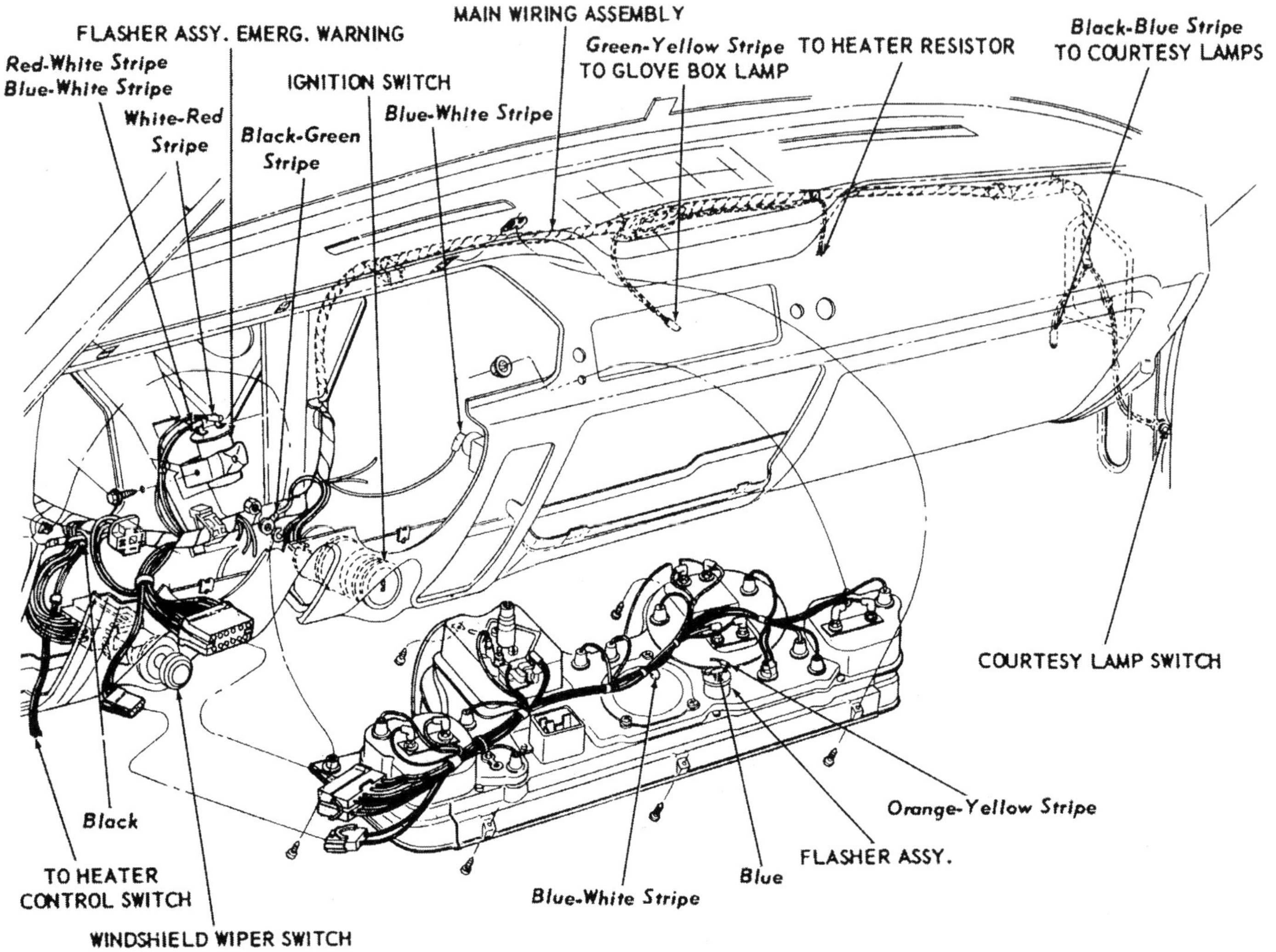

FIG 14:3 Instrument panel layout

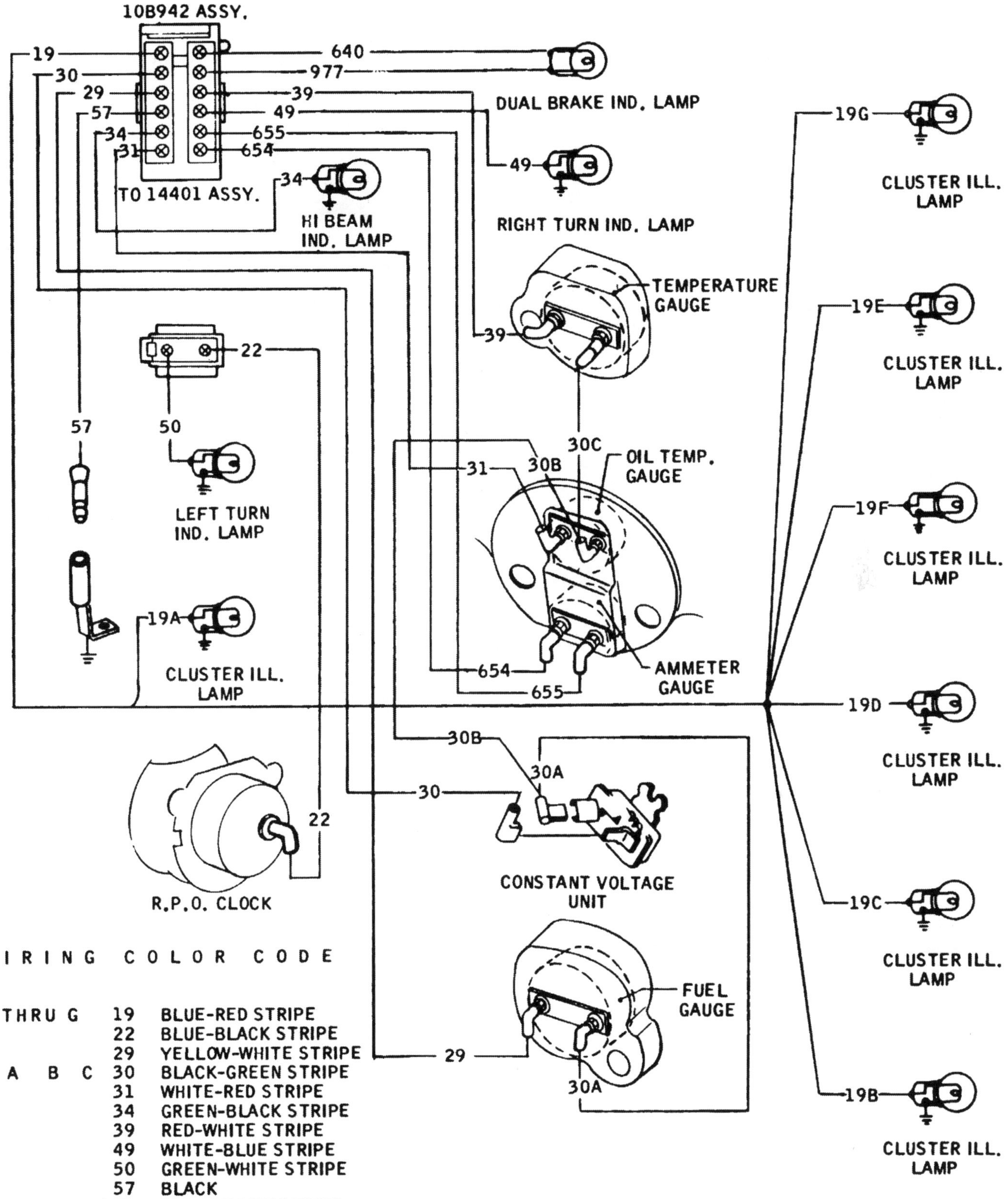

WIRING COLOR CODE

A THRU G	19	BLUE-RED STRIPE
	22	BLUE-BLACK STRIPE
	29	YELLOW-WHITE STRIPE
A B C	30	BLACK-GREEN STRIPE
	31	WHITE-RED STRIPE
	34	GREEN-BLACK STRIPE
	39	RED-WHITE STRIPE
	49	WHITE-BLUE STRIPE
	50	GREEN-WHITE STRIPE
	57	BLACK
	640	RED-YELLOW STRIPE
	654	VIOLET
	655	YELLOW
	977	VIOLET
	●	SPLICE
	⏚	GROUND

FIG 14:4 Instrument panel wiring

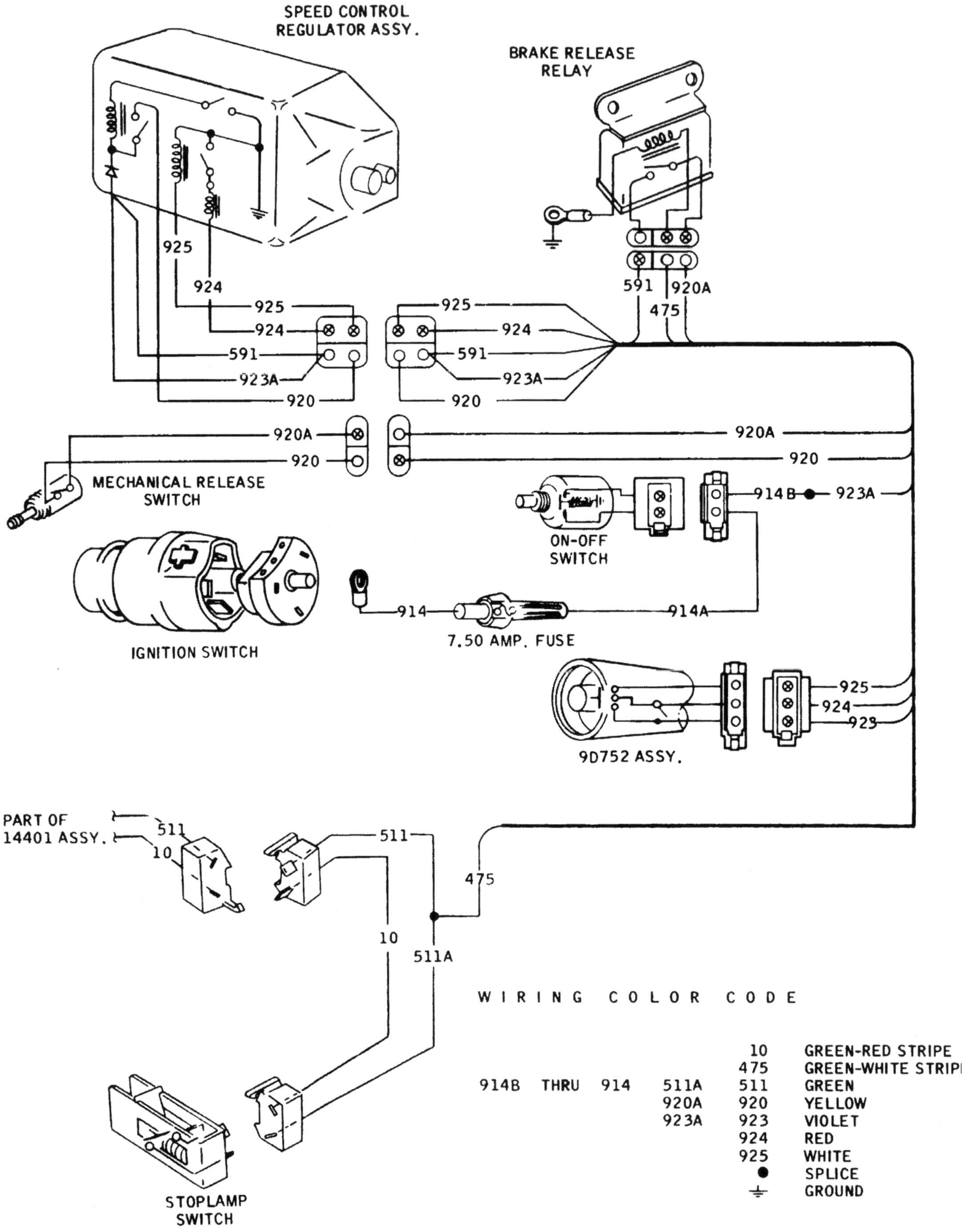

W I R I N G   C O L O R   C O D E

		10	GREEN-RED STRIPE
		475	GREEN-WHITE STRIPE
914B THRU 914	511A	511	GREEN
	920A	920	YELLOW
	923A	923	VIOLET
		924	RED
		925	WHITE
		●	SPLICE
		⏚	GROUND

FIG 14:5 Perfect Circle speed control

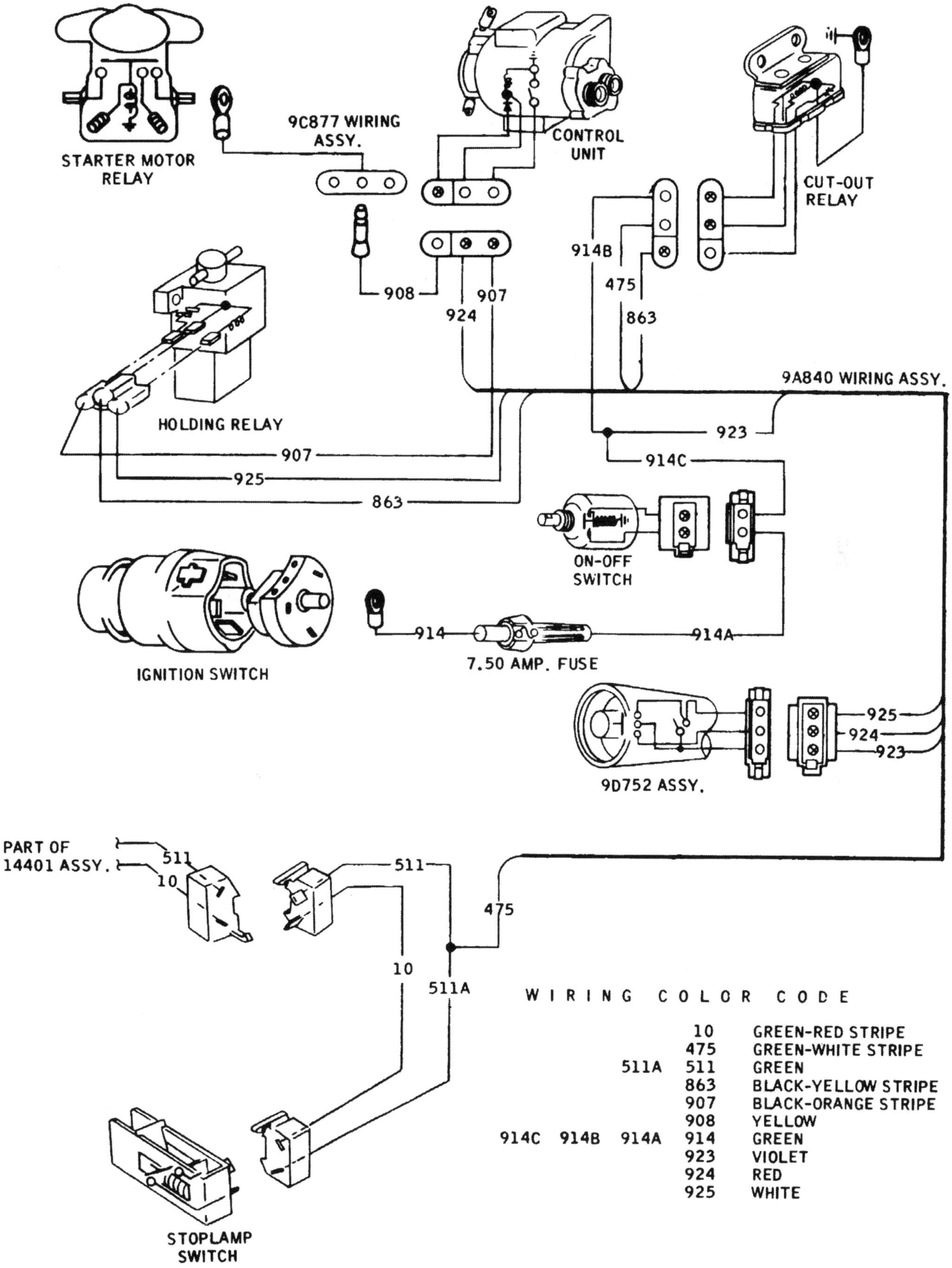

WIRING COLOR CODE

	10	GREEN-RED STRIPE
	475	GREEN-WHITE STRIPE
511A	511	GREEN
	863	BLACK-YELLOW STRIPE
	907	BLACK-ORANGE STRIPE
	908	YELLOW
914C 914B 914A	914	GREEN
	923	VIOLET
	924	RED
	925	WHITE

FIG 14:6 Curtiss-Wright speed control

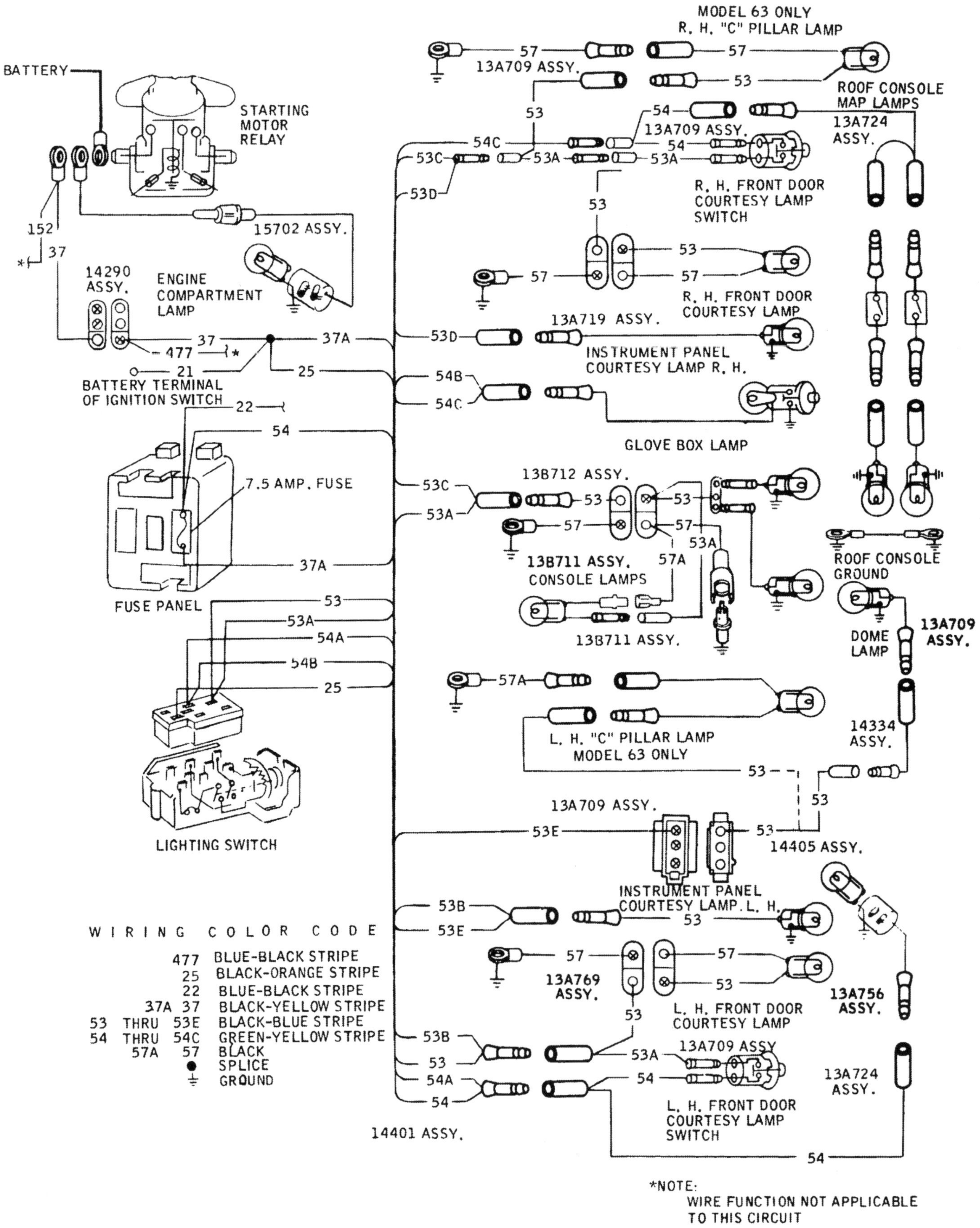

FIG 14:7 Interior lighting

Inches	Decimals	Millimetres	Inches to Millimetres: Inches	Inches to Millimetres: mm	Millimetres to Inches: mm	Millimetres to Inches: Inches
1/64	.015625	.3969	.001	.0254	.01	.00039
1/32	.03125	.7937	.002	.0508	.02	.00079
3/64	.046875	1.1906	.003	.0762	.03	.00118
1/16	.0625	1.5875	.004	.1016	.04	.00157
5/64	.078125	1.9844	.005	.1270	.05	.00197
3/32	.09375	2.3812	.006	.1524	.06	.00236
7/64	.109375	2.7781	.007	.1778	.07	.00276
1/8	.125	3.1750	.008	.2032	.08	.00315
9/64	.140625	3.5719	.009	.2286	.09	.00354
5/32	.15625	3.9687	.01	.254	.1	.00394
11/64	.171875	4.3656	.02	.508	.2	.00787
3/16	.1875	4.7625	.03	.762	.3	.01181
13/64	.203125	5·1594	.04	1.016	.4	.01575
7/32	.21875	5.5562	.05	1.270	.5	.01969
15/64	.234375	5.9531	.06	1.524	.6	.02362
1/4	.25	6.3500	.07	1.778	.7	.02756
17/64	.265625	6.7469	.08	2.032	.8	.03150
9/32	.28125	7.1437	.09	2.286	.9	.03543
19/64	.296875	7.5406	.1	2.54	1	.03937
5/16	.3125	7.9375	.2	5.08	2	.07874
21/64	.328125	8.3344	.3	7.62	3	.11811
11/32	.34375	8.7312	.4	10.16	4	.15748
23/64	.359375	9.1281	.5	12.70	5	.19685
3/8	.375	9.5250	.6	15.24	6	.23622
25/64	.390625	9.9219	.7	17.78	7	.27559
13/32	.40625	10.3187	.8	20.32	8	.31496
27/64	.421875	10.7156	.9	22.86	9	.35433
7/16	.4375	11.1125	1	25.4	10	.39370
29/64	.453125	11.5094	2	50.8	11	.43307
15/32	.46875	11.9062	3	76.2	12	.47244
31/64	.484375	12.3031	4	101.6	13	.51181
1/2	.5	12.7000	5	127.0	14	.55118
33/64	.515625	13.0969	6	152.4	15	.59055
17/32	.53125	13.4937	7	177.8	16	.62992
35/64	.546875	13.8906	8	203.2	17	.66929
9/16	.5625	14.2875	9	228.6	18	.70866
37/64	.578125	14.6844	10	254.0	19	.74803
19/32	.59375	15.0812	11	279.4	20	.78740
39/64	.609375	15.4781	12	304.8	21	.82677
5/8	.625	15.8750	13	330.2	22	.86614
41/64	.640625	16.2719	14	355.6	23	.90551
21/32	.65625	16.6687	15	381.0	24	.94488
43/64	.671875	17.0656	16	406.4	25	.98425
11/16	.6875	17.4625	17	431.8	26	1.02362
45/64	.703125	17.8594	18	457.2	27	1.06299
23/32	.71875	18.2562	19	482.6	28	1.10236
47/64	.734375	18.6531	20	508.0	29	1.14173
3/4	.75	19.0500	21	533.4	30	1.18110
49/64	.765625	19.4469	22	558.8	31	1.22047
25/32	.78125	19.8437	23	584.2	32	1.25984
51/64	.796875	20.2406	24	609.6	33	1.29921
13/16	.8125	20.6375	25	635.0	34	1.33858
53/64	.828125	21.0344	26	660.4	35	1.37795
27/32	.84375	21.4312	27	685.8	36	1.41732
55/64	.859375	21.8281	28	711.2	37	1.4567
7/8	.875	22.2250	29	736.6	38	1.4961
57/64	.890625	22.6219	30	762.0	39	1.5354
29/32	.90625	23.0187	31	787.4	40	1.5748
59/64	.921875	23.4156	32	812.8	41	1.6142
15/16	.9375	23.8125	33	838.2	42	1.6535
61/64	.953125	24.2094	34	863.6	43	1.6929
31/32	.96875	24.6062	35	889.0	44	1.7323
63/64	.984375	25.0031	36	914.4	45	1.7717

UNITS	Pints to Litres	Gallons to Litres	Litres to Pints	Litres to Gallons	Miles to Kilometres	Kilometres to Miles	Lbs. per sq. In. to Kg. per sq. Cm.	Kg. per sq. Cm. to Lbs. per sq. In.
1	.57	4.55	1.76	.22	1.61	.62	.07	14.22
2	1.14	9.09	3.52	.44	3.22	1.24	.14	28.50
3	1.70	13.64	5.28	.66	4.83	1.86	.21	42.67
4	2.27	18.18	7.04	.88	6.44	2.49	.28	56.89
5	2.84	22.73	8.80	1.10	8.05	3.11	.35	71.12
6	3.41	27.28	10.56	1.32	9.66	3.73	.42	85.34
7	3.98	31.82	12.32	1.54	11.27	4.35	.49	99.56
8	4.55	36.37	14.08	1.76	12.88	4.97	.56	113.79
9		40.91	15.84	1.98	14.48	5.59	.63	128.00
10		45.46	17.60	2.20	16.09	6.21	.70	142.23
20				4.40	32.19	12.43	1.41	284.47
30				6.60	48.28	18.64	2.11	426.70
40				8.80	64.37	24.85		
50					80.47	31.07		
60					96.56	37.28		
70					112.65	43.50		
80					128.75	49.71		
90					144.84	55.92		
100					160.93	62.14		

UNITS	Lb ft to kgm	Kgm to lb ft	UNITS	Lb ft to kgm	Kgm to lb ft
1	.138	7.233	7	.967	50.631
2	.276	14.466	8	1.106	57.864
3	.414	21.699	9	1.244	65.097
4	.553	28.932	10	1.382	72.330
5	.691	36.165	20	2.765	144.660
6	.829	43.398	30	4.147	216.990

INDEX

Made in United States
Orlando, FL
04 November 2026

90200821R00111